NEPA IN THE COURTS

A Legal Analysis of The National Environmental Policy Act

NEPA

Published by Resources for the Future, Inc.
Distributed by The Johns Hopkins University Press,
Baltimore and London

in the Courts

A LEGAL ANALYSIS OF THE NATIONAL ENVIRONMENTAL POLICY ACT

Frederick R. Anderson
assisted by Robert H. Daniels
ENVIRONMENTAL LAW INSTITUTE

RESOURCES FOR THE FUTURE, INC.
1755 Massachusetts Avenue, N.W., Washington, D.C. 20036

Resources for the Future is a nonprofit corporation for research and education in the development, conservation, and use of natural resources and the improvement of the quality of the environment. It was established in 1952 with the cooperation of the Ford Foundation. Part of the work of Resources for the Future is carried out by its resident staff; part is supported by grants to universities and other nonprofit organizations. Unless otherwise stated, interpretations and conclusions in RFF publications are those of the authors; the organization takes responsibility for the selection of significant subjects for study, the competence of the researchers, and their freedom of inquiry.

This book is one of RFF's policy studies, which are directed by Sterling Brubaker. It was edited by Ruth B. Haas. Cover art by Clare Ford.

RFF editors: Mark Reinsberg, Vera W. Dodds, Joan Tron, Ruth B. Haas.

Manufactured in the United States of America
Distributed by The Johns Hopkins University Press, Baltimore, Maryland 21218

Library of Congress Catalog Card Number 73-45
ISBN 0-8018-1510-X (cloth) $15.00
ISBN 0-8018-1559-2 (paper) $6.95

Foreword

THE NATIONAL Environmental Policy Act (NEPA) is the most comprehensive legislative statement of the nation's recently formed commitment to protect the environment. Its language is expansive in setting forth the need to achieve productive harmony between man and nature. The Act requires the government to take account of environmental considerations in all of its actions and establishes the Council on Environmental Quality and defines its functions.

The very scope and generality of the law's mandate threatened its effectiveness. This danger was in the minds of its drafters: late in the legislative process they added an "action-forcing" provision in §102(2) (C) requiring all federal agencies to prepare a detailed statement on all actions significantly affecting the quality of the human environment. This statement must describe the environmental impact, set out the unavoidable adverse effects, discuss alternatives to the proposed action, and deal with the long-term and irreversible effects that the action entails.

This rather specific procedural requirement, little debated in Congress, has become the heart of NEPA and has had a profound impact on agency decision making. At first glance it seems a rather fragile instrument to bear so much weight, but its effectiveness is owed in large part to the fact that it provides a point of entry for concerned citizens to challenge government actions and to invoke the spirit of the other sections of NEPA. Although the government has moved through guidelines and procedures to implement the law, legal points raised by hundreds of citizens—the private attorneys general—have given the courts a major new role in shaping its meaning. The process of legal challenge, court ruling, and agency accommodation is far from complete, but after three years of litigation the bare language of the Act has been subjected to a great deal of legal interpretation. The possibility that NEPA would become a mere noble expression of purpose has been turned aside and NEPA has become a vital, if still contentious, part of the decision process.

This study by Frederick R. Anderson of the Environmental Law Institute is an attempt to trace the way in which the courts have interpreted NEPA to date. The record is generated through the action of private individuals and groups who have had to establish their standing to intervene, to have access to the documentation, and have then challenged the quality of agency compliance. The principal points at issue form the structure of the study, with the analysis of cases organized to reveal the trend of decision.

This review does not pretend to be a complete evaluation of the operation of NEPA. No analysis is attempted of the completeness or quality of agency compliance as revealed in procedures, individual impact statements were not sampled for adequacy, nor is there an evaluation of the social costs and effectiveness of introducing the impact statement requirement into the decision process. The present work began as a background paper commissioned by Resources for the Future for a more comprehensive study dealing with some of these questions. When it became apparent that it was still somewhat early to attempt the broader study and that the legal review was both perishable and able to stand as an independent piece, the Environmental Law Institute was persuaded to expand its report into this more complete analysis. While the cutoff date for court cases discussed (March 1973) practically guarantees that the study will be incomplete in some respect before its publication, there is good reason to expect that the general interpretation of NEPA law presented here will remain valid for some time to come.

Originally undertaken as part of the RFF program of studies in public policy, this study of legal interpretation departs somewhat from the usual RFF focus on the economic and technical aspects of resource and environmental problems. While such studies may provide the principles for policy decisions, we recognize that the way in which principles are filtered through institutions and laws in the end determines their impact. Hence we are pleased to publish this account of how some innocuous-appearing procedural language can become such a powerful engine for change once the public and the courts have access to it.

This study will appeal first of all to environmental lawyers concerned with the implementation of NEPA. While meeting the standards of accuracy and completeness required for their use, it is nonetheless written in language accessible to all. It commends itself especially to administrators who deal with NEPA matters and to environmental activists who rely upon this law. Indeed, almost anyone concerned with environmental management will find much of interest in the way NEPA has been interpreted by the courts.

Joseph L. Fisher, President

March 1973 *Resources for the Future, Inc.*

Preface

CONGRESS through the National Environmental Policy Act of 1969 (NEPA) has established a national policy requiring all federal agencies to give full consideration to environmental effects in planning their programs. To ensure that the agencies implement this policy, NEPA prescribes specific "action-forcing" procedures which agencies must observe. This requirement has begun to bring about fundamental reform in the decision-making processes of federal agencies.

From the beginning the courts have played a central role in enforcing NEPA's requirements, especially the action-forcing procedures. In three years a NEPA cause of action has been included in 149 separate litigations, some of which have produced more than one reported opinion. This book is about the judicial interpretations of NEPA which have accumulated since it was signed into law on New Year's Day, 1970.

One action-forcing procedure in particular has received the lion's share of attention during NEPA's first three years. In the brief compass of §102(2) (C), Congress required each federal agency to prepare a detailed statement of environmental impact on every major federal action that might significantly affect environmental quality. By December 31, 1972, the agencies had filed 3,635 of these statements with the Council on Environmental Quality. The statement must discuss alternatives to the proposed action and must be circulated for comment to other federal agencies, to state and local governments, and to the public. The section's strict procedural duties, which courts can enforce, require that the agencies commit themselves—on paper and in advance—to the possible environmental consequences of their proposed actions. Court-enforced full disclosure, needless to say, promises basic change in the pattern of federal decision making in the environmental area. By focusing largely on the decisions that interpret §102(2) (C), we are not discounting the Act's other important provisions. So far, however, NEPA's goals have been realized primarily through this one key provision.

The courts have required about three years to establish the basic trends of interpretation for the first generation of NEPA issues, i.e., those primarily concerned with §102(2) (C): Must all agencies comply with NEPA? What level of federal involvement in an action requires preparation of a statement? What magnitude must a "major" federal action achieve before an impact statement must be prepared? Does the Act apply to the environmentally protective regulatory programs of federal agencies? How does §102(2) (C) apply to projects already in progress when NEPA was enacted? Must it always be a federal official who prepares the impact statement? How much and what kind of information must it contain? How is NEPA to be applied when several agencies are involved in a single action? To be sure, the courts have not completely answered these questions for which Congress provided only partial answers or, as is often enough the case, no answers at all. Much progress has been made, however, enough to permit a tentative first attempt to put the house of judge-made NEPA law in order.

We are aware that hairsplitting judicial interpretations imposed on a statute of broad scope may exalt form over substance and even prevent an agency from setting its own, better-conceived course toward the statutory goal. At this point we would do well to remember Gibbon's account of Diocletian's administrative reforms, which ultimately backfired and had the effect of hardening the arteries of the Roman bureaucracy. NEPA, after all, is an administrative reform statute.

We think, however, that in spite of some excesses in reading the Act too literally, NEPA's potential for lasting reform of federal government rests on the detailed judicial interpretations handed down in the first three years. These decisions lead the agency horses to water, even if they do not always drink. The courts have kept the agencies from straying from the careful, focused consideration of environmental values in day-to-day decision making that Congress intended.

It appears likely that the courts will continue to bring NEPA to bear even more directly on the substance of agency decision making. If this is to be the case, a host of new issues will arise for resolution. In chapter VII we discuss several district and circuit court opinions holding that NEPA imposes substantive duties on agencies to make environmentally sound decisions. These cases affirm that agency discretion to make the final decisions on federal actions is still quite wide, but the process of circumscribing it through further interpretation of NEPA has apparently begun.

Although it is too early to tell, a recent district court opinion from Texas may constitute something of a bench mark for NEPA in the courts. Just as the lengthy district court opinion two years ago in the

Gillham Dam case led the way in identifying and resolving key first-generation NEPA issues, the even more lengthy decision in the Trinity River–Wallisville Dam case[1] in early 1973 may point the way toward an eventual synthesis of NEPA's substantive and procedural provisions so that NEPA goes far beyond its role as a mere full disclosure statute. In the Trinity River–Wallisville Dam case, each step in the Army Corps of Engineers' entire decision-making process for a $1.3 billion river basin development project, and especially for a $29 million component dam, was minutely examined by the court in light of NEPA's provisions. Even the Corps' techniques for determining the cost-benefit ratio for the projects were subjected to intense scrutiny; the Corps was ordered to revise its methods of computation—and, one must unavoidably surmise, its results—in light of NEPA's requirements. If the courts require revision of existing agency review processes to ensure decisions consistent with NEPA, and if they actually begin to circumscribe the scope of allowable agency decision making under NEPA, then a second generation of NEPA issues will have been born.

The organization and approach of this book might have been considerably different had we not let the decided cases largely define its scope. For the most part, we have confined our analysis to the issues that actual controversies have raised. But where we thought it appropriate, we have expressed our views on the correctness of the courts' resolution of particular issues and on the direction which we think a developing line of litigation should take. Occasionally our remarks go beyond criticism of the cases to take up issues which the courts have only begun to consider.

As might be expected from so much litigation, NEPA fairly bristles with judicial glosses. All this grist for the lawyer's mill, however, does not mean that the agencies have in fact thoroughly implemented NEPA and have actually begun to make better decisions. This book sheds little light on the actual difference, if any, that NEPA is making in the final decisions of the federal agencies and in the quality of the environment. Most reports are decidedly pessimistic. Our effort here, however, is not totally irrelevant to the realities of federal decision making. Failure to comply with NEPA usually carries a serious penalty: an injunction restraining the agency from further action pending strict compliance with the Act. Routine implementation of NEPA through litigation may not be a very realistic or desirable prospect, but litigation does afford the courts both the

[1] Sierra Club v. Froehlke, Appendix B.

opportunity to castigate misuse of discretion and to point the way toward adequate administrative oversight of the NEPA process.

Furthermore, we have not discussed proposed amendments to NEPA nor the various bills patterned on NEPA which have been enacted by state legislatures. Although attempts to amend NEPA itself have failed, Congress has partially exempted the Environmental Protection Agency[2] and the Atomic Energy Commission from the burdens of full NEPA compliance.[3] The national pattern of strict judicial enforcement is being repeated for the "state NEPAs," which have now been enacted in at least eight states.[4] Both these matters lie somewhat to the side of our focus on NEPA litigation.

This study is intended for the use of attorneys—private practitioners, public interest lawyers, government counsel, law professors, and students. We have endeavored, however, to reach others with a primary interest in NEPA's application—political scientists, economists, government administrators, policy analysts, planners, and the like. We also intended that the analysis be comprehensible to the legally aware layman, of whom there are quite a few in citizens' groups, even if it is comprehensible only after a second reading. Legal terminology is usually as impenetrable as the user makes it; we would like to believe that difficulties with our analysis more often than not can be traced to a necessary economy in providing the background of the relevant legal doctrines and the facts of the cases.

The appearance of new decisions which we were not able to include during the last days of manuscript preparation serves as another reminder that the interpretation of NEPA is in midcourse. For this reason I would like to ask readers to communicate omissions, errors, and viewpoints to me before I prepare a second analysis as a chapter in a larger study of federal environmental law

---

[2] *See* the Federal Water Pollution Control Amendments of 1972, 33 U.S.C. §1151 *et seq.*, Pub. L. No. 92–500, 86 Stat. 816 (Oct. 18, 1972). The section modifying EPA's NEPA obligations is 33 U.S.C. §511(C)(1). *See* chapter IV, pp. 108 *ff*.

[3] 42 U.S.C. §2242, Pub. L. No. 92–307, 86 Stat. 191 (June 2, 1972). The authority granted to the AEC to issue temporary operating licenses on completion of expedited review procedures expires on October 30, 1973.

[4] The states are California, Delaware, Indiana, Montana, New Mexico, North Carolina, Washington, and Wisconsin. Puerto Rico also has a "little NEPA," and Hawaii has implemented a NEPA-like procedure by Executive Order. *See* Comment, *States Enact Environmental Protection Measures,* 2 ELR 10177 (August 1972), where full statutory citations to these provisions are given.

now in progress under the direction of the Environmental Law Institute. That study will appear in the winter of 1973.

As an aid to readers, we have included three appendices of NEPA materials. Appendix A includes the text of the Act. Appendix B contains full citations to the 149 NEPA litigations which have resulted in reported judicial dispositions—memorandum opinions, findings of fact and conclusions of law, orders, or even trial transcripts. In a sense Appendix B is a summary of all decided NEPA cases, and may be as useful to practitioners as portions of the analysis. It also serves as an index of page numbers in this book where the cases cited are discussed. Appendix C includes the text of the final guidelines of the Council on Environmental Quality.

We have generally followed *A Uniform System of Citation* (11th ed.), except in one important particular. Where a case is cited more than once in a specific chapter, and no particular page reference to the reports is required, we have referred the reader to Appendix B rather than to the earlier footnote containing the full citation. This system will avoid any errors which might occur because the footnotes are renumbered in each chapter. Each page containing a case reference is indexed in Appendix B, as mentioned above. For brevity, a large list of cases in a footnote may occasionally be cited to Appendix B, although particular cases have not yet been cited for the first time.

Over the past eight months my colleague, Robert Daniels, must have often felt that he was lured into this effort under false pretenses. Immediately after graduation from the Harvard Law School in June 1972, Bob came to the Environmental Law Institute to participate in what he then thought to be a three-month survey of the case law on NEPA. By the end of September, and after a careful scouring of the then-decided cases, we had promised a book to Resources for the Future and had plunged into the analysis. To Bob's great credit, he managed to carry both his burden in the preparation of this manuscript and his responsibilities as a Russell Sage Fellow at Stanford University, where he has been since September 1972. Chapter V and the section on the adequacy of impact statements in chapter VI are largely his work.

Many generous persons read our manuscript under considerable time pressures. We are not taking refuge in a polite convention when we credit their comments and suggestions with making the book much better than it could conceivably have been had we persisted in errors and mistakes even more outrageous to contemplate than those which we insisted on retaining against their better judgment. Edward Strohbehn of the Natural Resources Defense Council returned his detailed, perceptive comments within just a few days of receipt of the manuscript. James Moorman, Director of the Sierra Club Legal Defense Fund, must also have sacrificed a good portion of his Christmas holidays for the same purpose, as did Robert Kennan of the National Wildlife Federation and Gus Speth, also of the Natural Resources Defense Council. Timothy Atkeson, General Counsel to the Council on Environmental Quality, provided numerous helpful ideas and facts, as did William Lake, Charles Lettow, and Philip Soper, all formerly of the Council.

Over several months of writing, the staff of the Environmental Law Institute responded in good spirits and with outstanding effort to the exceptionally short time frame within which we worked. Many of the ideas developed here were born in *Environmental Law Reporter* Comments. Thomas Alder, the Institute's president; Grant Thompson, acting associate editor; Robert Funicello, assistant editor; and Durwood Zaelke helped sharpen these ideas over the two years of the *Reporter*'s existence. Thomas Guilbert and Erica Sims, both assisting in the Institute's study of federal environmental law, read the manuscript and provided numerous helpful comments. Neal Strauss assembled the table of NEPA cases. James Scott energetically checked numerous citations. I owe gratitude for a job so very well done to Jane Cantor who typed so many drafts of the manuscript, and to Anne Chacon, the Institute's editorial secretary, who pitched in to ease us through the last three months of preparation. Finally, there is Susanne, my wife, and Lindsey, Sally, and Kendall, who cooled their heels patiently while I wrote. The stern guardians of format tell me that I cannot have a page to dedicate this book to them, but if I could I would.

Environmental Law Institute Frederick R. Anderson

Washington, D. C.

March 1973

Contents

I

NEPA and Congress

THE NATIONAL Environmental Policy Act of 1969 (NEPA)[1] was signed into law on New Year's Day, 1970. Congressional discussion of the need for a national environmental policy had spanned the preceding decade, although only a few months elapsed between the introduction of NEPA's progenitors in the Senate and House and their enactment in modified form. With NEPA, Congress charted a new course for the federal agencies in environmental protection, and, as might be expected with innovative legislation, important problems were not fully considered before the bill became law. In particular, under the press of a closing session Congress had little time to speculate about the future of NEPA's unique action-forcing provisions, which were put in final form just before the bill went to conference.

In passing NEPA, Congress appeared to have five major purposes in mind. It wanted to enlarge the federal agencies' basic mandates through enactment of a national environmental policy, establish specific action-forcing procedures for the implementation of that policy, create the Council on Environmental Quality (CEQ), foster the development of information on and indices of environmental quality, and provide for an annual CEQ report of progress toward these goals.[2] In retrospect, the first two goals were by far the most important. Their interaction is what has made NEPA so innovative and controversial. Yet the largest portion of NEPA's legislative history is taken up with establishing the dynamics of environmental systems, diagnosing the extent of environmental harm insofar as it is known (and calling for the study and measurement of what is not yet known), identifying the federal institutional shortcomings which contribute to environmental deterioration, and endorsing the need for comprehensive federal planning, coordination, and decision making under a unified national policy. The subject of enforcement of such a

[1] See the full text and citations in Appendix A.

[2] SENATE COMM. ON INTERIOR AND INSULAR AFFAIRS, NATIONAL ENVIRONMENTAL POLICY ACT OF 1969, S. REP. No. 91–296, 91st Cong., 1st Sess. 3, 9, 10, 14 (July 9, 1969) [hereinafter cited as Senate Report].

policy on the working level in the federal agencies did not command Congress' full attention at any point.

The full text of the Act appears in Appendix A. Specifically, §101(a) instructs the federal government to protect and restore the environment in accordance with a general national policy, declared by the Act, that the government shall endeavor "to create and maintain conditions under which man and nature can exist in productive harmony." The national environmental policy is spelled out in §101(b) in six specific environmental mandates to the federal government. These give content to NEPA's substantive policy and ensure that NEPA's lengthy opening passages are more than a mere hortatory preamble.[3] Further, in §102(1) Congress stated that "to the fullest extent possible . . . the policies, regulations, and public laws of the United States shall be interpreted and administered in accordance with the policies set forth in this Act." In §102(2), in responding to fears that the new policy might become an empty utterance unless the statute included a means of ensuring that federal agencies would implement the policy, Congress addressed itself to the task of designing a system that could translate NEPA's goals into action. This system was also subject to compliance "to the fullest extent possible." Section 102(2) (C) was one of eight "action-forcing" provisions set up to ensure that the federal government bore these general goals and directives in mind in making specific decisions. As a Senate report put it:

> If goals and principles are to be effective, they must be capable of being applied in action. S. 1075 thus incorporates certain "action-forcing" provisions and procedures which are designed to assure that all Federal agencies plan and work towards meeting the challenge of a better environment.[4]

Section 102(2) (C) requires all federal agencies to prepare detailed, written statements of the environmental impacts which major proposed actions may cause, including alternative actions and their impacts. The detailed statement must be circulated to local, state, and federal agencies for comment, and it must accompany the proposed action throughout the agency's usual decision-making processes.

---

[3] *Hearings on S. 1075, S. 237 and S. 1752 Before the Senate Committee on Interior and Insular Affairs,* 91st Cong., 1st Sess. at 116 (April 1969) [hereinafter cited as Senate Hearings].

[4] Senate Report, *supra* note 2, at 9.

Sections 101 and 102(1) on the one hand, and § 102(2) (C) on the other, thus may be seen as interlocking; the former direct the implementation of the national environmental policy, while the latter relates the directive to specific legislation and actions, provides a mechanism for coordinating views from various specialized agencies, and assembles information on particular impacts and alternatives for use in "existing agency review processes."

Although most public discussion and litigation has involved the impact statement requirement, the other provisions of NEPA should not be overlooked. Section 102(2) (A) calls upon the agencies to approach environmental problems through the integrated use of natural and social sciences and "environmental design arts." Section 102(2) (B) asks that agencies develop methods for taking unquantified environmental values into account on a par with the usual economic and technical considerations. Section 102(2) (D) stresses and elaborates the critical requirement of § 102(2) (C) (iii) that less damaging alternatives to proposed action be sought out. Section 102 (2) (E) calls upon the agencies to help improve man's global environment.

In addition to the above action-forcing provisions, §103 directs agencies to review their policies and practices and bring them into line with the Act. This neglected but fertile provision seems to call upon the agencies for a great deal more than the *pro forma* "§103 statements" that were required to be filed by July 1, 1971.

Title II of NEPA creates the Council on Environmental Quality and defines its scope and responsibilities. The CEQ's role, as it has evolved, has been primarily to advise the White House on environmental policy and the President's environmental message. It has also prepared three excellent annual reports. Although CEQ's influence on the federal agencies has been exerted primarily through informal discussion and criticism, the agency has played an aggressive role in promulgating guidelines for agency implementation of NEPA and in seeing that they are recorded in the *Federal Register*.[5]

Congress began to consider some of NEPA's goals several years

---

[5] See Appendix C for the full text of CEQ Guidelines, 36 FED. REG. 7724, ELR 46049 (April 23, 1971). For a subsequent CEQ memorandum to agency counsel which defines in greater detail and widens considerably CEQ's interpretation of NEPA's requirements, *see Recommendations for Improving Agency NEPA Procedures*, ELR 46162 (May 16, 1972). Finally, *see* 38 FED. REG. 10856 (May 2, 1973) for proposed revisions in the April 1971 CEQ Guidelines.

before the legislation that became NEPA was introduced in 1969. Ten years earlier, Senator Murray had proposed "The Resources and Conservation Act," which called for a "unified statement of conservation, resource and environmental policy" and the creation of a high-level council of environmental advisors. Similar legislation was introduced by Senators Nelson, Jackson, and Kuchel in the mid-1960s.[6] Two documents prepared in 1968 so influenced NEPA's subsequent development that they are often included in its legislative history. In June 1968, the Subcommittee on Science, Research and Development of the House Committee on Science and Astronautics published a report entitled *Managing the Environment*[7] which linked fragmented governmental decision making and neglect of the environment to some of the nation's main environmental problems. In the same year, the full committee joined with the Senate Committee on Interior and Insular Affairs, chaired by Senator Henry Jackson, to sponsor a joint Senate–House "colloquium" to discuss a national policy on the environment. The colloquium led to publication of a document entitled *Congressional White Paper on a National Policy for the Environment.*[8] That paper contained the basic elements of the national environmental policy enacted in §101 of NEPA.

NEPA's legislative history began with a number of bills which hinted at the final form of the legislation but did not emerge from committee.[9] NEPA's immediate forebears were H.R. 6750, intro-

---

[6] *See* Richard N. L. Andrews, Environmental Policy and Administrative Change: The National Environmental Policy Act of 1969 (1970–71) at 76–109 (unpublished doctoral dissertation, Department of City and Regional Planning, University of North Carolina, 1972). A second unpublished doctoral dissertation focuses exclusively upon NEPA's legislative career. *See* Terence T. Finn, Conflict and Compromise: Congress Makes a Law, The Passage of the National Environmental Policy Act (Department of Government, Georgetown University, 1972) (683 pp.).

[7] SUBCOMM. ON SCIENCE, RESEARCH AND DEVELOPMENT OF THE HOUSE COMM. ON SCIENCE AND ASTRONAUTICS, MANAGING THE ENVIRONMENT, Serial S, 90th Cong., 2d Sess. (June 1968).

[8] *Joint House-Senate Colloquium to Discuss a National Policy for the Environment, in Hearings Before the Senate Comm. on Interior and Insular Affairs and the House Comm. on Science and Astronautics,* 90th Cong., 2d Sess. [No. 8] (July 17, 1968). The second document, the "white paper," is cited to the same two committees, same session, CONGRESSIONAL WHITE PAPER ON A NATIONAL POLICY FOR THE ENVIRONMENT, Serial T (Comm. Print, October 1968).

[9] *See* Senate Report, *supra* note 2, at 10–12; HOUSE COMM. ON MERCHANT MARINE AND FISHERIES, COUNCIL ON ENVIRONMENTAL QUALITY, H.R. REP. NO. 91–378, 91st Cong., 1st Sess. 2–3 (July 11, 1969) [hereinafter House Report].

duced February 17, 1969, and S. 1075, introduced the following day.[10] Interestingly, Senator Jackson's bill (S. 1075) did not mention a national environmental policy or action-forcing provisions when it was introduced, in spite of his earlier efforts to enlist congressional support for such a policy. The early version of S. 1075 would have authorized the secretary of the interior to conduct environmental research and would have created the CEQ. Congressman Dingell's bill likewise called for the creation of CEQ and supplied only a brief general statement of policy, but did not mention new research responsibilities for the secretary of the interior. Further, his bill was not intended to be a wholly new law, but rather was an amendment to the Fish and Wildlife Coordination Act.

Both Senator Jackson and Congressman Dingell may have been influenced to omit the policy provisions from their early proposed legislation in order to ensure that the bills, once introduced, would be referred to their respective committees, where they could then be accordingly altered. Senator Jackson may have been trying to avoid a jurisdictional controversy with Senator Muskie's Subcommittee on Air and Water Pollution of the Senate Committee on Public Works. Eventually this dispute led to the inclusion of §104, which has caused important problems in NEPA's interpretation. (See the discussion of the agencies to which NEPA applies in chapter IV, beginning at page 106.) Congressman Dingell may have wished to forestall a similar dispute between the House Merchant Marine and Fisheries Committee, sympathetic to NEPA, and the House Interior and Insular Affairs Committee, chaired by Congressman Wayne Aspinall, who became one of NEPA's strongest opponents in the House.

In the Senate, Senator Jackson's committee conducted a hearing on April 16, 1969.[11] Six weeks later he introduced an amendment to S. 1075 which contained "a declaration of national environmental policy."[12] The amendment stated the policy much as it exists in §101 today; however, two other important provisions did not survive subsequent review. First, his amendment further provided that "each person has a fundamental and inalienable right to a healthful environment." This language was changed in conference to "each

[10] H.R. 6750, 91st Cong., 1st Sess. (1969); S. 1075, 91st Cong., 1st Sess. (1969).

[11] Senate Hearings, *supra* note 2.

[12] *Id.* at 207 (Appendix 2).

person should enjoy a healthful environment," to avoid the creation of a court-enforceable right.[13] A second provision called for a "finding" by the responsible federal official as to the environmental impact of proposed agency action. Although the requirement of a "finding" did not survive into the language of the Act, it is the obvious forebear of the §102(2) (C) impact statement. It can be traced to the principal member of Senator Jackson's "brain trust," Professor Lynton K. Caldwell of the University of Indiana, whose testimony at the Senate hearing laid the groundwork for the inclusion of the action-forcing provisions.[14] This bill was reported unanimously to the Senate, which passed it by unanimous consent, without debate, on July 10, 1969.[15]

It was clear by this time that the legislation would add to and modify the existing mandates of federal agencies in a very important way. The legislation placed an environmental imperative upon those agencies which had earlier contended that they lacked authority to consider the environmental effects of their actions. Evidence for this position appears near the end of the Senate hearings on S. 1075, where Dr. Caldwell suggested:

> In the licensing procedures of the various agencies such as the Atomic Energy Commission or the Federal Power Commission or the Federal Aviation Agency there should also be, to the extent that there may not now exist fully or adequately, certain requirements with respect to environmental protection.[16]

Prompted by Dr. Caldwell's testimony, Committee Chairman Jackson indicated that instead of trying to revamp the operating statutes of the various agencies within the jurisdiction of a number of different committees, it might be better to "lay down a general requirement that would be applicable to all agencies that have responsibilities that affect the environment rather than trying to go through agency by agency."[17] This was done by blanketing the agencies with a common requirement that they affirmatively implement the national environmental policy and comply with the action-forcing requirements.

---

[13] CONFERENCE REPORT ON S. 1075, H.R. REP. NO. 91–765, 91st Cong., 1st Sess. 3 (Dec., 1969); *also* 115 CONG. REC. at 39701–04 (Dec. 17, 1969), and at H 12633 (daily ed. Dec. 17, 1969) [hereinafter cited as Conference Report].

[14] Senate Hearings, *supra* note 3, at 116.

[15] 115 CONG. REC. 19008–13 (1969).

[16] Senate Hearings, *supra* note 3, at 116.

[17] *Id.* at 116–17.

In the House, hearings had been held on Congressman Dingell's H.R. 6750 for seven days during the summer of 1969.[18] The committee reported out a "clean bill," H.R. 12549, which was practically identical to H.R. 6750. On September 23, H.R. 12549 was passed by the House by a vote of 372 to 15,[19] but not before Congressman Aspinall succeeded in obtaining two important amendments. The first applied the House bill to all environmental impacts, not just to fish and wildlife, which were the exclusive province of Congressman Dingell's subcommittee. This amendment probably was an attempt to maneuver subsequent jurisdiction and oversight into Congressman Aspinall's committee. The second amendment cut directly into the substance of the legislation. It stated that "nothing in this Act shall increase, decrease, or change any responsibility of any Federal official or agency."[20] The amendment would have negated the substantive effect of the Senate-passed legislation, had it not been dropped in conference.

In the meantime, S. 1075 had been sent to the House. Its friends in the House succeeded in having the Speaker refrain from the usual procedure of assigning it immediately to a House committee, which probably would have been Congressman Aspinall's Interior and Insular Affairs Committee, the House's "opposite number" to Senator Jackson's committee. The Senate-passed bill was held by the Speaker until passage of H.R. 12549, whose language was immediately substituted into the Senate bill, which was then returned to the Senate with a request for a Senate–House conference.[21]

However, before the Senate turned its version over to the conference committee, it made several very important additions and changes. In so doing, the Senate faced for the first time several matters of substance regarding the jurisdictional clash which had figured covertly in Senate consideration of the bill all along. Senator Muskie sought and obtained an amendment, the basic thrust of which apparently was to ensure that air and water standards set under legislation which had emerged from his subcommittee would not be affected by the new Act. Specifically, the jurisdictional dispute took the form of language in a new, ambiguously worded

[18] *Environmental Quality, Hearings on H.R. 6750 et al., Before the Subcomm. on Fisheries and Wildlife Conservation of the House Comm. on Merchant Marine and Fisheries,* 91st Cong., 1st Sess. (1969). *See also* Terence Finn's dissertation, *supra* note 6, at 311–75.

[19] 115 CONG. REC. 26568–91 (1969); *also in* 115 CONG. REC. H 8263–86 (daily ed. Sept. 23, 1969).

[20] *Id.* at 26589; daily ed. at H 8284.

[21] 115 CONG. REC. 26568–91 (Sept. 23, 1969) (vote 372-15).

§104 which attempted to bar the federal agencies from concerning themselves with water or air quality under NEPA when, for instance (to take the only example given), state certifications of water quality had already been obtained under water quality legislation.[22] The conference let the ambiguous language of §104 stand without change or further clarifying legislative history. The "explanation" of §104 provided in the legislative history, it must be said, shed more darkness than light. The legacy of the Muskie–Jackson compromise[23] continues to plague judicial interpretations of the Act in the form of two interrelated issues: (1) Can an agency not specifically empowered to set air and water standards rely on its NEPA authority to impose a higher standard than that set under air and water legislation, as a condition of that agency's grant of a federal license, permit, or other enablement?[24] (See the discussion of §511(C) (2) of the Federal Water Pollution Control Amendments of 1972 in chapter IV of this book at page 108.) (2) Must environmental protection agencies prepare impact statements and otherwise comply with NEPA? (See chapter IV, page 106 for a discussion of agencies to which NEPA applies.)

Further, under the agreement between Senators Jackson and Muskie, but with no apparent reason for a "compromise," the requirement for a "finding" was changed to a requirement for a "detailed statement" on major federal action, including a new provision for the description of alternatives to proposed action.[25] Moreover, the responsible federal official was required to consult with and obtain the comments of affected sister agencies before preparing the statement, which together with the comments and views of appropriate federal, state, and local agencies, was to be made available to the President, the CEQ, and the public, and was to accompany the original proposed action through existing agency review processes. Thus §102(2) (C) had reached its final form.

These changes and additions, together with a few less important

---

[22] *See* 115 CONG. REC. 29046–63 (Senate consideration of Water Quality Improvement Act of 1970); *id.* at 29066–89 (Senate discussion of position in conference committee on NEPA); *id.* at 40923–28 (House debate on conference report on NEPA); 116 CONG. REC. 8984 (Senate debate on conference report on Water Quality Improvement Act).

[23] *See* Comment, *Landmark Decision on the National Environmental Policy Act: Calvert Cliffs' Coordinating Committee, Inc. v. Atomic Energy Comm'n,* 1 ELR 10125, especially the subsection entitled "Novel Statutory Interpretation of Section 104: The Muskie-Jackson Compromise," at 10127.

[24] *See* Comment, *id.*

[25] 115 CONG. REC. S. 12117–18, 12146–47 (daily ed. Oct. 8, 1969).

ones, went to the Senate–House conference as part of S. 1075. The Senate conferees had been instructed to insist upon the recent modifications to their bill brought about by the Muskie–Jackson compromise. In conference, additional important changes occurred before NEPA took final form.[26] The most important modification was the inclusion in § 102 of the requirement that the agencies had to comply with the action-forcing provisions "to the fullest extent possible." The phrase was viewed as a compromise between Senator Jackson's desire to force action from the agencies and Congressman Aspinall's insistence that the Act should leave agency mandates entirely as they were before NEPA was enacted. The conference report stated that the phrase somehow "qualified" NEPA duties; however, an unusual second document, a statement by the bill's managers on the part of the House, gave a reading to the phrase which essentially adopted Senator Jackson's view. The statement said that the § 102 duties were to be complied with fully by every agency unless existing statutory law expressly prohibited full compliance or made it impossible.[27] As chapter III explains, the strict interpretation of compliance "to the fullest extent possible" has prevailed in the courts.

The conferees had reached agreement by December 17, 1969.[28] Both houses agreed to the conference report a few days later.[29] A week later, the President signed NEPA into law. The quick final vote on the bill, which had to be accepted or rejected *in toto* because congressional procedures forbade amendment to a conference-approved bill, has left a lingering impression that Congress had not fully considered the action-forcing procedures. Yet one might ask whether fuller consideration would have resulted in any better understanding of the potential effectiveness of the innovative action-forcing devices. The gist of the § 102(2) (C) requirement for a detailed statement was contained in the requirement for a "finding," which was introduced in NEPA's legislative history quite early as

---

[26] *See* Conference Report, *supra* note 13. *See also* Andrews' dissertation, *supra* note 6, at 97–99.

[27] 115 Cong. Rec. 39703 (Dec. 17, 1969) (conference report and appended statement); *id.* at 40418 (statement by Senator Jackson). Also in the daily edition, 115 Cong. Rec. H. 12633, with statement by House managers at 12635 (Congressman Aspinall did not sign).

[28] *See* Conference Report, *supra* note 13.

[29] 115 Cong. Rec. 40415–27 (Dec. 20, 1969) (Senate); *id.* at 40923–28 (Dec. 23, 1969) (House). These debates are in the daily editions at 115 Cong. Rec. S. 17450–69 (Dec. 20, 1969) (Senate); *id.* at H. 13091–96 (Dec. 23, 1969) (House).

the brainchild of Lynton Caldwell. Certainly neither Professor Caldwell nor the senators and congressmen and their staffs involved thought that the "findings" requirement would develop as §102(2)(C) has, although there was time to speculate about or even provide for implementation of the "findings" requirement. Moreover, the requirement of a "finding" should trigger analysis of the judicial role more readily than that for a "detailed statement," as has been noticed by courts which have subsequently interpreted NEPA.[30]

The analysis provided by the Senate report of the phrase "major federal action significantly affecting the quality of the human environment" did state that the section applied to "major actions, such as project proposals, proposals for new legislation, regulations, policy statements, or expansion or revision of ongoing programs."[31] The specific scope of the section's five subsections, however, went undiscussed. What is an "irreversible commitment" of resources? What is a viable "alternative"? What was meant by "the relationship between short-term uses and long-term productivity"? Judicial interpretation has had to rely more often than usual on the spirit of the Act or on hypothecations of congressional intent in order to flesh out the skeletal statutory provisions. Admittedly, the courts and agencies have had the benefit of CEQ Guidelines,[32] but these have been weighed by the courts and not always followed.[33]

Some direct light on the scope of §102(2) (C) is shed by the examples of degradation in NEPA's legislative history. These examples, coupled with the clear intent of Congress to lay a substantial share of the blame at the feet of ineffectual federal environmental efforts, argue implicitly for the widest possible application for the action-forcing clause. The Senate report, for example, details the environmental consequences of a wide variety of specific federal activities.[34]

Support for vigorous implementation of §102(2) (C) and for a wide reading of its provisions is contained in the introductory language of §102 requiring compliance "to the fullest extent possible."

[30] *E.g.*, City of New York v. United States, 344 F. Supp. 929, 936, 2 ELR 20688, 20690 (E.D.N.Y. June 7, 1972) (footnote 16).

[31] Senate Report, *supra* note 2, at 20.

[32] *See* text in Appendix C.

[33] *See, e.g.*, Greene County Planning Board v. Federal Power Comm'n; Calvert Cliffs' Coordinating Comm. v. Atomic Energy Comm'n; Anaconda Co. v. Ruckelshaus, all in Appendix B.

[34] Senate Report, *supra* note 2, at 4, 8.

Further support may be inferred from the requirement that *all* federal agencies must comply. These phrases have been the workhorses of NEPA interpretation over the past three years.

As has been mentioned, Congress had little to say about how the action-forcing requirements should be enforced. A colloquy between Senator Jackson and Dr. Caldwell late in the hearings suggests that the Office of Management and Budget (OMB) was the body that would primarily control the impact statement procedure, although judicial review was never ruled out. The senator indicated that what was then the Bureau of the Budget should "exercise prudence and discretion in requiring that environmental policies and standards be adhered to in connection with the responsibilities of the Federal Establishment."[35] As Dr. Caldwell said, "the Bureau of the Budget should be authorized and directed to particularly scrutinize administrative action and planning with respect to the impact of legislative proposals and particular public works projects on the environment."[36] This may explain why §102(2) (C) speaks specifically of "legislative proposals and recommendations," since these normally have to be cleared with OMB before submission to Congress.[37]

Be that as it may, NEPA as finally drafted makes no reference to OMB, nor has the office seemed particularly eager to take on the unfamiliar job of environmental clearance of legislative proposals, to say nothing of other actions. The General Accounting Office (GAO) reports:

> Only a limited number of the statements had been prepared on proposed legislation and OMB was not requiring the Federal agencies to furnish the statements as a prerequisite for legislative clearance, except for water resources projects.[38]

Although according to the GAO report OMB has not pressed the agencies to file statements, the OMB has ample authority to do so, and has traditionally exercised close review of agencies' legislative

[35] Senate Hearings, *supra* note 3, at 117.

[36] *Id.* at 116.

[37] *See* Office of Management and Budget, Legislative Coordination and Clearance, OMB Circular A–19 (Revised) (July 31, 1972).

[38] Comptroller General of the United States, REPORT TO THE SUBCOMM. ON FISHERIES AND WILDLIFE CONSERVATION, HOUSE COMM. ON MERCHANT MARINE AND FISHERIES, IMPROVEMENTS NEEDED IN FEDERAL EFFORTS TO IMPLEMENT THE NATIONAL ENVIRONMENTAL POLICY ACT OF 1969 51 (May 18, 1972).

proposals.[39] The guidelines of the Council on Environmental Quality state that OMB will supplement the general CEQ Guidelines with specific instructions to agencies about how to fit the NEPA §102 process into the OMB-controlled legislative clearance process.[40] However, in its Circular No. A–19 (Revised), dated July 31, 1972, OMB appears to avoid the issue by referring the agency back to the CEQ Guidelines. Circular A–19 states that the impact statement requirement "is dealt with generally in guidelines issued by the Council on Environmental Quality." The only additional requirement imposed by OMB is that "information copies of required §102 statements should be submitted to OMB if available at the time clearance is requested."[41] This requirement coincides with the requirements of a more recent OMB bulletin regarding NEPA's implementation.[42]

The CEQ has not filled in the vacuum left by OMB's lax enforcement of the 102 process. An executive order issued March 15, 1970 made the newly created Council on Environmental Quality responsible for issuing:

> Guidelines to Federal agencies for the preparation of detailed statements on proposals for legislation and other Federal actions affecting the environment, as required by section 102(2) (C) of the Act.[43]

Because the CEQ lacks the institutional power of OMB over funding, it is uncertain how CEQ can assure effective agency compliance with Executive Order 11514 and the CEQ Guidelines. It is significant that the council still does not officially approve or disapprove particular agency procedures, statements, or failure to prepare statements, relying instead on informal consultation, although it has played an active role in getting the agencies to

---

[39] *See* Legislative Coordination and Clearance, OMB Circular No. A–19 (Revised) (July 31, 1972); Evaluation, Review, and Coordination of Federal and Federally Assisted Programs and Projects, Circular No. A–95 (Revised) (February 9, 1971) (Transmitted Memoranda Nos. 1 and 2 containing revisions, July 26, 1971, and March 8, 1972, respectively); Memorandum for the Heads of Departments and Agencies, dated October 5, 1971, ELR 46001.

[40] CEQ Guidelines, §3(d). Text in Appendix C of this book.

[41] OMB Circular A–19, §7(d)(2).

[42] Proposed Federal Actions Affecting the Environment, OMB Bulletin 72–6, §3(a), ELR 46135 (Sept. 14, 1971). It should be pointed out that water resources projects are required to have final statements filed on them prior to OMB review.

[43] Protection and Enhancement of Environmental Quality, Exec. Order No. 11514, §3(h), 35 FED. REG. 4247, ELR 45003 (March 5, 1970).

prepare their own procedures for NEPA compliance, has guided them toward a broad reading of the Act in the CEQ Guidelines, and has occasionally criticized particular statements on a nonsystematic, *ex parte* basis. So far as administrative enforcement of NEPA is concerned, therefore, the responsibility for policing the agencies appears to have fallen between two stools. This omission is felt particularly with respect to legislative impact statements. (See chapter IV, pages 125 *ff.*). In these circumstances, responsibility has fallen back on the courts.

The legislative history is virtually silent on the possibility of judicial enforcement of the Act, although as a general principle judicial enforceability is assumed in the absence of a specific congressional directive either way.[44] At one point the Senate report does state that NEPA defined the changes which it was to bring about in terms intended to give guidance to the courts. It states:

> The [Senate Interior] committee believes that it is necessary to move ahead to define the "environmental" desires of the American people in *operational terms* that the President, Government agencies at all levels, *the courts*, private enterprise, and the public can consider and act upon [emphasis added].[45]

A further reference by Senator Jackson to institutional failure to cope with environmental problems drew attention to the interrelated components of government and implicitly to the place of the courts in ensuring that each component functions properly in its role. In the floor debate immediately preceding NEPA's passage by the Senate, Senator Jackson stressed the inadequacy of present policies and institutions in dealing with various forms of environmental degradation and then turned again to the obligations of the government in dealing with them.[46] His phrasing is important in light of the changes which litigation under NEPA has helped to bring about. He said that the enactment of NEPA was an exercise of a "primary function" of government, which was "to improve the institutional policy and the legal framework for dealing with these problems." It is going too far to suggest that the improvements in the "institutional policy" and the "legal" frameworks mentioned by Senator Jackson included a program of litigation for the enforcement of

---

[44] Administrative Procedure Act §10, 5 U.S.C. §§701–706 (1966). *See also* L. JAFFE, JUDICIAL CONTROL OF ADMINISTRATIVE ACTION 336 *ff.* (1965).

[45] Senate Report, *supra* note 2, at 13.

[46] 115 CONG. REC. S. 17452 (daily ed. Dec. 20, 1969).

NEPA. However, in these passages the senator does place much of the blame for environmental degradation on the unresponsiveness of governmental institutions. In seeking a remedy, he says that NEPA was intended to alter both the governmental policy-making framework and the legal framework within which environmental problems are solved. By identifying the importance of institutional procedures and constraints, he directs attention to the courts, which interpret and apply the law and which oversee the allocation of the work of government among its branches.

II

How a NEPA Suit Reaches Court

THREE YEARS after NEPA's passage, after scores of favorable judicial interpretations have been entered on the books, one may easily forget that active judicial implementation was not a foregone conclusion the day the Act became law. Instead of applying the standard of strict procedural compliance to every step in the "102 process" and requiring the impact statement to evince careful agency consideration of environmental impacts and values, the courts could have adopted a hands-off attitude and left to the executive branch the task of putting NEPA to work. Instead of agreeing that "it is hard to imagine a stronger mandate to the courts"[1] than NEPA, more federal judges might have concluded that the Act "would not seem to create any rights or impose any duties of which a court can take cognizance."[2]

The courts' leading role in requiring compliance with NEPA may be traced in large measure to their current willingness to review all agency action more closely than they did only a few years ago. In a parallel development, expanding notions of standing have enabled the courts to entertain "public interest" lawsuits brought by citizens and organizations which allege, not only that their environmental interests are injured when NEPA is improperly implemented, but that they suffer injury to special interests which NEPA created when an agency fails to provide information to the public or to guarantee an opportunity for comment and participation.

Other potential judicial barriers to NEPA litigation pose much less of a problem, although certainly if the courts had chosen to impose high bond requirements in "public interest" NEPA suits, some of

---

[1] Texas Comm. on Natural Resources v. United States, —— F. Supp. ——, ——, 2 ELR 20574, 20575 (W.D. Tex. 1970).

[2] Bucklein v. Volpe, —— F. Supp. ——, ——, 1 ELR 20043, 20044 (N.D. Cal. 1970). Early scholarly opinion was generally optimistic: E. Hanks and J. Hanks, *An Environmental Bill of Rights: The Citizen Suit and the National Environmental Policy Act of 1969,* 24 RUTGERS L. REV. 230 (1970); R. Peterson, *An Analysis of Title 1 of the National Environmental Policy Act of 1969,* 1 ELR 50035 (January 1971).

the plaintiffs would not have been able to provide security for preliminary relief. Finally, the timing of a NEPA challenge involves certain problems because, again, "public interest" litigants cannot always easily ascertain when the federal government is acting or which part of it is acting, in time to organize for litigation. The traditional equitable doctrines of laches, estoppel, and exhaustion of administrative remedies have consequently been altered somewhat for service in these special circumstances.

JUDICIAL REVIEW

The courts have been vigorous in reviewing agency compliance with NEPA. They have enforced strict standards of procedural compliance, and in instances where Congress failed to specify how the Act should be implemented, they have imposed judge-made requirements which give it a wider scope. As a result, the courts are thought of as the principal enforcers of NEPA. Through its procedures, they have expanded and enhanced their roles as overseers of the administrative process for a very large category of agency decision making.

Neither the Act nor its legislative history mentions judicial review. The courts themselves were left to decide whether the Act implicitly conferred jurisdiction to review, or whether review was only available with the assistance of the Administrative Procedure Act or other statutes. Most courts chose the conventional alternative and relied additionally upon the APA and other statutes. Two courts, however, found such a compelling case for review that they held that NEPA alone was sufficient to confer jurisdiction. In *Environmental Defense Fund v. Hardin*,[3] the court relied solely upon NEPA in reviewing the Department of Agriculture's Mirex fire ant control program. Defendants argued that the Federal Insecticide, Fungicide and Rodenticide Act (FIFRA)[4] provided exclusive circuit court review procedures, but the court found that independent review under NEPA could go forward even if the government were to withdraw the pesticide's registration pursuant to FIFRA hearings. The court in *Citizens for Clean Air v. Corps of Engineers*,[5] while relying on the APA, nevertheless cited the Mirex case approvingly and said, "the

[3] 325 F. Supp. 1401, 1 ELR 20207 (D.D.C. 1971).

[4] 7 U.S.C. §§135–135k, 61 Stat. 163, 2 ELR 41301. *Now amended and cited as* The Federal Environmental Pesticide Control Act of 1972, 7 U.S.C. §135 *ff.*, Pub. L. No. 92–516, 86 Stat. 973 (Oct. 21, 1972).

[5] 349 F. Supp. 696, 2 ELR 20650 (S.D.N.Y. 1972).

Act's structure and its interpretation by administrative agencies and courts firmly establish that judicial review is crucial to NEPA implementation."[6]

The second case to base jurisdiction exclusively upon NEPA, *SCRAP v. United States*,[7] did so in spite of a strong argument that other legislation forbade the court to tamper with an Interstate Commerce Commission ruling on a temporary freight rate increase pending a full hearing. The three-judge district court said:

> In our view, NEPA implicitly confers authority on the federal courts to enjoin *any* federal action taken in violation of NEPA's procedural requirements, even if jurisdiction to review this action is otherwise lacking.[8]

Citing *Committee for Nuclear Responsibility, Inc. v. Seaborg*,[9] *City of New York v. United States* and *Calvert Cliffs' Coordinating Committee v. Atomic Energy Comm'n* to buttress its position, the court appeared to rely upon the Act's novel imposition of "new and unusual" procedural duties, its direct effect upon each agency's processes, and its unquestioned application to a presidential decision to conduct a nuclear test in *Committee for Nuclear Responsibility.*

The Courts and Active Review in NEPA Cases

The reasons behind the willingness of the courts to review agency decisions under NEPA undoubtedly are numerous and complex. Factors as diffuse as the mood of the country about environmental quality, the rise of the "captive" agency, and the tendency of courts toward strict enforcement of statutory procedural requirements when they cannot reach substantive agency decisions,[10] all certainly have played some role in influencing the courts to accept NEPA jurisdiction. The Act's requirement of compliance "to the fullest extent possible" has made an important difference (see chapter III). But one particular reason that the courts did not adopt a hands-off attitude appears to outweigh the rest. This view suggests that NEPA has entered the lists at just the time the courts are generally tightening their review of agency decision making, and that NEPA's

---

[6] 349 F. Supp. at 703, 2 ELR 20653.

[7] 346 F. Supp. 189, 2 ELR 20486 (D.D.C.), *application for stay pending appeal denied sub nom.* Aberdeen & Rockfish Ry. Co. v. SCRAP, —— U.S. ——, 2 ELR 20491 (1972) (Burger, C. J., Circuit Judge).

[8] 346 F. Supp. at 197, 2 ELR at 20489.

[9] All in Appendix B.

[10] L. JAFFE, JUDICIAL CONTROL OF ADMINISTRATIVE ACTION 566 (abridged ed., 1965).

reform-minded provisions contribute to the courts' efforts in the same way that stricter judicial review does.

NEPA does not specify a role for the federal courts in its implementation, but NEPA's objectives and those of reviewing courts are in many respects similar. NEPA's ultimate purpose may be environmental protection, but the measures which it prescribes to reach this goal are similar to those provided by the APA and judge-made rules on review. Both NEPA and the evolving standards of judicial review call for the establishment by the agencies of procedures for principled decision making, for the articulation in the record of the reasoning which supports the decision taken, for the elaboration of the risks which proposed action entails, for discussion and consideration of alternatives as a test of the soundness of decisions taken, for a wider view of the public interest under long-standing agency missions, and for increased public participation. Thus it may not be wide of the mark to suggest that the courts have found a surrogate review mechanism in the §102 process.

To understand more fully why the courts did not adopt a hands-off approach to NEPA,[11] and instead turned it to their broader purposes, we must examine the recent history of relationships between reviewing courts and administrative decision makers in the environmental area. In recent years Congress has required a number of development-oriented agencies to take environmental factors into account, so that they do not sacrifice some public values while pursuing others contained in basic agency missions. The requirements include considering alternatives and consulting with sister agencies to obtain their opinion of the likely impact of a proposed action. The courts have read these statutes as broadly as possible to require agency consideration of environmental effects and have tightened the standards of judicial review through the APA and the judge-made standards governing review.

In §10(a) of the Federal Power Act,[12] Congress required the Federal Power Commission to consider "other beneficial public uses, including recreational purposes," in addition to commerce and water power, when it licenses a dam or related project. In *Scenic Hudson I*[13] the Second Circuit read §10(a) broadly and imposed on the

---

[11] *See* note 2 *supra* and text accompanying note 2.

[12] 16 U.S.C. §803(a).

[13] Scenic Hudson Preservation Conference v. Federal Power Comm'n (Scenic Hudson I), 354 F.2d 608, 1 ELR 20292 (2d Cir. 1965), *cert. denied sub nom.* Consolidated Edison Co. of New York v. Scenic Hudson Preservation Conference (Scenic Hudson II), 384 U.S. 941 (1966); 453 F.2d 463, 1 ELR 20496 (2d Cir. 1971), *cert. denied,* 407 U.S. 926, 2 ELR 20436 (1972).

commission an affirmative duty to develop and consider less environmentally damaging alternatives to its proposals. Two years later, in 1967, the Supreme Court approved this reading of the Federal Power Act.[14] Section 10(a) by itself did not explicitly require consideration of alternatives. The *Scenic Hudson I* rationale therefore may have a wider application, because it is based on a broad view of what an agency must reasonably do to justify its decision when competing public interests are at stake. *Environmental Defense Fund v. Environmental Protection Agency*,[15] discussed later, seems in accord. The duty to discuss alternatives may perhaps be viewed as an expansive and universally applicable interpretation of the APA's requirement for a "statement of basis and purpose" to support agency action. As a judge-made requirement that agencies justify their choices as the best among a range of choices, the interpretation illustrates how the courts are reaching for a means to improve administrative decision making.

In §4(f) of the Department of Transportation Act,[16] Congress specifically required the department to consider alternatives to proposed transportation projects if the projects encroach on public parks, wildlife refuges, or historic sites. In an interpretation of this section, the Supreme Court held that the secretary of transportation must specifically find that no feasible and prudent alternative to the project exists. More generally, the Court defined the kind of record which must support a decision when an agency is proceeding informally and is not bound by the APA's formal requirements regarding quasi-judicial and quasi-legislative agency decision making. In this instance interpretation of an environmental statute gave the Supreme Court the opportunity to establish principles of judicial review regarding a very large and important category of federal action.[17]

In other cases, the initiative appears to lie decisively with the courts rather than with Congress. In *Kennecott Copper Corp. v. Environmental Protection Agency*[18] the District of Columbia Circuit held that it was unable to decide whether the EPA Administrator had complied with the Clean Air Act[19] in setting the secondary ambient air quality standards for sulfur oxides, because EPA had

---

[14] Udall v. Federal Power Comm'n, 387 U.S. 428, 1 ELR 20117 (1967).

[15] 465 F.2d 528, 2 ELR 20228 (D.C. Cir. 1972).

[16] 49 U.S.C. §1653(f). *See also* 49 U.S.C. §1651(b)(2).

[17] Citizens to Preserve Overton Park v. Volpe, 401 U.S. 402, 1 ELR 20110 (1972). (Full citation to other decisions in the litigation in Appendix B.)

[18] 462 F.2d 846, 2 ELR 20116 (D.C. Cir. 1972).

[19] 42 U.S.C. §1857 *et seq., as amended.*

not adequately explained the basis of the standard set. The court remanded for a fuller articulation of the basis for decision. The APA simply required a "concise general statement" of the basis and purpose of the standards. In an era of greater leniency in applying the standards of judicial review, the statement submitted probably would have passed muster. In remanding, the court referred repeatedly to the difficult task facing EPA and the unwillingness of the court to interfere either by reviewing the substantive decision or by delaying implementation of the new Clean Air Act. But it did remand.

Some of the most important recent statements regarding judicial review have appeared in cases applying the Federal Insecticide, Fungicide and Rodenticide Act (FIFRA).[20] Under FIFRA the administrator of EPA may launch lengthy hearings to determine whether or not to cancel the registration of a pesticide. Pending completion of the cancellation proceedings, he may immediately suspend the registration if an imminent hazard to public health exists. The cases interpret the record which the administrator must establish before taking either action. In *Environmental Defense Fund v. Ruckelshaus*[21] the District of Columbia Circuit remanded a decision not to suspend the registration of DDT for further proceedings in which the administrator was to correct his failure to explain his decision and to articulate the criteria upon which it was based. The court concluded by remarking upon the change which was taking place in the relationship between reviewing courts and administrative decision makers.

> We stand on the threshold of a new era in the history of the long and fruitful collaboration of administrative agencies and reviewing courts. For many years, courts have treated administrative policy decisions with great deference, confining judicial attention primarily to matters of procedure. On matters of substance, the courts regularly upheld agency action, with a nod in the direction of the "substantial evidence" test, and a bow to the mysteries of administrative expertise. Courts occasionally asserted, but less often exercised, the power to set aside agency action on the ground that an impermissible factor had entered into the decision, or a crucial factor had not been considered. Gradually, however, that power has come into more frequent use, and with it, the requirement that administrators articulate the factors on which they base their decisions. . . . Courts should require administrative officers to articulate the standards and principles that govern their discretionary decisions in as much detail as possible. Rules and regulations should be freely formulated by administrators, and revised when necessary. Discretionary

[20] *Supra* note 4.
[21] 439 F.2d 584, 1 ELR 20059 (D.C. Cir. 1971).

decisions should more often be supported with findings of fact and reasoned opinions [court's footnotes omitted].[22]

In *Environmental Defense Fund v. Environmental Protection Agency*[23] the administrator again refused to suspend registration of two pesticides pending review of their status, and the same circuit court again remanded for a fuller explanation of the reasons for the refusal. The court summed up this and its additional grounds for remand by saying, "We cannot discharge our role adequately unless we hold EPA to a higher standard of articulation."[24] As part of its finding that the administrator had inadequately discussed the benefits of *not* suspending the registrations, the court observed that a discussion of benefits requires at least some consideration of possible alternative ways of accomplishing the benefits which the pesticides bring. Thus the court's interpretation of FIFRA is not unlike the Second Circuit's interpretation of the Federal Power Act. Both emphasize the importance of a discussion of alternatives in an agency's record of decision.

The cases just discussed offer a few instances of the changing judicial attitude regarding review. Although they were chosen from the field of environmental regulation, similar cases have been decided holding agencies in other areas accountable to higher standards.[25] Although a general trend may exist reflecting wide dissatisfaction with numerous aspects of administrative performance, there may be a special reason why the courts have so closely reviewed agency decision making in the environmental area. Agency decisions in this area more frequently involve vital personal interests such as life, health, and safety which, if offered inadequate protection or allowed to be abused, could conceivably have far more injurious consequences to the public than agency abuse of traditional functions of economic regulation. There is a great deal of difference between regulating the securities market, and establishing levels at which air pollution poses an imminent danger to health; between awarding broadcast licenses, and determining the hazardousness of a pesticide; between fixing maximum rates that can be charged for livestock, and setting human health tolerances for asbestos, beryllium, or mercury. The courts may have concluded that the principles of judicial review

[22] 439 F.2d at 597, 1 ELR at 20064.
[23] *Supra* note 15.
[24] *Supra* note 15, 465 F.2d at 541, 2 ELR at 20235.
[25] *See, e.g.,* cases cited in court's note 21, Calvert Cliffs' Coordinating Committee v. Atomic Energy Comm'n, 449 F.2d 1109, 1 ELR 20346 (D.C. Cir. 1971), *cert. denied,* 404 U.S. 942 (1972).

forged in the heyday of economic regulation are not adequate for today's agency decisions which vitally affect health and other personal interests. Such interests have always had a special claim to judicial protection, especially when balanced against economic interests.[26]

Further, the courts may also be searching for the special interest which is at stake in environmental controversies where life and limb are *not* threatened, but less tangible aesthetic and psychological experiences are nevertheless in need of protection. The view that a separate standard of review should prevail in such cases has been suggested at least once,[27] and at least once rejected.[28] Certainly the draftsmen of NEPA were acutely sensitive to such values, and undoubtedly the courts in reviewing NEPA cases have been affected by this concern.

With this background one may perhaps see more clearly why NEPA has been generally accorded a sympathetic reception by the courts. For several years the courts have been attempting to evolve better standards of review from existing statutory and case law. They have asked that agencies take a wider array of public interests into account in carrying out their statutory missions, that agencies consider alternatives to their proposals, that they articulate the grounds for their decisions, that they spell out procedures for principled decision making, and that they provide for public participation of various kinds. Congress' enactment of NEPA reinforced this trend by further defining the record that must be made in certain agency decision making, especially discussion of the five specific points in § 102(2) (C), by requiring the study of alternatives, by detailing procedures for obtaining advance criticism and comments from state and local governments and the public, by establishing routine, fair procedures applicable across the board to each agency, and by requiring more of a record in informal agency proceedings than the *Overton Park* minimum requires.[29]

Thus judicial review and the NEPA process share many of the same goals. By policing NEPA, the courts are able to achieve partial review of the quality of agency decision making by enforcing a statute which Congress enacted to bring about reforms in the administrative process. At the same time, they are protecting public

---

[26] Environmental Defense Fund v. Ruckelshaus, *supra* note 21, 439 F.2d at 598, 1 ELR 20065.

[27] D. Sive, *Some Thoughts of an Environmental Lawyer in the Wilderness of Administrative Law*, 70 Col. L. Rev. 612, 629 (1970).

[28] Scenic Hudson Preservation Conference v. Federal Power Comm'n (Scenic Hudson II), *supra* note 13.

[29] Citizens to Preserve Overton Park v. Volpe, *supra* note 17.

interests in life, health, and the environment from administrative arbitrariness.[30] Perhaps more importantly, they are requiring the administrative process itself to confine and control its exercise of general discretion.[31] These have been and continue to be the fundamental tasks of reviewing courts.

So far we have been concerned with establishing that the courts are more sympathetic toward NEPA than toward other legislation, because NEPA's purposes and the purposes of judicial review are quite similar. What the courts may be unwilling or unable to do through the APA and general principles of review, they may in fact be willing or able to do through a strict application of NEPA. Now we will look briefly at the related issue of how far the courts will actually interject themselves into the statutory process to ensure full compliance with the Act.

How Review of Key NEPA Requirements Has Been Conducted

The cases at first glance appear to impose the conventional standards of full review to ensure procedural compliance and of limited review to curtail abuse of discretionary decision-making powers. Within these wide boundaries, however, the courts have become deeply involved in NEPA's implementation. While avoiding "unreasonable extremes," the courts have held the agencies to each detailed procedural step mandated by the Act, have expanded the range of judicially enforceable NEPA duties, and have undertaken unusually close scrutiny of agency compliance. Furthermore, different kinds of NEPA duties have prompted the courts to apply "uniform" principles of review in different ways, depending on the need for oversight from the judicial branch.

The court in the Trinity River–Wallisville Dam case stated that "the role of district courts in reviewing decisions of federal agencies under NEPA has not yet been clearly enunciated, and it appears that differing standards have been used."[32] We agree with this view, but

---

[30] *Supra* note 26.

[31] Environmental Defense Fund v. Ruckelshaus, *supra* note 21, 439 F.2d at 598, 1 ELR at 20064–65.

[32] Sierra Club v. Froehlke, Appendix B. (See the discussion above the court's footnotes 167–75.) The court goes on to attempt to state what it believes should be the uniform standard governing review in NEPA cases. Adopting the Supreme Court's test in Overton Park, *supra* note 17, and perhaps inaccurately stating the circuit court view in Save Our Ten Acres v. Kreger, —— F.2d ——, 3 ELR 20041 (5th Cir. 1973), the court holds that district courts must conduct a "substantial inquiry" in "agency decisions" under NEPA. However, the court makes no distinction, *e.g.*, between the agency decision whether or not to prepare a statement and the decision whether or not to proceed with the challenged project.

would add that "differing standards" have resulted more from the individualized character of separate NEPA requirements than from varying judicial views about which single standard ought to apply across the board in all NEPA challenges. In this section we will briefly survey the standards which have been applied to review of agency compliance with several different key NEPA requirements, saving detailed coverage until later chapters where we will examine more closely judicial interpretation of particular statutory commands.

The important developments in judicial review under NEPA have been focused in four areas. First, the courts have given special attention to preliminary agency determinations on whether or not to prepare impact statements. Second, they have subjected the adequacy of impact statements to close review. Third, the courts have held that agencies must be able to show on review that they have actually considered the environmental information developed in the 102 process. Fourth, some courts have found that NEPA creates judicially reviewable substantive duties to which agency decisions must conform.

The courts have carefully reviewed preliminary agency determinations on whether §102(2) (C) applies at all to the agency action at issue. An agency may attempt to avoid preparing an impact statement by finding that the action is not "federal," is not "major," does not "significantly affect the environment" or even is not yet an "action." To satisfy their own interpretations of *Overton Park*,[33] some agencies issue written "negative declarations" explaining why they think they do not have to comply.[34] Judicial review of threshold determinations is so crucial to NEPA's implementation that we have included an extended discussion of it in chapter IV (see pages 96 *ff.*). That discussion shows that the cases tend to commit to the agencies the preliminary decision on whether NEPA applies to them. In so doing, the majority of courts pay lip service to the arbitrary and capricious standard but find in many instances that abuse has occurred, so that the frequency of the finding belies the standard applied. A few courts are more forthright. They openly apply a stricter standard that appears to amount to *de novo* review.

Review of the adequacy of statements under the five subsections of §102(2) (C) is subject to a different set of considerations (see chapter VI, pages 200 *ff.*). Once again the courts apply the tradi-

---

[33] *Supra* note 17. *See also* the Second Circuit's two opinions in Hanly v. Kleindienst, fully cited in Appendix B.

[34] Council on Environmental Quality, Environmental Quality, Third Annual Report 232 (August 1972).

tional word formulas in testing for abuse of discretion, but they are strongly influenced by the Act's policy of full disclosure. Review here is conditioned by the courts' awareness that if the level of thoroughness of statement preparation is left to the agencies to determine, cursory or self-serving statements may be submitted which do not enable federal decision makers to "consider" environmental values as required by the policy set out in the Act. Hence the courts have found that a very large number of questions of law arise in the application of the five detailed requirements of §102(2) (C). They appear also to review *de novo* the adequacy of the scientific and technical information presented. The courts draw aid whenever they can from agency guidelines which define the scope of the statement, because the application of an agency's guidelines to its own activities is a question of law.

The use to which information on environmental impacts must be put has been discussed in *Calvert Cliffs', Natural Resources Defense Council v. Morton,*[35] and their progeny. These cases hold that NEPA binds agencies actually to "consider" the information in the statement, taking a "hard look" at environmental factors, which must then be carefully balanced off against other relevant non-environmental factors. A few cases indicate that the courts' role is at an end once a properly prepared, detailed statement has been submitted, but these cases now lack the edge, both in logic and in weight of authority. The courts will review both the formality, and the reality, of NEPA compliance.

Calvert Cliffs' itself said that "by compelling a formal detailed statement and description of alternatives, NEPA provides evidence that the mandated decision making process has in fact taken place."[36] Subsequent circuit and district opinions support the view that the impact statement is part of the reviewable agency record of decision. These cases are discussed in chapter VII (see pages 252 *ff.*), and in chapter VI, where we consider the adequacy and contents of impact statements. The full use of the statement in final agency decision making completes the §102(2) (C) process. Without outside scrutiny by the court, in which the court examines the precise manner in which the agency has related its environmental assessment

---

[35] Calvert Cliffs' Coordinating Comm. v. Atomic Energy Comm'n, Appendix B; Natural Resources Defense Council v. Morton, 337 F. Supp. 165, 2 ELR 20028 (D.D.C.), 337 F. Supp. 167, 2 ELR 20089 (D.D.C. 1971), *motion for summary reversal denied,* 458 F.2d 827, 2 ELR 20029 (D.C. Cir.), *dismissed as moot,* 337 F. Supp. 170, 2 ELR 20071 (D.D.C. 1972).

[36] 449 F.2d at 1114, 1 ELR at 20348.

to the action proposed, the process would lack an all-important, final check on its efficacy.

Courts have reviewed agency consideration of the information developed in the 102 process to ensure optimum use of that information in carrying out the national environmental policy. Whether that policy also imposes court-enforceable duties requiring substantively correct decisions is a closely related and rapidly developing issue. The problem is whether NEPA's substantive requirements for environmental protection, which are set out in §§101 and 102(1), create duties that are detailed and explicit enough to be enforced by the courts. The Eighth and Fourth Circuits have now held that they are, as has the district court in the important Trinity River–Wallisville Dam case,[37] while several other courts appear implicitly to have reached essentially the same conclusion (see chapter VII, pages 258 *ff.*).

STANDING

The courts have been active in reviewing agency compliance with NEPA, in large part because private citizens interested in NEPA's enforcement have been granted standing to sue. As "proper parties to request adjudication of a particular issue,"[38] citizens' groups have provided the courts with numerous opportunities to rule on NEPA's implementation. Few challenges to citizens' standing have succeeded, even after the Supreme Court's recent decision in *Sierra Club v. Morton*,[39] in which the definition of the "injury" plaintiffs must allege was somewhat restricted.

Standing under NEPA is the same as standing under any other legislation; however, two special aspects of standing under NEPA deserve special mention. First, NEPA's enactment arguably creates new legal interests which enlarge the category of injuries which the public may sustain, so that harm to the public's right to know, to participate, and to have the interests of future generations protected may constitute injury in fact. Second, NEPA's broad scope has stimulated industry to bring suits under the Act which delay the implementation of the Environmental Protection Agency's air and pesticide programs. These suits raise the question whether the interests NEPA

[37] Sierra Club v. Froehlke, Appendix B.

[38] Flast v. Cohen, 392 U.S. 83, 100 (1968); Sierra Club v. Morton, Appendix B, footnote 3 of opinion.

[39] Appendix B.

was designed to protect are broad enough to encompass alleged corporate interests in the protection of the environment.

The doctrine of standing has undergone rapid expansion in the past seven years. In the wake of *Scenic Hudson I*,[40] plaintiffs alleging a variety of aesthetic, conservational, and recreational interests have convinced the courts of their right to bring suit. Three recent Supreme Court decisions have confirmed that right, while also attempting to provide a usable formulary test for lower courts to apply and while also attempting to refine the allegations of "injury in fact" necessary to support an action. A brief review of these developments is in order.

The notion of standing derives from the constitutional requirement that there be a "case or controversy" before the court, that is, that "the dispute sought to be adjudicated will be presented in an adversary context and in a form historically viewed as capable of judicial resolution."[41] Where the dispute is not between private parties, but involves the legality of administrative action, the Administrative Procedure Act indicates that the plaintiff must be "aggrieved by agency action within the meaning of a relevant statute" for suit to lie.[42] The three recent Supreme Court decisions attempted to enunciate formulas for determining whether a particular litigant has the requisite standing to seek review of agency action.[43]

The first rule is that the person bringing the suit must allege that he has suffered or will suffer "injury in fact" from the challenged action. "Injury in fact" is a legal term defined by the courts; it encompasses not only pecuniary loss, but also injury to "aesthetic, conservational, or recreational interests."[44] However, both the interest and the injury must be claimed, since the Court in *Sierra Club v. Morton* held that a "mere 'interest in a problem,' no matter how long-standing the interest and no matter how qualified the organization is in evaluating the problem,"[45] does not ensure that the organization is "adversely affected or aggrieved." There the Sierra

---

[40] Appendix B (2d Cir. 1967).

[41] Flast v. Cohen *supra* note 38, 392 U.S. at 204. *See also* Baker v. Carr, 369 U.S. 186, 204 (1962).

[42] Administrative Procedure Act, §10, 5 U.S.C. §702.

[43] Ass'n of Data Processing Organizations, Inc. v. Camp, 397 U.S. 150 (1970); Barlow v. Collins, 397 U.S. 157 (1970); Sierra Club v. Morton, Appendix B (1972).

[44] Ass'n of Data Processing Organizations, Inc. v. Camp, *supra* note 43, 397 U.S. at 154.

[45] Sierra Club v. Morton, Appendix B, 405 U.S. at 739, 2 ELR at 20195.

Club was faulted for failure "to allege that it or its members would be affected in any of their activities or pastimes by the . . . [proposed] development."[46] But the Court did seem to indicate that injury to even one member would have been enough to allow the Sierra Club to participate in the litigation. "It is clear that an organization whose members are injured may represent those members in a proceeding for judicial review."[47]

The second rule of standing, adopted in *Camp* and *Barlow* over Justice Brennan's dissent, is that the plaintiff's interest must be "arguably within the zone of interests to be protected by the statute."[48] "Arguably" must mean that the court is not to make a final determination on the merits of the claim, but is only to screen out those persons or interests which the statute clearly did not intend to protect. Plaintiff must pass this threshold test before the merits of his claim can be reached.

Before discussing the role of standing in NEPA cases, two further issues should be clarified. The first is that the *Sierra* rule appears to apply only when the proposed action would harm some citizens but not others. "The impact of the proposed changes in the environment of Mineral King will not fall indiscriminately upon every citizen."[49] The opinion in *Sierra* thus leaves some room for argument that the standing requirement may change in case of "generalized injury" (e.g., the effects on the general public of DDT use), although it would still be wise to claim some sort of potential, highly specific "injury in fact."[50]

A second issue concerns standing as a "representative of the public interest," or "private attorney general," where a party wishes to use injury to the general public to strengthen his case for review. *Sierra Club v. Morton* held that injury to the general public could be argued after standing to seek review had been established, even where such injury could not be used to establish standing initially.[51]

---

[46] 405 U.S. at 735, 2 ELR at 20194.

[47] 405 U.S. at 739, 2 ELR at 20195.

[48] Ass'n of Data Processing Organizations, Inc. v. Camp, *supra* note 43, 397 U.S. at 153.

[49] 405 U.S. at 735, 2 ELR 20194.

[50] Environmental Defense Fund v. Hardin, 428 F.2d 1093, 1096, 1 ELR 20050, 20051 (D.C. Cir. 1970). The possibility that Sierra Club v. Morton affects only those cases where "best plaintiffs" can be found is analyzed in an ENVIRONMENTAL LAW REPORTER Comment, *Sierra Club Decides the Mineral King Case*, 2 ELR 10034 (April 1972).

[51] 405 U.S. at 739, 2 ELR at 20195.

NEPA Cases on Standing

The courts have not encountered any particular difficulty in allowing standing to sue in NEPA cases, before or after *Sierra Club v. Morton.*[52] Several illustrative or problematic cases are discussed here. Because they are still relevant, the cases decided prior to the Supreme Court's opinion are discussed first.

Economic injury to property owners from proposed projects of course is still sufficient to confer standing. Hence the district court in the Gillham Dam case[53] ruled that a farmer who would no longer have rich alluvial soil deposited on his land because flooding would cease if the dam were built had a "personal, direct economic interest" sufficient to confer standing.[54] A similar analysis applied to the plaintiff in *Upper Pecos Ass'n v. Stans.*[55]

A second category of cognizable injury, however, consisting of injury to "aesthetic, conservational, or recreational interests," has given the courts somewhat more trouble, partly because *Sierra Club v. Morton* was pending in the Supreme Court. Plaintiffs in *West Virginia Highlands Conservancy v. Island Creek Coal Co.*[56] successfully claimed injury in this category, the Fourth Circuit distinguishing the holding of the Ninth Circuit in *Sierra Club v. Morton* on the grounds that the:

> Conservancy and its members have a special interest in the Otter Creek area. The area is one of the objects of their principal activities; they use it extensively, and they have studied it in detail. Their interest and the injury they would suffer are much more particularized and specific than those of Sierra Club and its members.[57]

Similarly, in the Cross-Florida Barge Canal case[58] the district court granted standing to individual plaintiffs as "Florida citizens and users of Florida recreational facilities . . . [who] will suffer real

---

[52] Appendix B. *But see* Kings County Economic Community v. Hardin, —— F. Supp. ——, 2 ELR 20151 (E.D. Cal. 1972).

[53] Environmental Defense Fund, Inc. v. Corps of Engineers, 325 F. Supp. 728, 1 ELR 20130 (E.D. Ark. 1970–71),—— F. Supp. ——, 2 ELR 20260 (E.D. Ark.), 342 F. Supp. 1211, 2 ELR 20353 (E.D. Ark.), *aff'd*, 470 F.2d 289, 2 ELR 20740 (8th Cir. 1972).

[54] 325 F. Supp. at 736, 1 ELR at 20133.

[55] 328 F. Supp. 332, 1 ELR 20228, 20229 (D.N.M.), *aff'd*, 452 F.2d 1233, 2 ELR 20085 (10th Cir. 1971), *vacated*, 93 S. Ct. 458 (1972).

[56] 441 F.2d 232, 1 ELR 20160 (4th Cir. 1971).

[57] 441 F.2d at 235, 1 ELR at 20161.

[58] Environmental Defense Fund, Inc. v. Corps of Engineers, 324 F. Supp. 878, 1 ELR 20079 (D.D.C. 1971).

injury if the anticipated environmental damage occurs."[59] The court in *Brooks v. Volpe*[60] also found such standing, since the individuals there "claim to use the campground for personal enjoyment and construction will allegedly interfere with their enjoyment."[61] Perhaps the most liberal case was *Kalur v. Resor*,[62] where plaintiffs had "direct contacts with non-navigable waters; they are conservationists who regularly engage in canoeing and other forms of outdoor water recreational activities," and accordingly had standing to sue to enjoin the *entire* Refuse Act Permit Program.[63] In none of these cases did the "zone of interests" part of the standing test pose any problems, it being generally assumed, in the words of *Kalur*, that "these interests are arguably within the zone of interest to be regulated or protected by the Refuse Act and NEPA."[64]

The result in *Kalur* may be contrasted with the fate of the second individual plaintiff in the Gillham Dam case, who had only "floated, fished, and otherwise derived benefit from the Cossatot in its natural state." The court deferred any determination, but remarked that "the standing of Mr. Remmell in his individual capacity is extremely doubtful under the law. The right of individual citizens to raise general issues affecting the public as a whole is not clear, although there is a tendency to recognize and encourage action by individual private citizens to vindicate certain declared public policies."[65] However, this ruling came before *Sierra Club v. Morton* and rested on the ability of organizational plaintiffs to sue. After that decision, Mr. Remmell may well be a "best plaintiff,"[66] although there is no indication in a recent district court ruling that it would now grant Mr. Remmell standing.[67]

An additional group of cases, also dealing with "aesthetic, recreational, or conservational interests," combines use with neighborhood

---

[59] 324 F. Supp. at 879, 1 ELR at 20079.

[60] 319 F. Supp. 90, 1 ELR 20045 (W.D. Wash. 1970), 329 F. Supp. 118, 1 ELR 20286 (W.D. Wash. 1971), *rev'd,* 450 F.2d 1193, 2 ELR 20139 (9th Cir.), 350 F. Supp. 269, 2 ELR 20704 (W.D. Wash.), 350 F. Supp. 287 (W.D. Wash.).

[61] 329 F. Supp. at 119, 1 ELR at 20286.

[62] 335 F. Supp. 1, 1 ELR 20637 (D.D.C. 1971).

[63] 335 F. Supp. at 6, 2 ELR at 20638.

[64] *Id.*

[65] Environmental Defense Fund v. Corps of Engineers, Appendix B, 325 F. Supp. at 736, 1 ELR at 20133.

[66] *See* Comment, 2 ELR 10134, especially at 10136.

[67] Environmental Defense Fund v. Corps of Engineers (Gillham Dam case), Appendix B, 342 F. Supp. at 1216, 2 ELR at 20355.

proximity. In *Nolop v. Volpe,*[68] which concerned a projected road through the middle of a college campus, the court said:

If the road project is executed the plaintiffs will be subjected to increased noise pollution and dangerous traffic-pedestrian conflicts. The plaintiffs assert their interest under a Congressional mandate requiring federal agencies to file an environmental impact statement.[69]

Thereafter, in *La Raza Unida v. Volpe,*[70] a California district court dealing with a road project held:

Certainly the plaintiffs residing in Hayward, Union City, and Fremont have standing, as they shall be affected directly by any activities adverse to the environment. It is *their* parklands and forests that may be damaged. The State implies that plaintiffs must show that they own property adjacent to the park areas; the court disagrees. A showing that they use the park and live nearby is enough to satisfy the standing requirement for these plaintiffs.[71]

The issue of standing for conservation organizations is also treated in the NEPA cases prior to *Sierra Club v. Morton,* but they are of mixed precedential value in the wake of the Supreme Court's ruling. Thus in the Gillham Dam case, Judge Eisele found the dissent more compelling in the Ninth Circuit's ruling in *Sierra Club v. Morton,* and concluded:

There can be no doubt that corporate [i.e., conservationist] plaintiffs are interested and antagonistic enough to present the issues vigorously and with the "concrete adversariness" referred to in *Baker v. Carr.* . . . NEPA makes clear the recognition by Congress of the important role of such private organizations [through the cooperation mandated in §102(2) (F)].[72]

A similar result was reached in *Izaak Walton League v. Macchia,*[73] where a local chapter of the Izaak Walton League had standing to sue to halt a private dredge-and-fill operation carried out under a Corps permit. The *Macchia* decision is notable for its bold language:

We do not share the fear of some earlier decisions that liberalized concepts of "standing to sue" will flood the courts with litigation. However, if that should be the price for the preservation and protection of our

---

68 333 F. Supp. 1364, 1 ELR 20617 (D.S. Dak. 1971).
69 333 F. Supp. at 1367, 1 ELR at 20618.
70 337 F. Supp. 221, 1 ELR 20642 (N.D. Cal. 1971), —— F. Supp. ——, 2 ELR 20691 (N.D. Cal. 1972).
71 337 F. Supp. at 232, 1 ELR at 20646.
72 325 F. Supp. at 735, 1 ELR at 20133.
73 329 F. Supp. 504, 1 ELR 20300 (D.N.J. 1971).

natural resources and environment against uncoordinated or irresponsible conduct, so be it. But such seems most improbable. Courts can always control the obviously frivolous suitor.[74]

Lower court decisions handed down after the *Sierra Club v. Morton* ruling tend to show that the Supreme Court has not drastically altered the requirements for standing in NEPA cases. In *Conservation Society of Southern Vermont v. Volpe*[75] individual residents and a conservation society with many local residents were permitted to challenge a highway undertaking. Only an association of railway passengers failed to show that their members had an individualized, personal stake in the outcome of the environmental lawsuit. In the Gillham Dam case, Judge Eisele permitted the Environmental Defense Fund to amend its complaint to allege that its members used the free-flowing Cossatot River for canoeing.[76] The circuit court in *Wilderness Society v. Morton*[77] ordered the district court to consider the status of Canadian intervenors in the trans-Alaskan pipeline case and to permit them to amend their complaint if necessary so that they might represent their members' viewpoints in court.[78] Similarly, the court in *National Forest Preservation Group v. Butz*[79] allowed standing for the limited purpose of challenging alleged administrative abuse of discretion.

In *Ward v. Ackroyd*[80] an individual plaintiff joined with the Sierra Club, the Natural Resources Defense Council, and a local highway opposition group in an effort to enjoin construction of part of the interstate system planned to be routed through two Baltimore city parks. The court found that the Sierra Club's allegations that it had 600 members in the Baltimore area and that its local chapter organized hikes and outings in the affected parks were sufficient to satisfy *Sierra Club v. Morton*. The local group likewise was granted standing because of similar activities.

The allegations of the Natural Resources Defense Council, in the court's view of *Sierra Club v. Morton*, however, were held insufficient to support its standing. The council's allegations were like those

[74] 329 F. Supp. at 513, 1 ELR at 20303.

[75] 343 F. Supp. 761, 2 ELR 20270 (D. Vt. 1972).

[76] 342 F. Supp. at 1216, 2 ELR at 20355.

[77] Wilderness Society v. Hickel, 325 F. Supp. 422, 1 ELR 20042 (D.D.C. 1970), *sub nom.* Wilderness Society v. Morton, 463 F.2d 1261, 2 ELR 20250 (D.C. Cir.), —— F. Supp. ——, 2 ELR 20583 (D.D.C.), —— F.2d ——, 3 ELR 20085 (D.C. Cir. 1973).

[78] 463 F.2d at 1262, 2 ELR at 20250 (D.C. Cir. 1972).

[79] 343 F. Supp. 696, 2 ELR 20571 (D. Mont. 1972).

[80] 351 F. Supp. 1002, 2 ELR 20405 (D. Md. 1972).

of the Sierra Club in *Sierra*, in that they lacked the specificity which the Supreme Court held was necessary to underpin organizational standing. The council's failure to spell out its interests, however, was an oversight only in retrospect, because at the time the complaint was filed, *Sierra* had not been decided and NRDC's minimal allegations were supported by impressive circuit and district opinion and by the weight of critical opinion on the direction which the law of standing would take. In line with the cases cited above, the court should have allowed NRDC to spell out its allegations of interest in greater detail, including allegations of its specific interest in the informational and participatory aspects of NEPA's implementation, discussed below.

Two additional cases involving NEPA have allowed standing while probing the margins of the Supreme Court's *Sierra* decision. The first, *Sierra Club v. Mason,*[81] contains an important dictum regarding the amount of detail which plaintiffs' allegations must supply in suggesting the harm which may flow from failure to comply with NEPA. In *Sierra Club v. Mason* the court allowed the Sierra Club to represent the interests of members who made recreational and commercial use of New York Harbor and Long Island Sound, where they challenged a Corps project to dredge the harbor and dump the spoil into the Sound. Defendants claimed that the *Sierra Club v. Morton* rule required a showing that the proposed dredging and dumping would adversely affect the club's members. The court held that the allegations necessary to show possible injury will necessarily vary from case to case. For example, on the facts in *Sierra Club v. Morton*, minimal allegations would have sufficed, the court reasoned, because commercial development of the Mineral King wilderness would obviously and necessarily affect plaintiffs' interests in the valley as wilderness. But the facts in *Mason* required more, the court said, because the potential injury to plaintiff's members from the dredging was less obvious. However, the court had no trouble holding that the club's members who swam and fished in the area could be injured, and warned against requiring too much when a NEPA suit was involved.

> It may well be that a connection sufficient to establish standing can be inferred once the pleadings locate a plaintiff . . . in a given environmental context through his recreational or economic activities or residence and allege that the proposed federal action threatens the quality of that environment. This test may be especially appropriate in instances where

[81] 351 F. Supp. 419, 2 ELR 20694 (D. Conn. 1972).

little is publicly known about the environmental impact of the proposed action and where plaintiff sues to require an impact statement. *To require him to allege more than a generalized, non-frivolous threat to the environment in which he lives, works or plays might, in some cases, require him to state what has not yet been determined but what may be detailed in the environmental impact statement he wants prepared* [emphasis added].[82]

The second case, *SCRAP v. United States*,[83] involved a challenge by Students Challenging Regulatory Agency Procedures to an Interstate Commerce Commission interim order allowing a temporary 2.5 percent rail freight rate surcharge that perpetuated the discriminatory rate structure favoring shipment of virgin ore over recyclable metal scrap. The government vigorously contested the students' standing, arguing that they had no more than a general interest in seeing NEPA enforced and that they were indistinguishable from other citizens equally concerned with environmental protection. The three-judge district court accepted as sufficient plaintiff's allegations that its members use the forests, streams, mountains, and other resources in the Washington, D.C. area and that this use is disturbed by nonuse of recyclable goods. Judge Wright for the panel distinguished *Sierra Club v. Morton* on the ground that the students did not fail to allege actual use of the affected areas, as did the Sierra Club in the Mineral King case.

Judge Wright squarely refused to accept defendant's argument that plaintiffs must somehow distinguish themselves from the "great many others" who share their interests and use natural resources, relying upon the Supreme Court's remark that "the fact that particular environmental interests are shared by the many rather than the few does not make them less deserving of legal protection through the judicial process."[84] Hence the panel in *SCRAP* appears to have adopted the "best plaintiff" interpretation of *Sierra* discussed above.

Other environmental cases not involving NEPA in which the Min-

---

[82] 351 F. Supp. at 424, 2 ELR at 20695. This problem is similar to the general problem of the burden of proof in NEPA cases, a topic we have not dealt with directly in our study. However, the district court in its comprehensive opinion in Sierra Club v. Froehlke (Trinity River–Wallisville Dam), Appendix B (above its footnotes 178–82), provides reasons why the burden of proof in NEPA cases should shift to the agency after a *prima facie* showing of noncompliance. Several cases are cited as supporting this view. *See also* International Harvester v. Ruckelshaus, ——F. Supp. ——, 3 ELR 20133 (D.C. Cir. 1973).

[83] Appendix B.

[84] 346 F. Supp. at 196, 2 ELR at 20488.

eral King decision is applied also broadly construe citizens' group standing and of course are of precedential value in NEPA suits. They lie outside the scope of the discussion here.[85]

Only a few NEPA cases decided since *Sierra Club v. Morton* deny standing. In *Maddox v. Bradley*[86] the court held that plaintiff landowners whose property had been condemned specifically lacked standing under NEPA. They had not alleged individualized harm "because of the lack of any environmental study," and admitted in court that the only harm suffered was in not being allowed to water their livestock on the land which they formerly owned. Although the court said it was applying *Sierra Club v. Morton* to reach this result, it would perhaps be more accurate to say that the court found no valid environmental interest of defendants lying even arguably within the zone of interests protected by NEPA. Defendants were much more like the businessmen denied standing in *Zlotnick*, discussed later.

In *Coalition for the Environment v. Linclay Development Corp.*,[87] three individual citizens and three citizens' groups sought to challenge the construction of Earth City, a 1,700-acre residential, commercial, and light-industrial development in St. Louis, Missouri. The requirements of several federal statutes, including NEPA, allegedly had not been followed. The court applied *Sierra Club v. Morton* and found that not even the three individual plaintiffs who lived near the site of Earth City, and would definitely be affected by the development, had alleged sufficiently individualized harm to support standing. On the facts, *Linclay Development* seems to have been wrongly decided.

A third case, *San Francisco Tomorrow v. Romney*,[88] also wrongly applied the Mineral King ruling. The district court in this case denied standing to conservation organizations and a neighborhood resident seeking an impact statement on an ongoing urban renewal project. The Ninth Circuit reversed with respect to the individual and San Francisco Tomorrow but indicated that the Sierra Club's standing in such a suit was questionable in light of specific provisions of the club's charter.

---

[85] *See* Comment, *Recent Cases on Standing*, 2 ELR 10194 (September 1972).
[86] 345 F. Supp. 1255, 2 ELR 20404 (N.D. Tex. 1972).
[87] 347 F. Supp. 634, 2 ELR 20555 (E.D. Mo. 1972).
[88] 342 F. Supp. 77, 2 ELR 20273 (N.D. Cal. 1972), *aff'd in part, rev'd in part*, —— F.2d ——, 3 ELR 20125 (9th Cir. 1973).

Two Special Aspects of Standing Under NEPA

Congress enacted a "new and unusual statute" in NEPA. The Act creates an important new public right to be informed of the possible environmental consequences of federal activities, to have alternatives considered, and to have the interests of future generations taken into account. By extension through administrative guidelines, NEPA also grants rights to the public to participate in the 102 process. Here we suggest that these new rights have expanded the category of injurable interests.

In another vein, industry has sought to participate in the implementation of NEPA by bringing suits against the Environmental Protection Agency where it has not prepared impact statements on its activities (see pages 113 *ff.*). These suits, which are vulnerable to the charge that they have been brought merely to delay the imposition of EPA standards which are costly for the firms to meet, also raise the issue whether the interests the plaintiffs seek to protect can meet the Supreme Court's requirements for standing.

New Injurable Interests. The courts usually treat NEPA cases as they do other administrative law cases. Plaintiffs have standing when the federal action to be taken might harm their interests. For example, in the Gillham Dam case harm was threatened to a farmer's economic interests and to the recreational, aesthetic, and conservational interests of the Environmental Defense Fund in the Cossatot as a free-flowing river. But several unconventional interests are also at stake in the NEPA process, including the public's right to be informed and its right to participate in the process of identifying environmental risks by commenting on the proposed action.

NEPA contains explicit directions to the federal government to involve the public in federal decisions affecting the environment. Section 101(a) says that "it is the continuing policy of the Federal Government, in cooperation with . . . concerned public and private organizations," to protect environmental quality. Section 102(2)(F) requires the agencies to make available to "institutions and individuals" information which will be useful in maintaining and improving environmental quality.

Executive Order 11514 reinforces these provisions. It directs federal agencies to "develop procedures to ensure the fullest practicable provision of timely public information . . . in order to obtain the views of interested parties."[89] The order indirectly refers to the

---

[89] Exec. Order No. 11514, §2(b), 35 FED. REG. 4247, ELR 45003 (March 5, 1970).

right of the public to "relevant information," which would include impact statements at a minimum. The CEQ Guidelines direct the agencies to submit draft statements to CEQ "together with all comments received thereon . . . from . . . private organizations and individuals."[90] The guidelines further state that public hearings should be provided for "whenever appropriate."[91]

Judicial opinions interpreting the public's role in the NEPA process are analyzed elsewhere (see pages 234 *ff.*). They confirm beyond any doubt that the public has established the right to be informed, to participate in the 102 process, and to have its viewpoints considered when the final agency decision is made. Especially interesting in this regard is the Second Circuit's second decision in *Hanly v. Kleindienst.*[92] The court placed heavy reliance on §102(2)(B)—which requires that agencies develop methods and procedures for taking nonquantifiable factors into account alongside economic factors—in remanding to the General Services Administration for a fresh threshold decision (whether to prepare a statement) that shows how the public received adequate notice and how the public was allowed to participate in the decision.[93] Accordingly, since the refusal to compile a statement or the preparation of an inadequate one denies the public both the basic information that NEPA attempts to provide and the opportunity to influence the final decision through the impact statement, it is at least arguable that "to satisfy the direct injury requirement . . . plaintiffs in NEPA cases need only allege injury to their right to know and participate."[94] Additional support for this view emerges from the Gillham Dam case's discussion of the standing of conservation organizations.[95]

In *Sierra Club v. Mason*[96] the court acknowledged that "in the narrow sense" the harm plaintiff was alleging was to the NEPA process alone, and that even if the original proposal survived the review, "public disclosure" would still have served "an important policy of the Act, and may well affect governmental decision-making in the future."[97] This argument also underlies the claim of the Scientists' Institute for Public Information to standing in their suit to

[90] *See* Appendix C, §3(a)(5).
[91] *Id.*, §10(e).
[92] Hanly v. Mitchell, "Hanly II," Appendix B.
[93] —— F.2d at ——, 2 ELR at 20722–24.
[94] Comment, *supra* note 48, at 2 ELR 10137.
[95] 342 F. Supp. at 1216, 2 ELR at 20355, note 3.
[96] Appendix B.
[97] 351 F. Supp. at 427, 2 ELR at 20697.

require an impact statement covering the Atomic Energy Commission's breeder reactor program as a whole.[98]

The extension of standing to cover "informational" harm or harm to the right to participate by having one's views taken into account, where carried out under a statute that makes public information and participation a prime objective, is in line with the incorporation of "aesthetic, conservational, or recreational" interests into those that could be considered "injured in fact," as is done in *Scenic Hudson I*[99] and *Sierra Club v. Morton.* The example of "aesthetics" indicates that the mere inability to quantify or measure a harm does not deprive the party who suffered it of standing. Moreover, should it be thought desirable to limit possible additional plaintiffs, the injury might be held to be limited to those individuals or groups whose regular practice it is to disseminate or analyze environmental information, and who thus lose the opportunity to participate or to carry out their role in the absence of an impact statement.

Still other harms might flow from agency failure to follow §102(2)(C). In the absence of a statement, there is no way to tell if the alternative chosen was indeed the best, or even if there were other alternatives. And since these alternatives could benefit citizens in other areas, the failure to consider them means that these hypothetical beneficiaries lose the procedural right to have *their* alternative considered as well. For example, in *Delaware v. Penn Central,*[100] where another governmental body happened to be the litigant, the court found that defendant's planned fill operation, carried out under a Corps permit, would probably be injurious in fact, as it would "make the land unusable for the purposes presently planned by the State and County."[101]

Finally, NEPA also speaks emphatically of the interests of future generations. Thus §101(b) (1) speaks of the "responsibilities of each generation as trustee of the environment for succeeding generations," and §102(2)(C)(iv) requires discussion of "the relationship between local short-term uses of man's environment and the maintenance and enhancement of long-term productivity." In this light, the essence of the standing question is to find suitable surrogates who may sue to protect the interests of future generations by

---

[98] Scientists' Institute for Public Information v. Atomic Energy Comm'n, —— F. Supp. ——, 2 ELR 20642 (D.D.C. 1972). *See* ELR Dig. [182], Doc. H, Brief for Appellant (filed May 3, 1972).

[99] Appendix B.

[100] 323 F. Supp. 487, 1 ELR 20105 (D. Del. 1971).

[101] 323 F. Supp. at 492, 1 ELR at 20107.

having the long-term consequences of environmental decisions focused and weighed through the impact statement.

Industry's Standing. Industry has recently taken an interest in suing to compel the federal agencies, especially the Environmental Protection Agency, to prepare impact statements.[102] The decided cases leave the standing of industrial plaintiffs in doubt, however, because the courts are as yet unconvinced that the interests which NEPA was designed to protect even arguably encompass alleged corporate interests in the environment. Underlying the courts' unease is the distinct possibility that industry's newfound interest in NEPA may be traced, not so much to its desire to see the federal government engage in environmentally sound decision making, but to a desire to see NEPA used to delay federal initiatives intended to protect the environment.

The leading case on the issue was decided well before the spate of recent actions aimed at EPA. In *National Helium Corp. v. Morton*[103] the district and circuit courts allowed the National Helium Corporation to sue for an impact statement on the secretary of the interior's action in exercising an option to cancel the company's helium supply contract. Plaintiff successfully blocked cancellation of its contract to extract helium from natural gas on the ground that no NEPA study had been made of the effect which termination would have on the "conservation of natural resources." The helium would be irretrievably lost to the atmosphere.

The district court relied on the theory of private attorneys general, "whereby one who feels aggrieved by administrative actions has standing in the public interest to see that governmental action is done legally and within the guidelines of due process."[104] This case would thus seem to fall within the dictum of *Sierra Club v. Morton.* Economic injury gets plaintiff into court, where he has further standing to argue the public interest in support of his claim. However, the decision of the Tenth Circuit in affirming presents a major problem, since it was there suggested that the remedy for plaintiff's private financial claim would be in the Court of Claims under the Tucker Act,[105] and that "it is their asserted representation of the public interest . . . which in final analysis justifies their seeking judicial

[102] *See, e.g.*, Anaconda Co. v. Ruckelshaus, —— F. Supp. ——, 3 ELR 20024 (D. Colo. 1972); *see also* 4 NATIONAL JOURNAL 1871 (Dec. 9, 1972).

[103] 326 F. Supp. 151, 1 ELR 20157 (D. Kan. 1971), *aff'd*, 455 F.2d 650, 1 ELR 20478 (10th Cir. 1971).

[104] 326 F. Supp. at 157, 1 ELR at 20159.

[105] 28 U.S.C. §1346.

review."[106] It appears that the case might have come out differently after *Sierra* distinguished an assertion of the public interest from the standing necessary to obtain review in the first place. The National Helium Corporation would have had to show that its *private* claim entitled it to judical review, and this is what the Tenth Circuit indicated had not been shown.

The public interest which National Helium alleged that it sought to protect was quite similar in important respects to the private interest advanced. Moreover, no governmental regulatory programs would be held up while the issue of NEPA compliance was resolved. National Helium argued persuasively that the public and the company would both be injured if a full study were not conducted of venting to the atmosphere a valuable natural resource—the helium contained in natural gas. A citizens' group might well bring a suit in similar circumstances alleging the wastage of a less polluting fossil fuel, such as natural gas, which is also vented to the atmosphere at some wells. Such a group might even have brought the *National Helium* suit, objecting to the wastage of helium as an "irreversible and irretrievable commitment of natural resources," in the words of §102 (2) (C) (v). On the facts of *National Helium*, no public interest in pollution control is frustrated by delays from suits brought by a private interest which directly causes the pollution, consumption, or destruction, rather than the preservation, of a natural resource or environmental amenity.

Although public and private interests merge in *National Helium*, the decision contains troubled language regarding plaintiff's *bona fides* as a protector of the environment; both courts remarked that it was indeed "passing strange" to see an oil company so garbed. But the Tenth Circuit refused to entertain the question.

> We are unable to say that the companies are motivated solely by protection of their own pecuniary interest and that the public interest aspect is so infinitesimal that it ought to be disregarded altogether. It is not part of our function to weigh or proportion these conflicting interests. Nor are we called upon to determine whether persons seeking to advance the public interest are indeed conscientious and sincere in their efforts. True, the plaintiffs are not primarily dedicated to ecological improvement, but they are not on this account disqualified from seeking to advance such an interest.[107]

---

[106] 455 F.2d at 654, 1 ELR at 20479.
[107] *Id.*

The *National Helium* case was cited without discussion in *Getty Oil v. Ruckelshaus*,[108] a Getty Oil Company suit challenging implementation of the Clean Air Act for, among other grounds, failure to prepare an impact statement on proposed regulations. The court reasoned that Getty's higher electricity costs provided it with "position and economic stake" under the Clean Air Act and that its allegation of "inefficient utilization of natural resources" might bring its claim within the ambit of NEPA. Of course, since virtually all regulatory actions may in some way affect resource use, this would seem to make the Act highly useful for those who wish to challenge or delay such measures; however, the authority of *Getty* is diluted by the district court's failure to explain the rationale of its citation of *National Helium* and by the Third Circuit's ruling that the district court lacked jurisdiction to hear the case.

Other cases attack the issue more directly. In *Zlotnick v. District of Columbia Redevelopment Land Agency*,[109] Judge Gerhard Gesell of the Federal District Court for the District of Columbia dismissed a suit by downtown property owners to enjoin condemnation of their land for urban renewal, holding that they could not compel the preparation of a more adequate impact statement, because they had only their own financial interests in mind. Their environmental concern was, to say the least, remote and speculative and beyond the zone of interests which NEPA protects. The court said:

> To date the Federal Courts have been extremely liberal in permitting almost anyone to interpose environmental objections [footnote omitted]. It would appear, however, that unless a modicum of common sense is interposed against this trend, the Act may well be misused by private commercial interests to obfuscate and delay essential federal projects to the real detriment of the very environmental and community interests the Act was designed to protect.[110]

And elsewhere:

> It is not enough that an inadequate environmental impact statement may have been filed for these very preliminary stages of a much larger project that will take years to complete. To have standing grounded on a federal statute plaintiffs must assert an interest "arguably within the zone of interests to be protected or regulated by the statute . . . in question." [citations omitted]. Plaintiffs can at best claim only a remote,

[108] 342 F. Supp. 1006, 2 ELR 20393 (D. Del.), *rev'd*, 467 F.2d 349, 2 ELR 20683 (3rd Cir. 1972), *cert. denied*, 41 U.S.L.W. 3389 (Jan. 16, 1973).

[109] —— F. Supp. ——, 2 ELR 20235 (D.D.C. 1972).

[110] —— F. Supp. at ——, 2 ELR at 20236.

insubstantial, highly speculative and ephemeral interest in the environment. They have nothing but their own financial interest to protect as their continuing efforts to achieve financial settlement emphasize. Obviously they would abandon their environment concerns in a moment if the price were right.[111]

A second district court opinion, *Pizitz v. Volpe*,[112] lends strong support to the theory of *Zlotnick*, but in dicta. In *Pizitz*, businessmen unsuccessfully sought to enjoin further construction of certain highway overpasses because the impact statement for the projects was allegedly inadequate. Although in the end the court held that the impact statement fully evaluated environmental effects and that defendants had complied with all of NEPA's procedural requirements, it did characterize the suit as a "spurious" NEPA case.

The Court has characterized plaintiffs' case as "spurious" because, as the pleadings reflect, plaintiffs' primary concern in filing and prosecuting this litigation was to avert a threatened loss of business. The environmental aspects of the case were brought in only to maintain the action. Of course, the National Environmental Policy Act was not designed to prevent loss of profits, but was intended only to promote governmental awareness of environmental problems. Nevertheless, plaintiffs have invoked the provisions of this Act seeking to enjoin future construction on the Huntsville Memorial Parkway, and this Court has considered the case accordingly.[113]

Having thus characterized the lawsuit, the court does not explain why it does not dismiss for lack of standing. One possible reason may be that the court felt that on appeal a finding that the impact statement was adequate would be more defensible than a finding that plaintiffs' case could not even be heard. (In fact, the circuit court did affirm.) Another possible reason may be that the court intended to convey through the cited passage, albeit ambiguously, that plaintiffs lacked standing (theirs was a "spurious" suit), but if they *did* have standing, they could not prevail on the merits. In any event, the case is similar to *National Helium* and *Zlotnick* in the court's distaste for the use to which plaintiffs attempted to put NEPA.

The cases just discussed reflect both differing interpretations of the law and judicial disquiet with the equities of industry-originated NEPA litigation. The problem can be traced to two entirely different

---

[111] *Id.*

[112] —— F. Supp. ——, 2 ELR 20378 (M.D. Ala.), *aff'd*, 467 F.2d 208, 2 ELR 20379 (5th Cir.), *modified on rehearing*, 467 F.2d 208, 2 ELR 20635 (5th Cir. 1972).

[113] ——F. Supp. at ——, 2 ELR at 20378.

types of interests which corporations have alleged. On the one hand, industry has economic interests to protect. These are traditional interests which may easily sustain "injury in fact." On the other hand, industry has begun to challenge whether NEPA has been effectively implemented, which suggests that it has an additional and entirely distinct set of interests. These are interests in sound federal environmental policy, the consideration of environmental side effects by EPA, etc.

Certainly, injury to economic interests is specific enough to meet the Supreme Court's "injury in fact" criterion. Further, interests in NEPA's effective implementation, such as having EPA consider the environmental side effects of its regulatory programs, clearly lie within the zone of interests NEPA was intended to protect. The question is whether economic injury to corporate interests can be used to meet the "injury in fact" criterion, while public interests are used to meet the "zone of interests" criterion. Although the Supreme Court said in *Sierra* that it was not interpreting the zone of interests criterion, its opinion definitely suggests that the interests used to meet the "injury in fact" criterion must be the same interests used to meet the "zone of interests" criterion.

This view of the requirements for standing creates a distinct Scylla and Charybdis situation for industries wishing to sue for review of agency noncompliance with NEPA. In order to have their NEPA claims heard, it would appear that such plaintiffs have either to argue that they themselves suffer specific injury in fact to the interests they share with the general public in having NEPA enforced (so that their interests can meet the zone of interests test), or they have to argue that the economic injury which they might suffer from regulation somehow lies within the zone of interests which NEPA protects. Either alternative will possibly create an insurmountable barrier to industry standing.

If industry chooses to rely on the specific harm or "injury in fact" that it supposedly suffers in not having NEPA implemented, then it has before it the difficult task of establishing that harm to a corporation may somehow be "individualized" and more specific than harm to the public at large. The Sierra Club should have alleged that members used Mineral King Valley for specific recreational and environmental activities; it could not rely upon its "corporate" general interest in the public lands. A similar burden rests on industry. Moreover, various possible plaintiffs exist who would suffer more direct harm from failure to implement NEPA than a regulated industry. If EPA's failure to prepare impact statements on point source emissions of air pollutants has caused harm, surely that harm falls first upon citizens living near the sources and second upon the organizations and their members who have as their *raison d'etre* gathering information on and participating in the setting of emissions standards adequate to protect the public. Such groups are interested, not only in whether EPA complies with NEPA in the manner sought by industry, but also in whether EPA's conduct of the NEPA process includes consideration of alternatives (such as stricter standards or different techniques) that would actually cause economic harm to the industry plaintiff. (This last problem may even be severe enough to raise the question whether industry has the necessary adversarial interest under Article III.)

If industry opts for the other argument and maintains that its economic interests are arguably within the zone of interests which NEPA protects, then it is immediately faced with the fact that NEPA is quintessentially an environmental protection statute, notwithstanding language in §101 about consideration of other national interests. Granted, both economic and technical data may appear in the impact statement where the reasonableness of an alternative to the proposed action is discussed. But the statement primarily supplies one component of the final agency decision-making process—an assessment of the environmental impact of proposed action. Industrial plaintiffs' economic interests may be advanced in the normal course of established agency decision-making processes.

LACHES, EXHAUSTION, RIPENESS, AND BONDS

The doctrines of judicial review and standing of course are not alone in limiting access to the courts for resolving disputes about NEPA's enforcement. Other traditional doctrines have come into play, but in less important or interesting ways. A few of them will be discussed briefly at this point.

It is always possible for plaintiffs not to press perfectly legitimate claims until the challenged action is well on its way to completion, thereby opening themselves to the defense of laches. Laches is an equitable doctrine, however, and the strong public interest component present in NEPA actions may outweigh competing considerations.[114] The Fourth Circuit in the *Arlington Coalition* case[115]

---

[114] *See* Pennsylvania Environmental Council v. Bartlett, 315 F. Supp. 238, 2 ELR 20752 (M.D. Pa. 1970), *aff'd*, 454 F.2d 613, 1 ELR 20622 (3rd Cir. 1971); Harrisburg Coalition Against Ruining the Environment v. Volpe, 330 F. Supp. 918, 1 ELR 20237 (M.D. Pa. 1971). In both these cases the

flatly stated, "we decline to invoke laches . . . because of the public interest status accorded ecology preservation by Congress."[116] This the court did although NEPA had been applicable to the ongoing highway project for over a year before suit was filed, and although courts (including the Fourth Circuit) had dismissed recent, similar actions involving highway hearings on the ground that laches was a bar to action. A Fourth Circuit district court has applied the *Arlington Coalition* rule,[117] while a New York three-judge district court apparently refused to entertain the argument that plaintiffs were too late in filing their action on the ground that primary responsibility for implementing NEPA rests with the agencies, whose failure cannot be excused by plaintiffs' tardiness.[118]

The argument of the New York three-judge court is similar to that advanced by plaintiffs in *Clark v. Volpe*,[119] where the Fifth Circuit specifically rejected the argument that laches can never apply to a suit brought by a private citizen asserting a public right.[120] The court went on to hold that the doctrine did bar the suit, on facts similar enough to those in the *Arlington Coalition* case to suggest that the Fifth and Fourth Circuits may have made inconsistent applications of the doctrine. The Fifth Circuit tried to distinguish cases where action (in this instance, construction) was substantially completed from those where action was still in the planning stages. The distinction is inadequate, however, at least as far as NEPA is concerned, as *Arlington Coalition* made plain by focusing upon the construction *remaining* to be completed and the shaping of further incremental federal action.

Failure to exhaust administrative remedies may also result in dismissal or remand of a NEPA suit. As NEPA procedures become

---

court refused to accept the argument that laches was present. On the facts, one might be persuaded that the only factor distinguishing these cases from more routine litigations where laches is accepted was the public interest character of the suits.

[115] Arlington Coalition on Transportation v. Volpe, 332 F. Supp. 1218, 1 ELR 20486 (E.D. Va. 1971), *rev'd*, 458 F.2d 1323, 2 ELR 20162 (4th Cir.), *cert. denied sub nom.* Fugate v. Arlington Coalition on Transportation, 41 U.S.L.W. 3249 (Nov. 7, 1972).

[116] 458 F.2d at 1329, 2 ELR at 20163.

[117] Ward v. Ackroyd, 344 F. Supp. 1202, 1212, 2 ELR 20405, 20409–10 (D. Md. 1972).

[118] City of New York v. United States, 337 F. Supp. 150, 160, 2 ELR 20275, 20277 (E.D.N.Y. 1972), *further proceedings* 344 F. Supp. 929, 2 ELR 20688 (E.D.N.Y. 1972).

[119] 342 F. Supp. 1324, 2 ELR 20459 (E.D. La.), *aff'd*, 461 F.2d 1266, 2 ELR 20459 (5th Cir. 1972).

[120] 461 F.2d at ——, 2 ELR at 20460–63.

more settled, the courts may be expected to rest decisions more frequently upon exhaustion or the related doctrine of estoppel. For example, in *Greene County Planning Board v. Federal Power Comm'n*[121] the Second Circuit refused to enjoin the completion of two of three power transmission lines because the Federal Power Commission had not prepared impact statements on them, suggesting that one of its reasons was that petitioners did not raise timely objections in the commission's proceedings.[122] In *National Forest Preservation Group v. Butz*,[123] the court, in refusing to enjoin the "Big Sky" land exchange for failure to obtain EPA's comments on the impact statement, pointed out that no objection had been made to the absence of EPA's statement in the course of Forest Service administrative proceedings.[124] Finally, in *City of New York v. United States*[125] the court refused to dissolve the Interstate Commerce Commission's railway abandonment order, in part because:

> the plaintiffs who now seek to benefit from the Commission's failure completely to perform the tasks imposed upon it by NEPA exacerbated the problem by waiting until after the hearings were completed to raise the environmental question.[126]

Where extensive administrative proceedings precede agency action, it makes great sense to require participants to make their NEPA objections known at that stage. A court would be justified in relying upon exhaustion or estoppel grounds in refusing to delay agency action where parties or intervenors in earlier quasi-judicial or quasi-legislative proceedings had waited until the proceedings were terminated to object that an impact statement had not been prepared. The primary responsibility for complying with NEPA rests with the agencies. Nevertheless, a party or intervenor who is entitled (or obligated) to present his views on environmental impacts at a hearing or through written comments should not be permitted to allow the agency to postpone its own environmental assessment so that it can be attacked in court.

While judicial review under NEPA may be sought too late, it may also be sought too early. In *Businessmen for the Public Interest*

---

[121] 455 F.2d 412, 2 ELR 20017 (2d Cir.), *cert. denied*, 41 U.S.L.W. 3184 (Oct. 10, 1972).

[122] 455 F.2d at 425, 2 ELR at 20023.

[123] Appendix B.

[124] 343 F. Supp. at 703, 2 ELR at 20574.

[125] Appendix B.

[126] 337 F. Supp. at 164, 2 ELR at 20277.

v. Resor,[127] plaintiffs sought to have the Refuse Act Permit Program (insofar as it affected Lake Michigan) enjoined for failure to fully implement NEPA. The court declined to issue the injunction, since the case was found not to be ripe for preenforcement review before actual issuance of the permits. *Lloyd Harbor Study Group v. Seaborg*[128] reached a similar result, because the agency review process had not been completed. The court held that the refusal of an Atomic Safety and Licensing Board to allow evidence on non-radiological environmental effects was not yet reviewable, since the proceedings would be reviewed later by the Atomic Energy Commission.[129]

A premature request for review may offer the court an opportunity to forewarn defendants that when the time comes, compliance with NEPA will be expected. In *Save the Dunes Council, Inc. v. Froehlke*[130] the court held that the action was "untimely," but said that "by reason of the instant action, defendant . . . has been put on notice that its proposed application will come under close scrutiny"[131] in connection with its NEPA obligations.

Finally, one of the largest potential threats to NEPA litigation simply has not materialized. While the courts could impose high bond requirements on NEPA plaintiffs, they have not chosen to do so because of the public interest nature of most NEPA lawsuits. While one $20,000 bond was imposed,[132] at least four cases where bond was requested imposed no bond requirement,[133] at least six imposed a $100 requirement,[134] and the court in the Tennessee–Tombigbee case imposed $1.[135] The issues involved in bond requirements were thoroughly aired first in *Natural Resources Defense Council v.*

---

[127] —— F. Supp. ——, 3 ERC 1216 (N.D. Ill. 1971).

[128] —— F. Supp. ——, 1 ELR 20188 (E.D.N.Y. 1971).

[129] Similar rulings were made in Coalition for Safe Nuclear Power v. Atomic Energy Comm'n, —— F.2d ——, 2 ELR 20150 (D.C. Cir. 1972), and Atlanta Gas Light Co. v. Southern Natural Gas Co., 338 F. Supp. 1039 (N.D. Ga. 1972).

[130] —— F. Supp. ——, 2 ELR 20356 (N.D. Ind. 1972).

[131] —— F. Supp at ——, 2 ELR at 20356.

[132] Sierra Club v. Laird, —— F. Supp. ——, 1 ELR 20085 (D. Ariz. 1970).

[133] Texas Comm. on Natural Resources v. Resor; Silva v. Romney; Scherr v. Volpe; Boston Waterfront Residents v. Romney (full citations in Appendix B).

[134] Natural Resources Defense Council v. Morton; West Virginia Highlands Conservancy v. Island Creek Coal Co.; Wilderness Society v. Hickel; Natural Resources Defense Council v. Grant; Stop H-3 Ass'n v. Volpe; Sierra Club v. Froehlke (Trinity River–Wallisville Dam) (full citations in Appendix B).

[135] Environmental Defense Fund v. Corps of Engineers, Appendix B, 331 F. Supp. 925, 1 ELR 20466 (D.D.C. 1971).

Morton,[136] and in *Natural Resources Defense Council v. Grant,*[137] where the Fourth Circuit held that the North Carolina district court's $75,000 bond requirement was too high and that a nominal bond should be imposed, because plaintiffs were acting as "private attorneys-general."[138]

---

[136] Appendix B, 337 F. Supp. 167, 2 ELR 20089 (D.D.C. 1971).

[137] 341 F. Supp. 356, 2 ELR 20185 (E.D. N. Car.), *remanded,* —— F.2d ——, 2 ELR 20555 (4th Cir.), —— F. Supp. ——, 2 ELR 20647 (E.D. N. Car.), —— F. Supp. ——, 2 ELR 20648 (E.D. N. Car. 1972), —— F. Supp. ——, 3 ELR 20176 (E.D. N. Car. 1973).

[138] —— F.2d at ——, 2 ELR at 20556.

III

The Requirement of Strict Compliance

AFTER SOME initial hesitation, the courts have accepted the view that Congress intended for NEPA's various requirements to be interpreted and applied in the strictest manner. The phrase, "to the fullest extent possible," has become the touchstone of NEPA interpretation. The key language is contained in the opening words of §102.

> The Congress authorizes and directs that, *to the fullest extent possible*: (1) the policies, regulations, and public laws of the United States shall be interpreted and administered in accordance with the policies set forth in this Act, and (2) all agencies of the Federal Government shall . . . [comply with eight action-forcing provisions] [emphasis added].

The requirement of compliance "to the fullest extent possible" has been applied with special vigor to the provisions of §102(2) after the forceful interpretation given to the phrase in *Calvert Cliffs' Coordinating Committee v. Atomic Energy Comm'n.*[1] The requirement has set the tone for implementation of all aspects of NEPA's action-forcing provisions. While the requirement has been frequently cited in connection with the duties owed once it has been established that a statement must be prepared (see chapter VI), it has also helped ensure that the Act is applied to the broadest range of federal agencies and actions (see chapter IV) and that it is applied to as many projects and programs as possible which were in progress when NEPA was enacted (see chapter V). Thus, at least as far as the action-forcing sections are concerned, the requirement of implementation "to the fullest extent possible" has established a simple judicial philosophy of NEPA implementation. This guiding precept is fully supported by NEPA's legislative history, as chapter I has shown. Only a clear conflict of statutory authority can alter the requirement.

An important aspect of the strict compliance standard has been

---

[1] 449 F.2d 1109, 1 ELR 20346 (D.C. Cir. 1971), *cert. denied*, 404 U.S. 942 (1972). Calvert Cliffs' is discussed more fully in chapter VII. For scholarly discussion of this case, see note 4 in chapter VII.

somewhat overlooked in NEPA's interpretation by the courts. Section 102(1) requires that existing federal laws, regulations, and policies be interpreted and administered in accordance with NEPA's policy "to the fullest extent possible." The inference is that the specific substantive requirements of §101(b) must be accorded a larger influence than the courts have yet acknowledged. Dicta in *Calvert Cliffs'* notwithstanding, and in spite of certain less stringent language in §101, NEPA's substantive requirements may also be subject to a strict standard of compliance.[2] (See discussion of substantive requirements in chapter VII, page 258.)

As in many matters of NEPA interpretation, the decision in *Environmental Defense Fund v. Corps of Engineers*[3] indicated the direction subsequent judicial opinion would take. Decided a year after NEPA was enacted, the Gillham Dam case held that the Corps of Engineers had failed in ten specific ways to comply "to the fullest extent possible" with the detailed requirements of §102.[4] Six months later in *Calvert Cliffs'*, the District of Columbia Circuit stated the rule which has been most frequently relied upon by the majority of courts in requiring strict procedural compliance.

> The Act . . . contains very important "procedural" provisions—provisions which are designed to see that federal agencies do in fact exercise the substantive discretion given them. These provisions are not highly flexible. Indeed, they establish a strict standard of compliance.[5]

By NEPA's second anniversary courts in seven important NEPA litigations had applied the *Calvert Cliffs'* holding on strict compliance with NEPA's procedures.[6] The decision of the Fourth Circuit in *Ely v. Velde*[7] cited language from the *Calvert Cliffs'* opinion that explicitly based the requirement of strict compliance

---

[2] *See also* B. Cohen and J. Warren, *Judicial Recognition of the Substantive Requirements of the National Environmental Policy Act of 1969*, 13 B.C. Ind. & Com. L. Rev. 685, 697 (March 1972).

[3] 325 F. Supp. 728, 1 ELR 20130 (E.D. Ark. 1970–71), —— F. Supp. ——, 2 ELR 20260 (E.D. Ark.), 342 F. Supp. 1211, 2 ELR 20353 (E.D. Ark.), *aff'd*, 470 F.2d 289, 2 ELR 20740 (8th Cir. 1972).

[4] 325 F. Supp. at 758, 1 ELR at 20141.

[5] 449 F.2d at 1112, 1 ELR at 20347.

[6] National Helium Corp. v. Morton; Comm. for Nuclear Responsibility v. Seaborg; Ely v. Velde; Morningside–Lenox Park Ass'n v. Volpe; Kalur v. Resor; Natural Resources Defense Council v. Morton; City of New York v. United States (full citations in Appendix B). *See also* Silva v. Romney; Daly v. Volpe (second district court opinion); Scherr v. Volpe (7th Cir.); Harrisburg Coalition Against Ruining the Environment v. Volpe; Lee v. Resor; Citizens for Clean Air v. Corps of Engineers (full citations also in Appendix B).

[7] 321 F. Supp. 1088, 1 ELR 20082 (E.D. Va.), *rev'd*, 451 F.2d 1130, 1 ELR 20612 (4th Cir. 1971).

upon NEPA language requiring compliance "to the fullest extent possible."[8] Judge Wright said in the quoted portion of *Calvert Cliffs'*:

> We must stress as forcefully as possible that the language does not provide an escape hatch for foot-dragging agencies; it does not make NEPA's procedural requirements somehow "discretionary." Congress did not intend the Act to be such a paper tiger. Indeed, the requirement of environmental consideration "to the fullest extent possible" sets a high standard for the agencies, a standard which must be rigorously enforced by the agencies.[9]

In *Morningside–Lenox Park Ass'n, Inc. v. Volpe*[10] the district court cited the *Calvert Cliffs'* formulation of the standard of strict compliance, but discussed the earlier Gillham Dam case in detail for its actual application of the standard, citing with approval the Arkansas court's interpretation of the phrase, "to the fullest extent possible."[11]

The Second Circuit and a New York federal district court have suggested an extension of the concept beyond its interpretation in *Calvert Cliffs'* to require strict compliance even when the agency is reluctant or perhaps powerless to alter its proposed course of action. The Second Circuit gave "preservation of the integrity of the new Act" as a reason for this proposed interpretation,[12] citing the Act's legislative history, *Calvert Cliffs'*, and the district court opinion which is authority for this view, *City of New York v. United States.*[13] This interpretation of the usefulness of strict compliance even when the agency may not freely "consider" the facts and options set out in the impact statement has considerable merit. While the agency itself may be bound, Congress, the Executive, and the public can use the information thus developed to propose legislative action or action that may be taken by other agencies.

The strict compliance rule appears to resolve the doubt created by a few early decisions which found that consultation or studies which did not follow the plan of §102 nevertheless were acceptable as being in substantial compliance with the Act. Thus in *Bucklein v. Volpe*[14] the court found that the county applying for emergency road repair

---

[8] 451 F.2d at 1138, 1 ELR at 20614.

[9] Appendix B, 449 F.2d at 1114, 1 ELR at 20348.

[10] 334 F. Supp. 132, 1 ELR 20629 (N.D. Ga. 1971).

[11] 334 F. Supp. at 141, 1 ELR at 20631.

[12] Hanly v. Kleindienst (Hanly I), 460 F.2d 640, 648, 2 ELR 20216, 20220 (2d Cir. 1972) (full citation in Appendix B).

[13] 337 F. Supp. 150, 160, 2 ELR 20275, 20277 (E.D.N.Y.), *further proceedings* 334 F. Supp. 929, 2 ELR 20688 (E.D.N.Y. 1972).

[14] —— F. Supp. ——, 1 ELR 20043 (N.D. Cal. 1970).

funds had "amply considered" environmental factors and allowed the project to proceed without a formal statement. In *Citizens to Preserve Foster Park v. Volpe*[15] the court held, over objection, that although the statement filed there was not organized around the five factors specified by the Act, the defendant had complied "to the extent possible" with the Federal Highway Administration's Interim Procedures.[16] This decision was upheld, virtually without discussion, by the Seventh Circuit.[17] It may be distinguished on the ground that guidelines governing the preparation and adequacy of impact statements were not yet final. Presumably, now that final CEQ and Federal Highway Administration guidelines are fully approved and implemented, the Seventh Circuit will require strict compliance. Finally, in *Sierra Club v. Hardin,*[18] the Forest Service was allowed to rely on the informal environmental studies made by the applicant, although the court warned:

> Nothing in this opinion should be construed as implying that the procedures followed by the Forest Service in its efforts to comply with NEPA in this case will be found acceptable in the future under circumstances where it is fair to impute notice of the Act's provisions to all parties at or before the time a major federal project is conceived.[19]

These cases may best be seen as exemplifying the uncertainty that surrounded early efforts at implementation and, as *Sierra Club v. Hardin* indicates, provide little guidance now that the initial throes of implementation are past and the strict compliance standard has been widely accepted.[20] Further, there appears to have been a shift

[15] —— F. Supp. ——, 1 ELR 20389 (N.D. Ind. 1971), *aff'd*, 466 F.2d 991, 2 ELR 20560 (7th Cir. 1972).

[16] —— F. Supp. at ——, 1 ELR at 20391.

[17] 466 F.2d 991, 2 ELR 20560 (7th Cir. 1972).

[18] 325 F. Supp. 99, 1 ELR 20161 (D. Alas. 1971).

[19] 325 F. Supp. at 127, 1 ELR at 20171.

[20] *Quaere* whether Hiram Clarke Civic Club, Inc. v. Romney, —— F. Supp. ——, 2 ELR 20362 (S.D. Tex. 1971), does in fact approve substantial compliance to prevent "a confiscatory result against a private developer." As the court actually uses the phrase, "substantial compliance" seems to mean the same thing as "strict compliance." In New York v. Department of the Army, —— F. Supp. ——, 2 ELR 20507 (S.D.N.Y. 1972), the court appears to have allowed the Corps of Engineers to avoid preparing impact statements on each ocean dumping permit issued, since the Corps had prepared an adequate statement on the entire ocean dumping permit program. Yet the court did not explicitly state that it was allowing a form of substantial compliance and made plain that the only statement at issue was the comprehensive program statement.

in judicial attitude, as *Daly v. Volpe*[21] strikingly demonstrates. In *Daly* Judge Beeks adopted a substantial compliance view at the hearing on the preliminary injunction and denied relief. One year later in his decision after trial, he said:

I denied a preliminary injunction because I believed that there had been substantial compliance . . . [with NEPA]. I am now satisfied, however, that the law as it has since evolved mandates strict compliance with NEPA and that the position heretofore taken by me has been rejected. The provisions of NEPA are not highly flexible, but establish a strict standard of procedure.[22]

One last, special case which may be thought of as allowing limited substantial compliance with NEPA is the second decision of the Second Circuit in the now-famous litigation in *Scenic Hudson Preservation Conference v. Federal Power Comm'n.*[23] The court accepted the environmental findings made in the commission's final decision as the final impact statement.[24] The propriety of late compliance with NEPA was not discussed, although the Second Circuit's view on the inadequacy of the commission's guidelines for NEPA compliance was made clear in its later decision in *Greene County* (see discussion at pages 186 *ff.*). The majority opinion was exhaustively criticized in Circuit Judge Oakes' dissent and in Mr. Justice Douglas' dissent from the Supreme Court's denial of certiorari. Both judges detailed the loss of affirmative commission research, development of alternatives, focused review, etc., because NEPA had been partially avoided.[25] Yet the majority's refusal to be persuaded by these considerations may be partly explained by the circumstances surrounding this exceptional litigation. The environmental aspects of the project had been under review by the commission and the court since the mid-1960s; the legislation under which the commission had conducted lengthy environmental reviews had been interpreted

---

[21] 326 F. Supp. 868, 1 ELR 20242 (W.D. Wash. 1971), 350 F. Supp. 252, 2 ELR 20443 (W.D. Wash.), 350 F. Supp. 252, 3 ELR 20032 (W.D. Wash. 1972).

[22] 350 F. Supp. at 257, 2 ELR at 20444 (W.D. Wash. March 31, 1972).

[23] Scenic Hudson I, 354 F.2d 608, 1 ELR 20292 (2d Cir. 1965), *cert. denied sub nom.* Consolidated Edison Co. of New York v. Scenic Hudson Preservation Conference, 384 U.S. 941 (1966), Scenic Hudson II, 453 F.2d 463, 1 ELR 20496 (2d Cir. 1971), *cert. denied,* 407 U.S. 926, 2 ELR 20436 (1972).

[24] Scenic Hudson II, 453 F.2d at 481, 1 ELR at 20506.

[25] 453 F.2d at 491, 1 ELR at 20512 (Judge Oakes); 407 U.S. at 926, 2 ELR at 20436 (Mr. Justice Douglas).

so expansively by the Second Circuit court in *Scenic Hudson I* that the legislation has been frequently mentioned as a precursor of NEPA itself.[26] The court, perhaps weary of the litigation, over vigorous dissent allowed an exception to the requirement of strict compliance in view of the extensive environmental review that had already taken place. Perhaps no other litigation in environmental law can compare to *Scenic Hudson*; realistically, it affords little guidance as to how the courts—especially the Second Circuit—will decide subsequent NEPA cases.

In his discussion in *Calvert Cliffs'* of what is meant by compliance with NEPA "to the fullest extent possible," Judge Wright made the following statement, which has been repeated subsequently in several NEPA opinions:

> Section 102 duties are not inherently flexible. They must be complied with to the fullest extent, unless there is a clear conflict of *statutory* authority. Considerations of administrative difficulty, delay and economic cost will not suffice to strip the section of its fundamental importance [court's footnote omitted].[27]

NEPA has not lacked critics who believe that the administrative difficulty, delay, and costs involved in implementing NEPA are ensnarling the agencies in paperwork, halting the progress of key federal programs, and causing considerable economic loss both to government and to the private parties who must await governmental action before they themselves can act.[28] Various amendments have been proposed to ease the burden of NEPA compliance, and the Environmental Protection Agency recently became the first agency to obtain a partial exemption from preparing impact statements.[29]

The response that is most often given to NEPA's critics is that the reform of federal decision making which NEPA will bring about will eventually save more time, money, and bother than the NEPA procedures waste. In short, NEPA will be "cost-effective."[30] Fur-

---

[26] Council on Environmental Quality, Environmental Quality, Third Annual Report 223 (August 1972).

[27] Appendix B, 449 F.2d at 1115, 1 ELR at 20349.

[28] *See* Joint Comm. on Atomic Energy, Selected Materials on the Calvert Cliffs' Decision, its Origin and Aftermath, 92d Cong., 1st Sess. *passim* (Comm. print February 1972); H. Voight, *The National Environmental Policy Act and the Independent Regulatory Agency*, 5 Nat. Res. Lawyer 13 (1972); R. Gillette, *National Environmental Policy Act: Signs of Backlash are Evident*, 176 Science 30 (April 7, 1972).

[29] *See* Introduction, notes 2 and 3, and accompanying text.

[30] [NEPA's] . . . benefits have not been without costs to the government. The initial uncertainties about NEPA's meaning have spawned a large amount

thermore, the strict compliance standard requires immediate implementation, even if other considerations are also present which could delay indefinitely the implementation of the Act, because Congress placed emphasis on putting the overdue NEPA reform to work as soon as possible.

Judge Wright's remarks appear not only to apply to the actual direct burden to government of implementing NEPA's procedures, but also to the indirect costs, both to government and to private parties, of delaying construction projects, cancelling contracts, laying off workmen, etc. Injunctions granted in NEPA cases have in some instances exacted rather heavy indirect costs in order to ensure that agencies first complied strictly with the Act before undertaking major federal action. Such burdens of compliance, however, actually have had more to do with the relief granted in specific cases than with the courts' interpretation of the Act's requirement that agencies comply "to the fullest extent possible." Cases which discuss how the strict standard of compliance relates to the traditional test for preliminary injunctive relief are discussed in the final section of chapter VI (see pages 239 *ff.*).

---

of litigation, which is always costly in money and time. As NEPA principles become clearer, this problem should decline. The need to study environmental effects and to hire new personnel carries budgetary costs. These costs may run as high as $65 million a year when NEPA is fully underway. However, much larger amounts can be wasted on any one ill-advised federal project—for example, the Cross-Florida barge canal had cost $50 million when the President stopped it and would have cost $130 million more to complete. Moreover, careful analysis of the effects of government action is a logical component of good public administration. Much of the cost attributed to NEPA is going for studies that should be performed in any event.

Private investment decision making in many areas also has been touched by NEPA and the 102 process. Businesses subject to federal regulation or which receive federal funding are having to adjust to the agencies' new environmental awareness. Private planners for new power facilities, for federally assisted housing, and for development of the resources of federal lands must now consider the environmental issues spelled out in section 102(2)(C). The costs to business have in some instances been substantial. *Supra* note 26 at 258.

IV

Circumstances Requiring an Impact Statement

THE NEXT three chapters, which constitute the major portion of this book, focus on the requirement that an impact statement must be prepared under the conditions prescribed in §102(2) (C). This one action-forcing provision has received the most attention from the courts and consequently is treated in detail in our analysis. However, throughout the next three chapters it should be kept in mind that §102(2) (C) is not an end in itself; as one of the first steps in the NEPA process, it requires that the federal agency develop—and then disclose—a thorough estimate of the environmental impacts which its actions may cause. But preparation of such an assessment is an exercise of limited value unless it is used along with NEPA's other provisions to bring about actual environmental improvement through better federal decision making (see chapter VII).

In this chapter we consider the extent to which the courts have applied the impact statement requirement to the range of federal activities. The matter is one of degree; Congress obviously did not intend impact statements to be prepared each time an agency proposed a trivial action, nor did it intend to excuse all but the largest and most important actions from compliance. On balance, the courts have tended toward the widest possible scope of application. They have done so for a variety of reasons, the most frequent being that Congress intended NEPA to apply to all federal agencies and at the "lowest levels, where . . . most decisions are formulated and even finalized."[1]

Congress' rather general language necessarily places on agencies and courts the burden of spelling out the precise scope of the Act. Specifically, §102(2) (C) requires "all agencies" to include in "every recommendation or report on proposals for legislation and other major Federal actions significantly affecting the quality of the

---

[1] 115 CONG. REC. S. 12117 (daily ed. Oct. 8, 1969) (remarks of Mr. Church, Senate Conferee on NEPA).

human environment a detailed statement" on five specific environmental factors.[2] The phrase, "major Federal action significantly affecting the quality of the human environment," has engendered the bulk of litigation under NEPA. The courts have interpreted the phrase, "proposals for legislation," much less frequently.

In this chapter we first discuss the cases which define the kind and extent of agency participation necessary before an action can be called "federal." Second, we explore how large or how important the subject matter of a proposed agency action must be before it becomes "major" and "significantly affects" the environment. Third, we discuss whether "*all* agencies of the Federal government" must comply. Fourth, we consider the Act's requirement that agencies must prepare statements on "legislative proposals and recommendations." We take up fifth and last several possible limits on the scope of the requirement.

WHEN AND WHETHER AN ACTION IS "FEDERAL"

By its terms NEPA applies only to the federal agencies. Congress did not attempt to pull within the Act's impact statement requirement all the state, local, and private actions which may cause environmental degradation. Thus an important threshold issue in NEPA's implementation is whether a particular action is sufficiently "federal" for NEPA to apply. The question may be whether the action will ever be federal at any time, or whether the federal role has as yet become sufficiently manifest for NEPA to apply. The latter situation arises where state, local, or private activities which take place prior to federal funding or approval threaten to cause environmental impacts which might constrict federal options once the government becomes involved.

Whether an action which has admittedly been taken by a federal official is of a type or kind to which NEPA applies is a related but distinct issue. Before dealing with the degree and timing of federal actions, we should point out that the types or kinds of actions to which NEPA has been applied have been virtually unlimited. Because we do not treat this "nonissue" elsewhere, we will briefly discuss it first.

Types of Federal Action

When the federal agencies provide services and carry out direct construction programs, they must prepare impact statements. Thus

---

[2] *See* full text of §102(2) (C) in Appendix A.

the activities of federal agencies such as the Post Office, the Army Corps of Engineers, the Forest Service, the Bureau of Reclamation, and the armed services are covered. In deciding cases in this area, courts do not normally discuss whether the federal presence is sufficient; it is assumed to exist. Similarly, federal regulation of private conduct poses few problems to NEPA's application. A federal permit, license, certificate, or other regulatory activity that significantly affects the environment and qualifies as a major federal action is subject to §102 (2) (C). The regulatory activities of several independent regulatory commissions and of most departments have already been the subject of NEPA litigation. Again, in most of these cases the federal presence is assumed to exist without discussion. A problem could arise, however, where the ultimate authority of the federal regulator to control is a close question. There, it may be argued, no federal action takes place within the definition of NEPA.[3]

A few cases have been decided involving a private request to an executive department for permission to do something in federally controlled areas. In five of the six cases falling in this category, the courts either implicitly assumed that §102(2) (C) applied, or passed over the issue with the briefest affirmation that such federal actions were covered.[4] However, in *Davis v. Morton*[5] the Tenth Circuit had to reverse the district court below to hold that the Interior Department's loose guardianship over Indian lands sufficiently federalized a lease to a developer.

NEPA has been routinely applied to federal contracts, grants, and loans to private parties. A few litigated cases concern such funds, but do not discuss the issue in any great detail. Indeed, the very first case decided held that the Act applied to a projected Farmers Home Administration loan to be used for construction of a golf course and park. The district court in *Texas Committee on*

[3] *See* Davis v. Morton, 335 F. Supp. 1258, 2 ELR 20003 (D.N.M. 1971), *rev'd*, 469 F.2d 593, 2 ELR 20758 (10th Cir. 1972); McLean Gardens Residents Ass'n v. National Capital Planning Comm'n, —— F. Supp. ——, 2 ELR 20659 (D.D.C.), *motion for stay of injunction and summary of reversal denied*, —— F.2d ——, 2 ELR 20662 (D.C. Cir. 1972). Regulatory actions include the preparation of "policy, regulations and procedure-making." CEQ Guidelines, §5(a)(iii), Appendix C.

[4] Wilderness Society v. Hickel; West Virginia Highlands Conservancy v. Island Creek Coal Co.; Sierra Club v. Hardin; Natural Resources Defense Council v. Morton; Jicarilla Apache Tribe of Indians v. Morton (full citations in Appendix B).

[5] Appendix B.

Natural Resources v. United States[6] held there is "little doubt that in the future the type of activity involved here would be covered by the statute."[7] The court viewed the federal action as more than the decision to grant the loan; the entire project was the "action."[8]

Harder questions arise when the federal government underwrites part or all of the cost of projects planned and executed by the states. Perhaps the easiest case is that in which the federal government not only provides funds to the state but participates in some other substantial manner as well. *Environmental Defense Fund v. Hardin*[9] concerned a challenge to the Department of Agriculture's plan to spray the pesticide Mirex in order to control the spread of the imported fire ant. In applying NEPA the court characterized the plan as a "cooperative federal-state program"; plaintiff's brief reveals that the program was to spend $200 million over 12 years, and to be two-thirds federally funded. The research which led to the proposal was carried out at the federal level.

The application of NEPA to the federal matching program for funding of state-federal highways has been the subject of rather extensive litigation. However, there is no doubt that all the types of highways involved are subject to NEPA's requirements. This is true for the interstate system,[10] the "primary" road system,[11] road

[6] —— F. Supp. ——, 2 ELR 20574 (W.D. Tex. 1970), *vacated* 430 F.2d 1315 (5th Cir. 1970).

[7] —— F. Supp. at ——, 2 ELR at 20575.

[8] Four reported cases deal with development loans and mortgage guarantees to private builders by the Department of Housing and Urban Development. The court in Goose Hollow Foothills League v. Romney, 334 F. Supp. 877, 1 ELR 20492 (D. Ore. 1971), held that the department had erroneously determined that a $3 million grant for a student high-rise apartment was not within the Act. Silva v. Romney, 342 F. Supp. 783, 2 ELR 20385 (D. Mass. 1972), *rev'd*, —— F.2d ——, 3 ELR 20082 (1st Cir. 1973), did the same in connection with a 138-unit housing development. Echo Park Residents Committee v. Romney, —— F. Supp. ——, 2 ELR 20337 (C.D. Cal. 1971), assumed that § 102(2)(C) might be applicable to such grants, but upheld the departmental determination that the 66-unit project involved there would be without significant impact. In Hiram Clarke Civic Club v. Romney, —— F. Supp. ——, 2 ELR 20362 (S.D. Tex. 1971), the central issue appeared to be whether the impact of neighborhood dislocation was within the scope of § 102(2)(C), the court ruling that it was not. The court did not discuss whether such projects were "major," nor did it deny that statements might be required should project effects be found to be "environmental."

[9] 325 F. Supp. 1401, 1 ELR 20207 (D.D.C. 1971).

[10] *E.g.*, Lathan v. Volpe, and Arlington Coalition on Transportation v. Volpe, both cited fully in Appendix B, are circuit court decisions applying NEPA to the interstate system.

[11] *E.g.*, NEPA's application in Named Individual Members of the San Antonio Conservation Society v. Volpe, a protracted litigation cited fully in Appendix B.

upgradings,[12] and probably repair grants, although the latter have been involved in litigation in which the outcome was adverse to plaintiffs.[13] Difficulties arise only when state and federal officials seek to avoid the impact statement requirement either by arguing that the highways at issue will be built solely with state funds, or that the request that a statement be prepared is premature because federal funding and approval have not yet been sought (see discussion, pages 64 *ff.*).

Block grant and revenue-sharing programs in some ways pose the extreme test of NEPA's applicability. The federal role is theoretically as minimal and automatic as possible; one of the purposes of such programs is to get funds to the states with few federal conditions attached.[14] A leading NEPA case out of the Fourth Circuit, however, is strong precedent for NEPA's application to the federal actions involved. In *Ely v. Velde*[15] the circuit court overruled a district court determination that NEPA did not apply to a block grant by the Law Enforcement Assistance Administration, part of which was to pay 20 percent of the construction cost of a prison medical facility capable of handling 400 patients. The lower court had held that the Organized Crime Control and Safe Streets Act,[16] which provided that LEAA "shall approve" state requests, overrode the provisions of the later NEPA statute. The circuit court, however, looked to the legislative history of the earlier statute and found that it was intended to forestall creation of a "national police force," not to preclude the environmental review mandated by NEPA. The circuit court next concluded that the magnitude of federal involvement was sufficient for NEPA to apply, even if only 20 percent of the project was federally financed. It took account of "LEAA's overall involvement in the promotion and planning of the Center, as well as the cumulative impact of the proposed federal action."[17]

---

[12] *E.g.*, NEPA's application in Scherr v. Volpe, 336 F. Supp. 882, 2 ELR 20068 (W.D. Wisc.), 336 F. Supp. 886, 2 ELR 20068 (W.D. Wisc. 1971), 466 F.2d 1027, 2 ELR 20473 (7th Cir. 1972).

[13] Bucklein v. Volpe, —— F. Supp. ——, 1 ELR 20043 (N.D. Cal. 1970); Pennsylvania Environmental Council v. Bartlett, 315 F. Supp. 238, 2 ELR 20752 (N.D. Pa. 1970), *aff'd*, 454 F.2d 613, 1 ELR 20622 (3rd Cir. 1971). For further authority on NEPA's application to highways, *see* Daly v. Volpe, 350 F. Supp. 252, 2 ELR 20443 (W.D. Wash. 1972), text above note 14 and cases cited in footnote.

[14] *See* R. Strauss, *Revitalizing Our Federal System: The Rationale for Revenue Sharing*, 21 De Paul L. Rev. 889 (1972); A. Lynn and D. McGirr, *Revenue Sharing—An Idea Whose Time Has Come—Almost!*, *id.*, at 902.

[15] 321 F. Supp. 1088, 1 ELR 20082 (E.D. Va.), *rev'd* 451 F.2d 1130, 1 ELR 20612 (4th Cir. 1971).

[16] 42 U.S.C. §3701 *et seq.*

The real significance of this case extends beyond the LEAA grant program. Should revenue-sharing and other "no strings" proposals be adopted, the degree of federal participation, using the *Ely* "overall involvement" test, might be significant enough to warrant applying NEPA, particularly if the federal instrumentality for funding retains any responsibilities for promotion and planning. The basic theory of the block grant approach would be no more compromised by the application of NEPA than it is by similar federal constraints and requirements already applicable to such grants (e.g., civil rights legislation, the Hatch Act, the planning and reporting requirements of OMB, etc.).[18] It is true that application of NEPA would require the submission of enough information to enable the agencies to prepare impact statements. These submissions might be combined with OMB's requirements without undue cost in time. Another possible solution is suggested in the House committee report on the 1972 Housing Bill, which failed of passage.[19] The committee contemplated full NEPA compliance (the bill explicitly required it[20]) and suggested that the Department of Housing and Urban Development should first make a limited environmental review and then require the unit of local government administering the funds to prepare a full report on the environmental impact of any proposed major action as a condition of the grant. "Any local failure to discharge such responsibilities would be subject to possible withholding or reductions of further grant funding by the Department."[21]

Necessary Degree of Federal Participation

In some cases a close question may exist whether an action involves a federal agency to a degree meriting preparation of an impact

[17] 451 F.2d at 1137 note 22, 1 ELR at 20614 note 22.

[18] State agencies that receive block grants for law enforcement must "pass through" at least 40 percent of the planning money and 75 percent of action money to local government. 42 U.S.C. §3723(c) (1970). In addition, no more than one-third of any block grant may be expended for personnel. 42 U.S.C. §3731(d) (1970). *See* Tomlinson and Mashaw, *The Enforcement of Federal Standards in Grant-in-Aid Programs: Suggestions for Beneficiary Involvement,* 58 VA. L. REV. 600, 606 (1972).

[19] Housing and Urban Development Act of 1972, H.R. 16704, 92d Cong., 2d Sess. (1972).

[20] *Id.,* §1145.

[21] H.R. REP. NO. 92–1429, 92d Cong., 2d Sess. 60 (1972). For a less optimistic analysis *see* Note, *Ely v. Velde: The Application of Federal Environmental Policy to Revenue Sharing Programs,* 1972 DUKE L. J. 667 (1972).

statement. The case of *Kitchen v. Federal Communications Comm'n*[22] illustrates the point well. Plaintiffs sought to have NEPA applied to a telephone exchange building which the Bell Telephone Company of Pennsylvania proposed to construct in petitioner's residential neighborhood in the Schuylkill area near Philadelphia. They alleged that such construction required prior approval by the Federal Communications Commission under the Communications Act of 1934.[23] The court found that no *federal* action would take place, because under long-standing judicial interpretation the Communications Act specifically excludes such facilities from FCC jurisdiction.

A harder case involved Interior Department approval of a lease of Indian land for construction of a new city. In *Davis v. Morton*[24] the district court decided that no statement had to be prepared on the ground that there was no "federal" involvement in the lease. "The United States did not initiate the lease, participate financially in it, or benefit from it. Its further responsibilities under the lease consist only of subsequent approvals in its role as guardian."[25] As an alternate ground for its ruling, the court interpreted a statute passed after the effective dates of both NEPA and the lease which called for consideration of the environmental effects of further Indian lease approvals. The court concluded from this that Congress did not intend for NEPA to cover such leases, because it had passed the later legislation specifically to provide for special study of the environmental effects of Indian leases.

The Tenth Circuit recently reversed the district court holdings. With respect to whether granting leases on Indian lands constitutes "federal" action, the court said:

> Upon review of the lease and relevant case law, we feel the lower court erred in holding the lease did not constitute major federal action. The lease refers to the United States government countless times. All notices and approvals must be made by the Pueblo and the United States. The Secretary is required to give written approval before encumbrances can be made on the leased land. The lease protects the United States government against damage or injury to people or property on the leased premises. Certainly the fact the United States government might be held liable for injury or damages incurred on the Indian land unless the lease

[22] 464 F.2d 801, 2 ELR 20534 (D.C. Cir. 1972).
[23] 47 U.S.C. §§151–609 (1970).
[24] Appendix B.
[25] 335 F. Supp. at 1260, 2 ELR at 20004

provides otherwise makes the government more than an impartial, disinterested party to the contract.[26]

With respect to the alternate ground for the district court's ruling, the Tenth Circuit distinguished the later statute as being only peripherally and generally concerned with the environment and not in conflict with NEPA, which is a comprehensive statute detailing specific procedures for environmental protection. "The [later statute] . . . merely reaffirms congressional intent that environmental considerations are to play a factor [sic] in any Bureau of Indian Affairs decisions."[27]

An even harder case is *McLean Gardens Residents Ass'n v. National Capital Planning Comm'n.*[28] Plaintiffs successfully asserted that the role of the planning commission in reviewing and "approving" a rezoning request was a major federal action, although the local zoning commission had the final authority to grant or deny the request. The planning commission apparently was limited to an advisory role and had no veto power over the application.

Litigation over NEPA's applicability to highway projects eligible for federal aid has frequently involved the issue of adequate federal participation. NEPA has generally been held applicable, because even if short sections of each state's highways are supported entirely with state funds, each state's overall highway program is very much a joint federal-state venture. In highway cases the court's determination whether a sufficient degree of federal involvement does or does not exist at a certain time is heavily influenced by the likelihood that federal support or approval will eventually be necessary. Some courts require impact statements, although the federal role is at the time almost nonexistent, because later federal support and approvals are likely to be sought. Other courts are not so concerned about interim impacts and are willing to allow statement preparation to be delayed until the federal role becomes more explicit. Because relevant considerations are so closely interrelated, the highway cases will be discussed together in the next section, which focuses on the timing of federal involvement.

Davis, McLean, and the highway cases indicate that the courts will broadly construe the requirement of §102(2) (C) that an action must be sufficiently "federal" before a statement has to be prepared. In *Davis* and *McLean,* the federal role was somewhat

[26] 469 F.2d at 596, 2 ELR at 20760.
[27] 469 F.2d at 598, 2 ELR at 20760.
[28] Appendix B.

peripheral to the exercise of power—what one might normally take to mean "action." Yet NEPA applied. A possible explanation for the liberal reading given by the courts is that the requirements prescribed for the agencies by preexisting statutes are similar to the requirements of NEPA: consultation, planning, "guardianship," review, etc. By enforcing NEPA, the purposes of the statutes authorizing the limited federal roles are somewhat better achieved. A more likely explanation is that the requirement of compliance "to the fullest extent possible" has been interpreted by the courts to mean that the federal government should anticipate environmental impacts at every available opportunity, even when its own part in launching or authorizing the action is slight.

Timing of Federal Involvement

Private parties or state governments obviously may cause environmental impacts and narrow eventual federal options while readying a project for federal funding or approval. These impacts are problematic because they are not preceded by a federal action which triggers the preparation of an impact statement. Nevertheless, the federal agency must make an assessment of them when it is eventually asked to participate, including an assessment of the possibility that the private or state party deliberately attempted to narrow the agency's options before funding or approval was sought.

The key cases in this area are highway cases. As we mentioned earlier, the courts have played a low degree of current federal involvement off against the likelihood of eventual federal support or approval that would clearly require compliance with §102(2)(C). As a result, the two questions are closely intertwined.

Circumstances were favorable to NEPA's application in *Thompson v. Fugate.*[29] There the state of Virginia wanted to proceed with the condemnation of a portion of historic Tuckahoe Plantation in order to complete the circumferential highway around Richmond. The district court renewed its injunction against condemnation and federal approval or funding on the ground, among others, that an impact statement had not been prepared. The state's argument that it had not yet requested federal approval or funding and that therefore the particular segment of highway in question was not yet "federal" was rejected. The court pointed both to the state's admission

[29] —— F. Supp. ——, 1 ELR 20369 (E.D. Va.), *injunction expanded,* —— F.2d ——, 1 ELR 20370 (4th Cir.), *aff'd in part and rev'd in part,* 452 F.2d 57, 1 ELR 20599 (4th Cir. 1971), *further proceedings,* 347 F. Supp. 120, 2 ELR 20612 (E.D. Va. 1972).

that the segment was an integral part of the completed beltway and to prior federal approvals of the segment as part of the interstate system. The 8-mile Tuckahoe section could not be exempted from federal environmental requirements merely because it was part of a 29-mile state highway portion of the 75-mile beltway, even if federal approval and funding had not yet been sought. The court would not approve such a "subversion of the announced Congressional policy with respect to our national environment" and labeled such segmentation a "bureaucratic exercise . . . doomed to failure."[30]

In *Lathan v. Volpe*[31] the Ninth Circuit ordered a statement prepared prior to final funding approval, where federal officials had already approved the state's location plans and right-of-way acquisitions.

> Given the purpose of NEPA to insure that actions by federal agencies be taken with due consideration of environmental effects and with a minimum of such adverse effects, it is especially important with regard to federal-aid highway projects that the section 102(2)(C) statement be prepared early.[32]

Echoing *Thompson, Lathan* further suggested that it would be impermissible for a state to begin a project, make decisions that foreclosed future options, and then turn around and request federal aid.

This problem was specifically considered by Judge Peckham of the Northern District of California in *La Raza Unida v. Volpe.*[33] There a highway had received location approval from the FHWA, but not specific funding approval, and the state had not yet requested funds. Over objection that the FHWA was not yet "committed" to the project, the court found that "Group III" projects (those for which funds "may eventually" be provided) became federal upon location approval, though defendants had maintained that the state would "probably not" request funds. (The court distinguished *Northeast Welfare Rights Organization v. Volpe,*[34] discussed below as a "Group II" case, since location approval had not been sought there.[35])

---

[30] 347 F. Supp. at 124, 2 ELR at 20613.

[31] 455 F.2d 1111, 1 ELR 20602 (9th Cir. 1971), *modified on rehearing,* 455 F.2d 1122, 2 ELR 20090 (9th Cir.), 350 F. Supp. 262, 2 ELR 20545 (W.D. Wash. 1972).

[32] 455 F.2d at 1121, 1 ELR at 20605.

[33] 337 F. Supp. 221, 1 ELR 20642 (N.D. Cal. 1971), —— F. Supp. ——, 2 ELR 20691 (N.D. Cal. 1972).

[34] —— F. Supp. ——, 1 ELR 20186 (E.D. Wash. 1970).

[35] 337 F. Supp. at 227, 1 ELR at 20644.

Judge Peckham appeared mindful of two considerations in arriving at his decision. One was that the action undertaken by the state was irrevocable; over 30 percent of the necessary land had been acquired and the residents asked to become "voluntary relocatees." The second factor motivating application of NEPA was the court's perception of congressional intent: "It is obvious then, that strong federal policies exist towards adequate relocation assistance and protection of the environment."[36]

Judge Sweigert of the same district court had an opportunity to apply the *La Raza Unida* rationale only a few months after Judge Peckham's decision. In *Sierra Club v. Volpe*[37] federal and state defendants argued that by advertising for bids on construction of a bypass around Devil's Slide on coastal Highway 1, in contravention of FHWA guidelines, the state defendants had precluded themselves from receiving federal aid funds. Before rejecting this argument the court said:

> In *La Raza Unida* . . . this court . . . held that there was no merit to defendants' contention that federal statutes do not become applicable until the state actively seeks federal funds for a project; that for purposes of applying the various federal environmental statutes and regulations, a highway project becomes a federal aid highway when the state has obtained *location* approval and, that any project for which federal aid thereafter remains an open option falls within the federal statutes and regulations.
>
> The rationale of *La Raza Unida* is that Congressional policy statements in federal environmental and similar statutes, together with the legislative history of these enactments, indicate a great concern of Congress with problems of environmental protection, particularly in the area of highway construction; that common sense suggests that all the protection which the Congress has sought to provide would be futile gestures were the states and federal agencies allowed to ignore federal statutes and regulations until deleterious effects upon the environment have actually occurred while the option for receiving federal funds still remains open.[38]

In rejecting defendants' claim that the state was now ineligible for further federal aid, the court said that to all appearances the federal defendants had acquiesced in letting the project depart from FHWA guidelines in order to escape NEPA's application. To buttress its reasoning, the court cited the numerous steps taken by the state over several years prior to NEPA's passage to ensure that the project would qualify for aid.

---

[36] *Id.*
[37] 351 F. Supp. 1002, 2 ELR 20760 (N.D. Cal. 1972).
[38] 351 F. Supp. at 1007, 2 ELR at 20762.

The question whether a state may discontinue federal involvement by withdrawing requests for funds and continuing with a project alone specifically in order to avoid the application of the Act has been litigated. *San Antonio Conservation Society v. Texas Highway Dept.*[39] argues strongly that it may not. There the Texas Highway Department had maintained that it was "absolutely committed" to construction of the questioned freeway and that "the highway will be built with 100% state money if necessary." The state refunded the federal funds already received for highway construction. The state also tipped its hand, however, by admitting that if federal participation in the particular undertaking should cease, "other eligible projects will be submitted to take up available federal funding." The Fifth Circuit characterized this argument as an attempt to circumvent Congress.

> The North Expressway is now a federal project, and it has been a federal project since the Secretary of Transportation authorized federal participation. . . . As such, the North Expressway is subject to the laws of Congress, and the State as a partner in the construction of the project is bound by those laws. . . . The State may not subvert that principle [of the supremacy of federal law] by a mere change in bookkeeping or by shifting funds from one project to another.[40]

The circuit court also stressed that the decision to seek federal assistance had been voluntary and was made in full awareness "of the controversial nature of the project and of the applicable law." The court stated in conclusion that "while this marriage between the federal and state defendants seems to have been an unhappy one, it has produced an already huge concrete offspring whose existence it is impossible for us to ignore."[41]

The necessity of avoiding state action which would undercut the options open to federal decision makers also figured in the decision in *Arlington Coalition on Transportation v. Volpe.*[42] With regard to an ongoing highway undertaking, the Fourth Circuit held that the state could not proceed with construction while the Department of Transportation was carrying out the reconsideration mandated by NEPA. *Arlington Coalition* extended the *San Antonio* rule to cover cases where a state sought to proceed with a challenged project

---

[39] Appendix B.
[40] 446 F.2d at 1027, 1 ELR at 20388.
[41] 446 F.2d at 1028, 1 ELR at 20389.
[42] 332 F. Supp. 1218, 1 ELR 20486 (E.D. Va. 1971), *rev'd*, 458 F.2d 1323, 2 ELR 20162 (4th Cir.), *cert. denied sub nom.* Fugate v. Arlington Coalition on Transportation, 41 U.S.L.W. 3249 (Nov. 7, 1972).

while the federal agency was involved in reconsideration and preparation of an impact statement.

If we were to find—as we do—that federal law requires that the proposed route for Arlington I-66 be reconsidered, acquisition by the state of right-of-way along the proposed route during the reconsideration would make proceeding with the proposed route increasingly easier and, therefore, a decision to alter or abandon the route increasingly undesirable. Thus the challenged activities of the state highway department would make a sham of the reconsideration required by federal law. Action of a state highway department, challenged because furthering a project that under federal law allegedly must be reconsidered, is a matter in controversy arising under the laws of the United States. Federal jurisdiction over such state action is essential to preserve federal question jurisdiction in the application of federal statutes.[43]

Four cases have ruled NEPA inapplicable where federal involvement was not alleged or shown, or where planning or construction was carried out exclusively with state funding before any federal commitment was manifest. *Northeast Welfare Rights Organization v. Volpe*[44] held that a proposed freeway connector was not "federal," because "*at this time* the . . . Project is proceeding with state funds only, no final approval has been sought . . . and there is no immediate plan to seek federal financial participation."[45] Federal funding of the Spokane Metropolitan Area Transportation Study, which had proposed the connector, apparently did not constitute sufficient involvement in that particular project.

A recent case is in accord. In *Indian Lookout Alliance v. Volpe*[46] plaintiffs contended that Iowa's planned statewide 1,900-mile freeway system, as well as a specific 272-mile component freeway (Freeway 518) should be covered by separate impact statements. State and federal authorities had conceded that segments of Freeway 518 should be covered in impact statements, and a statement had actually been prepared for one 7-mile segment for which final federal approvals were pending. The court held that plans for the entire Iowa freeway system and for Freeway 518 were "tentative plans for future highway construction for at least 20 years," the maps for which showed "wide general corridors, which are subject to change."

Federal approval has not been requested and is not needed for such planning. No federal money has been involved. Neither has reached the

[43] 458 F.2d at 1329, 2 ELR at 20163.
[44] Appendix B.
[45] —— F. Supp. at ——, 1 ELR at 20187.
[46] 345 F. Supp. 1167, 3 ELR 20051 (S.D. Ia. 1972).

stage of being a "major Federal action." It would be impossible to prepare an E.I.S. on such indefinite proposals.[47]

The virtual inevitability of federal involvement, coupled with NEPA's clear command to include environmental considerations in the earliest planning of federally supported projects, especially before alternative sitings and possible palliative actions are foreclosed, strongly suggests that the *Northeast Welfare Rights* and *Indian Lookout Alliance* cases were wrongly decided. The court in *Indian Lookout Alliance* identified the indefiniteness of planning proposals as reason for finding that impact statement preparation would be impractical. Yet Congress intended for NEPA to have its primary impact in the agencies' earliest deliberations about proposed actions. By enacting a national environmental policy, Congress equipped the agencies with the authority they needed to make changes of the very kind singled out by the court as causing so much indefiniteness that the impact statements were "impractical" to prepare.

A third case reached a similar result, but with an important difference. In *Civic Improvement Committee v. Volpe*,[48] the Fourth Circuit held that a state-funded project was not subject to the impact statement requirement, *unless* plaintiffs could show the eventuality of federal participation. Moreover, the dissent would have remanded to allow further discovery by plaintiff, since the road widening there proposed was projected to become part of a circumferential route, other portions of which had already received over $460,000 in federal money.

The dismissal for failure to state a federal cause of action in *Bradford Township v. Illinois State Toll Highway Authority*[49] appears to be premised on a similar finding. The Seventh Circuit held that the district court correctly found that no federal claim had been stated or federal involvement shown in a proposed extension of the East-West Toll Highway. "There is no claim that the federal government or any federal agency is engaged in any manner in the construction of the Tollway extension. Nor is it alleged that any federal funds are to be used."[50]

---

[47] 345 F. Supp. at 1169–70, 3 ELR at 20052.

[48] —— F. Supp. ——, 2 ELR 20170 (W.D. N. Car.), *motion for partial preliminary injunction granted*, —— F.2d ——, 2 ELR 20249 (4th Cir.) (Craven, J.), *partial preliminary injunction vacated*, 459 F.2d 957, 2 ELR 20249 (4th Cir. 1972).

[49] 463 F.2d 537, 2 ELR 20322 (7th Cir.), *cert. denied*, 41 U.S.L.W. 3313 (Dec. 5, 1972).

[50] 463 F.2d at 540, 2 ELR at 20324.

The six cases extending NEPA's reach to highway actions that lack federal funding or immediate federal approval, although some federal involvement is present, seem to undercut the logic behind the four adverse decisions. This is true even though a close comparison of the facts in the ten cases might allow one to unwind a weak thread of continuity. The problem with the Fourth Circuit's view in the *Civic Improvement Committee* case is that the burden regarding the degree of federal participation was placed on plaintiffs, instead of on those state officials who might later seek federal aid. At the least, the *Lathan, La Raza, Sierra Club,* and *San Antonio* cases suggest that initial disavowal coupled with a later change of heart should be scrutinized most carefully and subjected to a requirement of good faith.

A few relevant cases also exist outside the highway area. In *Gage v. Commonwealth Edison*[51] plaintiffs sought to compel the Atomic Energy Commission to prepare an impact statement on a proposed nuclear facility, although the license applicant, Commonwealth Edison, was still in the process of acquiring the land necessary for the facility, and although its construction permit was still pending before the AEC. Plaintiffs argued that NEPA compliance was necessary before Edison committed major financial resources and before improper land use could occur. The court refused relief against both the AEC and Edison on the grounds that the claim against the AEC was premature and that private parties could not be enjoined under NEPA. Acquisition by a private party of land for a facility that eventually had to receive federal approval before construction could occur on the site in question did not trigger a "clear legal duty" under NEPA, although Edison had already filed its application for a construction permit.

The recent First Circuit decision in *Silva v. Romney*[52] shows how *Gage* may have been wrongly decided. It also negates the effect of an earlier First Circuit decision on the same point which was cited with approval in *Gage. Silva* also allows a private party to be enjoined pending completion of the NEPA review. Furthermore, dicta in *Silva* point the way toward a reasonable federal administrative policy toward prejudicial "prefederal" actions. Consequently, discussion of *Silva* is an appropriate way to conclude our discussion of the federal presence requirement of §102(2)(C).

---

[51] —— F. Supp. ——, 3 ELR 20068 (N.D. Ill. 1972).

[52] 342 F. Supp. 783, 2 ELR 20385 (D. Mass. 1972), *rev'd,* —— F.2d ——, 3 ELR 20082 (1st Cir. 1973).

In *Silva v. Romney* the district court had enjoined the Department of Housing and Urban Development from approving a $4,000,000 mortgage guarantee and $156,000 interest grant to the developer of a housing project until HUD had prepared a proper impact statement. In subsequent litigation the district court refused to enjoin the developer from cutting trees on the site, apparently believing that it lacked jurisdiction over the private party. The First Circuit held that the developer could be enjoined and remanded for the district court to determine if a preliminary injunction should issue.

The circuit court reasoned that NEPA applied because HUD and the developer had concluded a binding contract when the agency issued its "180-day commitment." Citing a 1958 Supreme Court decision,[53] the court held that the contract cemented a "partnership" with the federal government on which basis the private partner could also be enjoined.

To buttress its decision, the court also cited other NEPA cases in which private actions had been enjoined. In particular it stated that "we subscribe fully to the rationale of the highway cases . . . that once the partnership stage has been reached . . ., all parties . . . are subject to injunctive process."[54] Presumably the court would put the *Arlington Coalition* and *San Antonio* cases at the top of its list.

To reach its holding the court was hard pressed to distinguish its earlier ruling in *City of Boston v. Volpe*.[55] In *City of Boston* the district court held that the filing of an application requesting federal assistance for the construction of an airport runway could not be used to prevent the state agency from proceeding on its own without preparing an impact statement. On appeal, the First Circuit affirmed. Judge Coffin noted the stress of the CEQ Guidelines on early evaluation, but refocused the inquiry on whether a preliminary injunction should be issued, not against the federal agencies, but against the airport authority. The court rejected the proffered analogy to highway funding, since the "tentative allocation" made by the Federal Aviation Administration was an "administrative device," "preliminary and tentative in nature," and denied relief.

Nevertheless, the circuit court in *City of Boston* was aware of the possibility that applicants would attempt to time funding requests to avoid early environmental review. Judge Coffin suggested:

---

[53] Ivanhoe Irrigation District v. McCracken, 357 U.S. 275, 295 (1958).

[54] —— F.2d at ——, 3 ELR at 20083 note 5.

[55] —— F. Supp. ——, 2 ELR 20169 (D. Mass.), *aff'd* 464 F.2d 254, 2 ELR 20501 (1st Cir. 1972).

Where the state authority does rely on the expectation of federal aid, it goes ahead with construction prior to approval only at great risk to the prospects for funding, since the options of the federal agency become increasingly limited to bald approval or rejection with no opportunity for modification.[56]

He also indicated that noncompliance should be viewed with a jaundiced eye by the funding agency.

In most cases a state or community would be sensitive to its environmental obligations, not only to avoid jeopardizing its chances of obtaining assistance for the specific project, but also to avoid a negative report on future projects associated with the same facility. For, as we have noted, the federal agencies cannot close their eyes to ill-advised actions of the past as they assess a project in the present.[57]

The artificial distinction which the court had to make between the "preliminary and tentative" allocation of funds in *City of Boston* and the "180-day commitment" in *Silva* prompted it to suggest in *Silva* that the "crucial action" approach which it had adopted should be abandoned. In dictum the court urged the federal agencies to adopt guidelines for a "prepartnership regulatory scheme" which would preserve the status quo even before the agency became fully involved.

While this case, as we have noted, is not controlled by *City of Boston*, we confess to a sense of growing uneasiness in seeing decisions determining the obligations of federal and non-federal parties under NEPA turn on any one interim step in the development of the partnership between the parties. Such an approach unrealistically stresses adventitious factors which bear little relationship to either the broad concerns of NEPA or the interests of the potential grantee, private or public. Hence, in the present case, the mere fact that a binding contract has been entered into between HUD and the developer is but one manifestation of and quite irrelevant to an ongoing planning process by all parties to the project which must provide for the reasonable expectations of the parties. . . .[58] . . . We see no insuperable barrier to a regulatory scheme which would require applicants for federal aid to a potential "major federal action" to maintain the status quo of the project pending federal environmental review in accordance with NEPA. Such a requirement, though obviously hindering the freedom of the private developer, would recognize that the developer has voluntarily submitted to some degree of federal regulation as private in nature but a project which is intended to and which can go ahead on the same scale generally only with public assistance.

---

[56] 464 F.2d at 259, 2 ELR at 20503.
[57] 464 F.2d at 260, 2 ELR at 20504.
[58] —— F.2d at ——, 3 ELR at 20084.

> The coexistence of this regulatory gap and the strength of the Congressional and Presidential directives might well justify a court in requiring an agency to formulate status quo regulations [court's footnote omitted].[59]

This undoubtedly is the best alternative for avoiding the constriction of federal choices that may occur if private parties or states can commit resources before a careful prognosis of environmental impact is made. As the First Circuit pointed out, it is unfair to ask both "partners" to accept the regime of all-or-nothing, the total acceptance or rejection of which may occur if options are gradually narrowed before the federal partner steps in. Moreover, one may doubt that an applicant's peril before a development-oriented agency is real. Such agencies are not likely to make it a practice to "punish" applicants who attempt to avoid NEPA through *faits accomplis.* Hence judicial intervention would again be necessary. Here, as in many areas, a viable administrative solution would be best.

HOW LARGE A "MAJOR" FEDERAL ACTION MUST BE

While NEPA's sweeping policy applies to every governmental action, not every such action must be accompanied by a formal written impact statement compiled in the course of the §102 process. Although an agency must strive to implement NEPA's policy in all its various activities,[60] it will have to prepare an impact statement only for "major" actions which "significantly" affect environmental quality.

The case of *Citizens for Reid State Park v. Laird*[61] illustrates the point. The Department of the Navy made environmental studies and took environmental precautions from the very beginning in planning a mock amphibious assault upon an oceanside state park. The navy consulted with a variety of scientists and sister agencies. Although the navy consciously strove to implement NEPA's policy, it also determined that an impact statement did not have to be filed, because in its opinion the environmental impact of the landing was too insignificant to merit full review. Clearly influenced by the measures taken without the necessity of applying §102(2) (C), the court agreed with the navy's determination.

---

[59] —— F.2d at ——, 3 ELR at 20085.

[60] R. Peterson, *An Analysis of Title I of the National Environmental Policy Act of 1969*, 1 ELR 50035, 50038 (January 1971).

[61] 336 F. Supp. 783, 2 ELR 20122 (D. Me. 1972).

The following three sections examine the trends and evolution of judge-made law on the issue of the threshold of "major" federal action "significantly" affecting the environment. The first looks at how the courts have ruled over the past three years to put the congressional mandate to work. No clear litmus test for major federal action has emerged, but several trends are discernible. The second examines the decisions which shed light on whether Congress intended for one or two tests to emerge from the phrase, "*major* federal action *significantly affecting* the quality of the human environment." While the cases favor a two-test standard, they have never held that the small size of a federal project will exempt it from NEPA if its environmental effects are significant. Finally, the third section takes up the important issue of the standard of review to be applied in deciding if an agency's decision not to prepare a statement will be upheld. While a bare majority of the courts have used the language of the standard of arbitrary and capricious abuse of discretion, they nevertheless have looked closely at agency determinations, frequently reversing the agencies. Other courts have openly applied a stricter standard.

Trends in Judicial Interpretation

As NEPA has evolved, the courts have not resisted reviewing agency determinations and applying the Act when an agency decision not to prepare a statement is challenged. Because Congress failed to define the lower limits of major actions and significant effects, the courts have had either to rely upon the bare words of §102(2)(C), or to accept the interpretations supplied by the CEQ and individual agencies. While agency guidelines have played an important part in shaping judicial decision making on this issue, the courts have also defined the key terms independently so that they encompass a variety of large and small federal actions.

That the courts should be active in this aspect of NEPA's implementation is not hard to understand. Determinations of the magnitude of federal projects to be covered by §102(2) (C) affect NEPA's applicability in the most fundamental way. If low-level federal activities were exempted, one of Congress' main purposes in enacting the statute would fail to be achieved. Yet NEPA's legislative history shows that Congress intended to interrupt business-as-usual and affect decision making at the lowest agency levels.

Judicial failure to implement this purpose could have meant that the Act would have hardly deflected the wheels of government from their accustomed ruts. For this reason the courts have policed the lower boundaries of NEPA's application with greater than ordinary vigilance and have worked in concert with the CEQ guideline-setting process to keep the threshold low.

Understandably, the courts have shied away from developing ready-to-use formulas for applying NEPA. As the court in *Citizens for Reid State Park v. Laird* remarked, "the statutory language 'significantly affecting the quality of the human environment' is extremely broad and not susceptible to precise definition."[62] The court in *Goose Hollow Foothills League v. Romney*[63] made the same observation, but less charitably. "The court recognizes the infirmity of the phrase, 'significantly affecting the quality of the human environment.' Legislative direction couched in such terms can hardly be expected to produce clarity, decisiveness or predictability in administrative decisions."[64]

Nevertheless, a court has occasionally attempted to utilize a formulary approach. For example, in *Natural Resources Defense Council v. Grant*[65] the court interpreted major action as any action which "requires substantial planning, time, resources and expenditure," but until the court explained that a plan to channelize 66 miles of stream in one watershed at a total cost of $1.5 million ($750,000 state funds) was "certainly" federal, it was impossible to know the relative magnitudes in which the court was thinking.

Perhaps the closest a court has come to a formulary approach occurred recently in the Second Circuit's second decision in *Hanly v. Kleindienst*.[66] The government had already conceded that the construction of the jail in question was a "major" federal action. The court held that the meaning of the term "significantly" in the phrase "significantly affecting the quality of the human environment" could be isolated and interpreted as a question of law. Conceding that "significantly" was an "amorphous term," and remarking that Congress, CEQ, and the agencies had all failed to give it specificity, the court held:

> In the absence of any Congressional or administrative interpretation of the term, we are persuaded that in deciding whether a major federal

---

[62] *Id.*, 336 F. Supp. at 789, 2 ELR at 20125.
[63] 334 F. Supp. 877, 1 ELR 20492 (D. Ore. 1971).
[64] 334 F. Supp. at 880, 1 ELR at 20492.
[65] 341 F. Supp. 356, 2 ELR 20185 (E.D. N. Car.), *remanded*, —— F.2d ——, 2 ELR 20555 (4th Cir.), —— F. Supp. ——, 2 ELR 20647 (E.D. N. Car.), —— F. Supp. ——, 2 ELR 20648 (E.D. N. Car. 1972), —— F. Supp. ——, 3 ELR 20176 (E.D. N. Car. 1973).
[66] —— F.2d ——, 2 ELR 20717 (2d Cir. 1972) (complete citation to the litigation in Appendix B under Hanly v. Mitchell).

action will "significantly" affect the quality of the human environment, the agency in charge, although vested with broad discretion, should normally be required to review the proposed action in the light of at least two relevant factors: (1) the extent to which the action will cause adverse environmental effects in excess of those created by existing uses in the area affected by it, and (2) the absolute quantitative adverse environmental effects of the action itself, including the cumulative harm that results from its contribution to existing adverse conditions or uses in the affected area.[67]

There is less to the court's test than meets the eye. As Chief Judge Friendly's dissent points out, the test may "raise the floor" to permit agencies to escape the impact statement requirement.[68] "Comparative" and "cumulative" effects may be very difficult to distinguish. The "comparative" effect, it would seem, should be great whenever the "cumulative" effect is great, while the "absolute" effect, which "may represent the straw that breaks the back of the environmental camel," seems in the court's mind again to be a comparative factor.

Perhaps the best way to understand how low the threshold of major federal action has been set is to examine the facts of the cases. Admittedly, such a review overlooks other legal issues and equities which may figure more centrally in the cases. Nor does it cover the vast array of federal activities for which statements are prepared that never involve judicial challenge. But it does confirm, in a way which general discussion of trends cannot, that the level of federal action to which §102(2) (C) applies has been pushed quite low.

Of course, the impact statement requirement applies to the trans-Alaskan oil pipeline,[69] to the Cross-Florida Barge Canal,[70] to the detonation of a 5-megaton nuclear warhead under Amchitka Island, Alaska,[71] to the leasing of 380,000 acres of submerged lands for oil development in the Gulf of Mexico,[72] to the cancellation of a

---

[67] —— F.2d ——, 2 ELR at 20720.

[68] —— F. 2d at ——, 2 ELR at 20724.

[69] Wilderness Society v. Hickel, 325 F. Supp. 422, 1 ELR 20042 (D.D.C. 1970), *sub nom.* Wilderness Society v. Morton, 463 F.2d 1261, 2 ELR 20250 (D.C. Cir.), —— F. Supp. ——, 2 ELR 20583 (D.D.C. 1972), *rev'd,* —— F.2d ——, 3 ELR 20085 (D.C. Cir. 1973).

[70] Environmental Defense Fund v. Corps of Engineers, 324 F. Supp. 878, 1 ELR 20079 (D.D.C. 1971).

[71] Committee for Nuclear Responsibility v. Seaborg, 463 F.2d 783, 1 ELR 20469 (D.C. Cir.), 463 F.2d 788, 1 ELR 20529 (D.C. Cir.), 463 F.2d 796, 1 ELR 20532 (D.C. Cir.), *application for injunction in aid of jurisdiction denied sub nom.* Committee for Nuclear Responsibility v. Schlesinger, 404 U.S. 917, 1 ELR 20534 (1971).

[72] Natural Resources Defense Council v. Morton, 337 F. Supp. 165, 2 ELR 20028 (D.D.C.), 337 F. Supp. 167, 2 ELR 20089 (D.D.C. 1971), *motion for*

national program to purchase irreplaceable helium,[73] and to a nine-state program for spraying colonies of fire ants with insecticide.[74] But courts have also confirmed that NEPA applies to:

A federal block grant of $775,000 for a prison medical facility and prisoner reception center in a rural, historic area of Virginia[75]

A Corps of Engineers project to clear 3,000 acres of oxygen-consuming vegetation from 55 miles of the Gila River[76]

A permit which the Forest Service proposed to issue for the construction of a pulp mill on 200 acres within the Tongass National Forest, Alaska[77]

A grant for construction of a 26-mile highway (with 7.5 miles of additional upgrading of an existing road) to traverse part of the Santa Fe National Forest[78]

A HUD loan of $3.1 million for the construction of a 16-story high-rise apartment building in Portland, Oregon[79]

Corps of Engineers condemnation of 250 acres of grazing land for a flood control project[80]

Corps of Engineers permits to be issued to water-polluting facilities under the Refuse Act Permit Program[81]

The construction of an incinerator at the Walter Reed Medical Center[82]

A downtown Washington, D.C., urban renewal project covering five blocks[83]

---

summary reversal denied, 458 F.2d 827, 2 ELR 20029 (D.C. Cir.), *dismissed as moot*, 337 F. Supp. 170, 2 ELR 20071 (D.D.C. 1972).

[73] National Helium Corp. v. Morton, 326 F. Supp. 151, 1 ELR 20157 (D. Kan. 1971), *aff'd*, 455 F.2d 650, 1 ELR 20478 (10th Cir. 1971).

[74] Environmental Defense Fund v. Hardin, 325 F. Supp. 1401, 1 ELR 20207 (D.D.C. 1971).

[75] Ely v. Velde, Appendix B.

[76] Sierra Club v. Laird, —— F. Supp. ——, 1 ELR 20085 (D. Ariz. 1970).

[77] Sierra Club v. Hardin, 325 F. Supp. 99, 1 ELR 20161 (D. Alas. 1971).

[78] Upper Pecos Ass'n v. Stans, 328 F. Supp. 332, 1 ELR 20228 (D.N.M.), *aff'd*, 452 F.2d 1233, 2 ELR 20085 (10th Cir. 1971), *vacated*, 93 S. Ct. 458 (1972).

[79] Goose Hollow Foothills League v. Romney, Appendix B.

[80] United States v. 247.37 Acres of Land, —— F. Supp. ——, 1 ELR 20513 (S.D. Ohio 1971).

[81] Kalur v. Resor, 335 F. Supp. 1, 1 ELR 20637 (D.D.C. 1971).

[82] Montgomery County v. Richardson, —— F. Supp. ——, 2 ELR 20140 (D.D.C. 1972).

[83] Businessmen Affected Severely by the Yearly Action Plans, Inc. v. District of Columbia City Council, 339 F. Supp. 793, 2 ELR 20237 (D.D.C. 1972). *See also* Zlotnick v. Redevelopment Land Agency, —— F. Supp. ——, 2 ELR 20235 (D.D.C. 1972).

The comptroller of the currency's approval of an application to open a branch bank office in Woodstock, Vermont[84]

Temporary ICC suspension of rail freight rates to allow a 2.5 percent surcharge affecting the shipment of recyclable metal scrap[85]

The exchange of 10,200 acres of national forest land for 20,500 acres held by Big Sky of Montana, Inc.[86]

The review by the National Capital Planning Commission of plans for construction of a private commercial mall and housing complex in a Washington, D.C., neighborhood[87]

An FHA loan for the construction of a golf course and park in Texas[88]

Repair and expansion by the Park Service of the towpath along segments of the historic C&O Barge Canal near Washington, D.C.[89]

The cases which deny that NEPA is applicable because a major federal action with significant environmental effects was not involved also confirm that the courts have given NEPA a very broad reading. As with the cases just cited, other legal issues and equities than the issue of the action's magnitude may figure centrally in the final decision. The courts have held that NEPA does *not* apply to:

The construction by the U.S. Postal Service of a bulk mail facility on a 63-acre tract within a 350-acre industrial park[90]

The introduction of stretch jets into Washington National Airport in 1968[91]

HUD insurance assistance for a proposed 66-unit apartment building in Los Angeles[92]

The navy's plan to conduct a mock amphibious assault of 900

---

[84] Billings v. Camp, —— F. Supp. ——, 2 ELR 20687 (D.D.C. 1972).

[85] City of New York v. United States, 337 F. Supp. 150, 2 ELR 20275 (E.D.N.Y.), 344 F. Supp. 929, 2 ELR 20688 (E.D.N.Y. 1972) (three-judge panels).

[86] National Forest Preservation Group v. Butz, 343 F. Supp. 696, 2 ELR 20571 (D. Mont. 1972).

[87] McLean Gardens Residents Ass'n v. National Capital Planning Comm'n, Appendix B.

[88] Texas Committee on Natural Resources v. United States, Appendix B.

[89] Berkson v. Morton, —— F. Supp. ——, 2 ELR 20659 (D. Md. 1971).

[90] Maryland-National Capital Park and Planning Comm'n v. United States Postal Service, C.A., 349 F. Supp. 1212, 2 ELR 20656 (D.D.C. 1972).

[91] Virginians for Dulles v. Volpe, 334 F. Supp. 573, 2 ELR 20359 (E.D. Va. 1972).

[92] Echo Park Residents' Committee v. Romney, Appendix B.

marines for a three-day bivouac in an ocean-front state park in Maine[93]

HUD's intention to insure a $3.8 million loan for the construction of a 272-unit apartment complex in a semirural neighborhood 15 miles south of Houston, Texas[94]

The Interior Department's plan to fence 6,800 acres of grazing land acquired by condemnation to protect a proposed reservoir[95]

The Interior Department's plan to lease an additional 120 acres adjacent to acreage in an original 1961 lease for coal extraction for the San Juan Project in the Four Corners Energy Project[96]

A Federal Power Commission permit to add a 7-acre addition to an existing 17-acre liquid natural gas facility on the company's 500-acre holding in the New Jersey meadowlands[97]

U.S. Forest Service construction of a final 4.3-mile segment of a 16-mile-long gravel-surfaced one-lane road begun 30 years ago[98]

An Environmental Protection Agency grant for construction of a regional sewage treatment plant (no significant environmental impact)[99]

Several of these cases, however, cannot be fairly stated to be "major federal action" cases only. *Echo Park Residents' Committee v. Romney*,[100] involving a 66-unit housing project in Los Angeles, is widely cited as involving a project that did not meet the statutory standard; however, the court's conclusory determination that NEPA would not apply simply does not reveal the grounds on which the court acted.

The refusal of the court in *Jicarilla Apache Tribe of Indians v. Morton*[101] to apply NEPA to leasing 120 additional acres of strip-minable coal is perhaps the only defensible holding in a decision

[93] Citizens for Reid State Park v. Laird, Appendix B.
[94] Hiram Clarke Civic Club, Inc. v. Romney, Appendix B.
[95] Maddox v. Bradley, 345 F. Supp. 1255, 2 ELR 20404 (N.D. Tex. 1972).
[96] Jicarilla Apache Tribe of Indians v. Morton, —— F. Supp. ——, 2 ELR 20287 (D. Ariz. 1972), *aff'd*, —— F. 2d ——, 3 ELR 20045 (9th Cir. 1973).
[97] Transcontinental Gas Pipeline Co. v. Hackensack Meadowlands Development Comm'n, 464 F.2d 1358, 2 ELR 20495 (3rd Cir. 1972).
[98] Kisner v. Butz, 350 F. Supp. 310, 2 ELR 20709 (N.D. W.Va. 1972).
[99] Howard v. Environmental Protection Agency, —— F. Supp. ——, 2 ELR 20745 (W.D. Va. 1972).
[100] Appendix B.
[101] Appendix B.

which otherwise utterly fails to apply NEPA in the manner that Congress intended. The question of whether NEPA applies to the particular tract of 120 acres would be better treated as part of the larger question of NEPA's applicability to the San Juan Project and ultimately the entire Four Corners energy complex.

In *Virginians for Dulles v. Volpe*,[102] for reasons that are not at all clear, the court chose to find NEPA inapplicable to the introduction of stretch jets into Washington National Airport in 1968, not because NEPA cannot be applied retroactively, but because the action was not a major federal action. If regarded as part of a continuing federal action, rather than a completed one, introduction of the larger jets nevertheless did involve substantial front-end costs that make the action more like a one-time action than a continuing federal "program."

The court in *Citizens for Reid State Park v. Laird*[103] was strongly impressed with the elaborate environmental precautions taken by the navy in preparation for its mock amphibious assault. Had the precautions not been so thorough, and the study of potential adverse effects so complete, it is possible that the case would have had a different outcome. When it became clear to the court that known adverse impacts would be small, and especially that preparation and study had substantially reduced uncertainty about possible impacts, the court was reluctant to require more.

Finally, *Transcontinental Gas Pipeline Corp. v. Hackensack Meadowlands Development Comm'n*[104] may actually be viewed as authority that under present guidelines, §102(2) (C) would apply to the permit to add 7 acres to the existing 17-acre liquid natural gas facility. In *Transco* the Third Circuit held only that at the time the Federal Power Commission made its decision to permit the expansion, neither the agency nor the CEQ had published guidelines, and a "reasonable man" would probably not have thought the action to be either a "federal action" or "major."[105]

Turning away now from the facts of the cases to trends in interpretation, it appears that in some instances the courts have used NEPA's broad language and the CEQ Guidelines to reinforce their determinations that §102(2) (C) applies to particular projects of modest size. If there is any question about the Act's applicability, the court may look to a challenged project's long-range effects, or to its

[102] Appendix B.
[103] Appendix B.
[104] Appendix B.
[105] 464 F.2d at 1366, 2 ELR at 20499.

impacts when cumulated with those of allied activities, to find that potential direct and indirect environmental impacts in a larger context are indeed significant. The more consequences and alternatives one considers, the more important the narrow "triggering" action which the plaintiff has challenged may seem. For example, in *Ely v. Velde*[106] the Law Enforcement Assistance Administration argued that since the federal government was providing only 20 percent of the construction costs of a prison facility, the action was not a major federal action. The court said, "We think that in view of LEAA's overall involvement in the promotion and planning of the Center, as well as the cumulative impact of the proposed federal action, the NEPA definition of 'major federal action' has been satisfied," citing CEQ Guidelines.[107] In *Goose Hollow Foothills League v. Romney*,[108] Judge Goodwin reinforced his finding that an impact statement had to be prepared for a loan guarantee by HUD on a high-rise apartment building by stating, "I believe the defendants have ignored cumulative effects on the quality of the human environment which are by no means insignificant."[109] In *United States v. 247.37 Acres of Land*[110] the court pointed out that the reservoir for which land was being condemned was "but a small cog in an overall big plan," the overall flood control project having major environmental consequences.[111]

A second important trend in the NEPA cases on major federal action reveals that, consistent with the usual judicial role in reviewing administrative action, the courts have first examined what the agencies have required of themselves. The CEQ Guidelines go beyond the words of the Act, and to some extent fill in the gaps left by Congress. In addition, §3(a) (1) of the CEQ Guidelines calls upon the agencies themselves to develop their own guidelines defining agency actions to which NEPA will apply. Most agencies have done so, and the courts have had frequent recourse to agency and CEQ Guidelines in further defining the size and importance of the federal projects to which §102(2) (C) applies.

If agency guidelines cover the proposed action, or if the agency

---

[106] Appendix B.

[107] 451 F.2d at 1137 note 22, 1 ELR at 20614 note 22.

[108] Appendix B.

[109] 334 F. Supp. at 879, 1 ELR at 20492.

[110] Appendix B.

[111] —— F. Supp. at ——, 1 ELR at 20513. *See also* Sierra Club v. Froehlke (Trinity River–Wallisville Dam), Appendix B, where the court refused to consider the Wallisville Dam apart from the larger Trinity River Project, of which it was a small part.

has begun or completed an impact statement, or otherwise conceded that a statement should be prepared, then courts may view these actions as "interpretative" of NEPA, embodying the best administrative judgment of one of the agencies charged with putting the Act into effect. The surprising result is that in some instances the agencies themselves are responsible for the courts' application of the §102(2)(C) requirement to actions that some have considered too trivial, too routine, or too remote in consequences to merit an impact statement.

In several cases an agency has been, as it were, hoist with its own petard because its guidelines explicitly called for impact statement preparation: "A government agency is bound by its own regulations."[112] In *Scherr v. Volpe*[113] the Federal Highway Administration tried to argue that federal funding of the expansion of a 12-mile, two-lane state highway to four lanes was not a major federal action significantly affecting environmental quality. In upholding the district court, the Seventh Circuit tripped the FHWA over both a guideline defining major federal action which by its "express terms" applied to the project and over a second guideline that listed "significant effects" which the district court had identified as applicable to the disputed project.[114]

In *Silva v. Romney*[115] it was HUD's turn. There, the Department of Housing and Urban Development disputed the preparation of an impact statement for a $4 million mortgage guarantee and $156,000 interest loan for a 138-unit housing project. The district court held that "in light of the policy considerations behind NEPA and the fact that the . . . project appears likely to be found to be a major federal action, as defined by HUD's own guideline,"[116] a "special environmental clearance worksheet" prepared in lieu of an impact statement did not satisfy NEPA. Of the two relevant HUD guidelines, one required statements for controversial projects and the other denominated projects of this kind of more than 100 units as "major federal actions." Other NEPA cases appear to be in accord with *Scherr* and *Silva*.[117]

---

[112] Silva v. Romney, Appendix B, 342 F. Supp. at 785, 2 ELR at 20386.

[113] Appendix B.

[114] 466 F.2d at 1032, 2 ELR at 20455.

[115] Appendix B.

[116] 342 F. Supp. at 785, 2 ELR at 20386.

[117] *See* SCRAP v. United States, note 14; National Helium Corp. v. Morton; Hiram Clarke Civic Club v. Romney; Daly v. Volpe (second district court decision, on application of guidelines to statement adequacy), all cited in Appendix B.

Although agencies are not estopped to deny that they must prepare impact statements just because they have taken steps which suggest that they think they should, such measures are strongly indicative of accepted agency practices and may influence courts to require that complete statements be prepared. In the Gillham Dam case,[118] the defendants had already prepared an impact statement, but this did not improve their challenge to compliance and did not help them convince the district court that plaintiffs sought a retroactive application of the Act. In *Sierra Club v. Hardin*[119] defendants argued that the granting of a permit was not a major federal action, but the Department of Agriculture had forwarded a hastily compiled statement to CEQ one day before the permit was issued. Because of this action, the court simply assumed that a statement had to be prepared, although it also found substantial compliance. Finally, in *City of New York v. United States*[120] the court found that NEPA applied to the action of the ICC in approving a petition for railroad abandonment. The three-judge panel ruled:

> The Commission has gone a long way to resolving whatever doubt there was on the question by its proposed rules which include railroad abandonment proceedings among those actions which may have a significant effect on the quality of the environment.[121]

Agency practices lowering the threshold of "major" federal action have also occurred on a more subtle and perhaps more pervasive level. The agencies have by no means uniformly resisted preparing impact statements; many have prepared statements for a wide range of agency activities. By preparing the statement in a case which ultimately is litigated on another point ("retroactivity," adequacy, timing, etc.) the agency may have inadvertently contributed an example of a "concededly" major action which in time will be cited to show that a similar action falls within NEPA's requirements. For example, the defendants in *Upper Pecos Ass'n v. Stans*[122] conceded that a relatively modest highway undertaking required an impact statement, but disputed who should prepare it and when it had to be submitted. In *Greene County Planning Board v. Federal*

---

[118] Environmental Defense Fund v. Corps of Engineers, 325 F. Supp. 728, 1 ELR 20130 (E.D. Ark. 1970–71), —— F. Supp. ——, 2 ELR 20260 (E.D Ark.), 342 F. Supp. 1211, 2 ELR 20353 (E.D. Ark.), *aff'd* 470 F.2d 289, 2 ELR 20740 (8th Cir. 1972).

[119] Appendix B.

[120] Appendix B.

[121] 337 F. Supp. at 158, 2 ELR at 20276.

[122] Appendix B.

Power Comm'n[123] the commission conceded that the authorization for a 32-mile transmission line, one of three leading from a pumped storage project, was a major federal action, but chose instead to dispute when the commission had to prepare its statement. In *Hanly v. Mitchell*[124] defendants conceded that the construction of a jail in downtown Manhattan was a major federal action, although they contested the filing of a statement on the ground that the building would not significantly affect the local environment.

A third trend in judicial determinations regarding "major" actions concerns the evolution of *per se* categories to which the §102(2) (C) requirement will automatically apply. Judge Plummer in *Sierra Club v. Hardin*[125] expressed qualified optimism that "it may be possible in the future to develop some *per se* categories of major federal actions," but went on to say that so far as the construction of pulp mills was concerned, "for the present complete investigation of the impact of individual mills will continue to be appropriate."[126] The Third Circuit, however, did not even consider that the development of *per se* categories was a serious possibility. "Whether a project is a 'major federal action' is, of course, a question which can only be resolved through a careful case-by-case analysis."[127]

While the Third Circuit view probably represents the prevailing judicial attitude, *per se* categories have nevertheless begun to evolve in fact, either because NEPA's legislative history singled out certain obvious activities as stimulating NEPA's passage, or because agency practices and guidelines have established them. The trend is clearest for highway projects, water resources projects, and electrical generating facilities. It is less clear for housing programs, such as loan and mortgage guarantees, interest grants, and the like.

The Federal Highway Administration (FHWA) has adopted the policy of preparing impact statements for practically every "project" which it approves or funds.[128] The thoroughness with which FHWA has made highway projects a *per se* category of NEPA implementation has even been challenged on environmental grounds.

---

[123] 455 F.2d 412, 2 ELR 20017 (2d Cir.), *cert. denied* 41 U.S.L.W. 3148 (Oct. 10, 1972).

[124] Appendix B.

[125] Appendix B.

[126] 325 F. Supp. at 126 note 52, 1 ELR at 20171 note 52.

[127] Transcontinental Gas Pipeline Co. v. Hackensack Meadowlands Development Comm'n, 464 F.2d 1358, 1366, 2 ELR 20495, 20499 (3rd Cir. 1972).

[128] R. Peterson and R. Kennan, *The Federal-Aid Highway Program: Administrative Procedures and Judicial Interpretation*, 2 ELR 50001, 50014, 50017 (May 1972).

The criticism calls for more of a case-by-case approach, so that small and environmentally insignificant projects are bypassed in order to leave time for comprehensive treatment of larger projects with greater environmental impacts.

In *Scherr v. Volpe*,[129] the exception proved the rule. As already discussed, in *Scherr* the FHWA attempted to go back on its policy of impact statement preparation for small segments of highway. The court refused to allow FHWA to make an exception, citing its guidelines. In *Nolop v. Volpe*,[130] a case with facts very similar to those in *Scherr* (which was decided a month later), the FHWA did not challenge the project, choosing to concede that it was major federal action. Whether proven by the exception or by the rule, a *per se* category of federal action to which §102(2) (C) automatically applies appears to have evolved for highway projects.[131] Sometimes preparation takes place even where environmental impacts are practically nonexistent.

The Corps of Engineers has also been thorough in implementing the §102(2) (C) requirement. However, while assuming that NEPA applies to its projects and programs,[132] the Corps has not in general been criticized for overpreparation in the manner of the FHWA. Nevertheless, one may ask whether in *United States v. 247.37 Acres of Land*,[133] where defendants argued successfully that NEPA should apply to the condemnation of 250 acres of land for a Corps reservoir, the court's finding that NEPA had to be complied with was not a result of the evolution of a *per se* approach where Corps projects are concerned.

Likewise, the projects which the Atomic Energy Commission licenses or constructs are not usually capable of segmentation. Any action that the commission takes is likely to involve a facility which may cause highly significant environmental impacts. Thus while the siting and operation of nuclear power plants may constitute a *per se* category of NEPA implementation, no cases exist which illustrate how a *per se* approach resulted in preparation of an impact statement for an otherwise questionable federal action.

Programs involving the urban environment and housing, such as Farmers Home Administration loans, construction loans, interest

[129] Appendix B.

[130] 333 F. Supp. 1364, 1 ELR 20617 (D.S. Dak. 1971).

[131] *See also* Indian Lookout Alliance v. Volpe, Appendix B.

[132] *See, e.g.*, Texas Committee on Natural Resources v. Resor, —— F. Supp. ——, 1 ELR 20466 (E.D. Tex. 1971).

[133] Appendix B.

loans, mortgage guarantees, and other activities of the Department of Housing and Urban Development, and federally approved zoning or construction, appear to be moving toward *per se* treatment. The court in the first NEPA case, *Texas Committee on Natural Resources v. United States,*[134] said with regard to an FHA loan for construction of a golf course and recreational facility that "there is little doubt that in the future the type of activity involved here would be covered by the statute."[135] Recently, the court in *McLean Gardens Civic Ass'n v. National Capital Planning Comm'n*[136] held without discussion that the commission's review of the District of Columbia decision to allow construction of an intensively developed shopping mall and housing complex in a Washington neighborhood was major federal action. Major federal action was found on similar facts in *Businessmen Affected Severely by the Yearly Action Plans v. District of Columbia.*[137]

HUD programs present a mixed picture. In *Silva v. Romney*[138] the court held the department to an agency guideline which required the preparation of impact statements for multifamily projects of 100 units or more. Thus as far as this particular type of project is concerned, the agency itself established the category. For a similar project, CEQ Guidelines were held explicitly to govern.[139] Finally, the court in *Boston Waterfront Residents Ass'n, Inc. v. Romney*[140] found that demolition of buildings in the Fulton Street waterfront urban renewal project was a major federal action without citing agency or CEQ guidelines, choosing to rely directly on NEPA itself.

Two cases involving HUD projects, however, go the other way. In *Echo Park Residents' Committee v. Romney*[141] the court held that HUD's determination that NEPA did not apply to a 66-unit housing project was not an abuse of discretion. The basis of the court's reasoning does not appear in its terse Findings of Fact and Conclusions of Law. In *Hiram Clarke Civic Club v. Romney*[142] the court refused to disturb a "negative declaration" by HUD that an impact statement did not have to be prepared for a 272-unit apartment complex.

---

134 Appendix B.
135 —— F. Supp. at ——, 2 ELR at 20574.
136 Appendix B.
137 Appendix B.
138 Appendix B.
139 Goose Hollow Foothills League v. Romney, Appendix B.
140 343 F. Supp. 89, 2 ELR 20359 (D. Mass. 1972).
141 Appendix B.
142 Appendix B.

These two cases cast doubt on the evolution of a *per se* category of NEPA implementation as far as HUD programs are concerned. Until the agency and the courts have further interpreted HUD's guidelines and the Act itself, HUD projects are more likely to be judged on a case-by-case basis. This is especially true because HUD's implementation of NEPA has been far from enthusiastic[143] and because HUD, like the FPC and the ICC, has adopted a policy of issuing "negative declarations."[144]

A fourth trend in NEPA interpretation can be traced to the suggestion in the CEQ Guidelines that highly controversial actions should always be covered in impact statements.[145] This suggestion is subject to conflicting interpretations. On its face it is premised on the notion that controversy automatically makes an action "major" and that the public is an acceptable judge of significant environmental impacts. The Second Circuit disagrees, stating in what we believe to be a strained interpretation that the term "apparently refers to cases where a substantial dispute exists as to the size, nature or effect of the major federal action rather than to the existence of opposition to a use, the effect of which is relatively undisputed."[146] According to this view, neighborhood opposition, for example, is not enough to trigger the §102 process.

There exists a further dispute about the weight to be given the CEQ "requirement." In both *Save Our Ten Acres v. Kreger*[147] and the first district court opinion in *Hanly v. Mitchell*,[148] the courts held that the CEQ Guidelines were "advisory only," citing *Greene County Planning Board v. Federal Power Comm'n.*[149] Perhaps the unquoted part of the key sentence from *Greene County* provides the clue why three other courts accepted the CEQ point of view.

We would not lightly suggest that the Council, entrusted with the responsibility of developing and recommending national policies "to

---

[143] Of the total of 3,635 draft or final impact statements filed with CEQ by December 31, 1972, only 52 were prepared by the Department of Housing and Urban Development. COUNCIL ON ENVIRONMENTAL QUALITY, 102 MONITOR, vol. 2, no. 12, p. 82 (January 1973).

[144] *See, e.g.*, SCRAP v. United States (ICC) note 16, Appendix B. *See also* COUNCIL ON ENVIRONMENTAL QUALITY, ENVIRONMENTAL QUALITY, THIRD ANNUAL REPORT 232 (August 1972).

[145] CEQ Guidelines, §5(b), Appendix C.

[146] Hanly v. Kleindienst, Appendix B. *See* Hanly II, —— F.2d at ——, 2 ELR at 20720.

[147] Appendix B.

[148] Appendix B.

[149] Appendix B.

foster and promote the improvement of environmental quality," has misconstrued NEPA.[150]

When an agency has incorporated a test of "controversy" into its *own* procedures for deciding when to prepare statements, however, it is thereafter bound by such a rule. The court in *Nolop v. Volpe*[151] turned to the applicable FHWA procedures which "further require environmental impact statements 'where organized opposition has occurred or is anticipated to occur,' " and found that "the mere existence of this lawsuit shows that organized opposition has occurred. Polls showed that over 80 percent of the 5,625 University of South Dakota students oppose this project."[152] *Silva v. Romney*[153] involved a similar question. There the district court turned to HUD regulations which referred to "controversy" and found: "The regulations provide no guidance as to the definition of 'controversial.' However, evidence presented to the court showed there is considerable opposition to the project. . . . A government agency is bound by its own regulations."[154]

The fifth and last trend seems on reflection almost inevitable. What is surprising is that the trend has been so little in evidence during the three years the courts have been involved in NEPA's implementation. In the absence of rules or *per se* categories of major federal action, the courts might be expected to engage rather frequently in comparisons of the relative magnitudes of projects which have already been held to be important enough to require a statement. This the courts have not done frequently, although a few examples exist.

In *Sierra Club v. Hardin*[155] the court favorably compared the granting of a permit to construct a pulp mill in a national forest to the permit for the Alaskan oil pipeline and to a much smaller project for clearing vegetation from a portion of the Gila River in Arizona.[156] In *Environmental Defense Fund v. Corps of Engineers*[157]

---

[150] 455 F.2d at 421, 2 ELR at 20021. *See* Goose Hollow Foothills League v. Romney, Appendix B, 334 F. Supp. at 879, 1 ELR at 20492, Izaak Walton League v. Schlesinger, 337 F. Supp. 287, 294, 2 ELR 20039, 20042 (D.D.C.), *court-approved settlement*, 2 ELR 20039 (D.D.C. 1971); SCRAP v. United States, Appendix B, 346 F. Supp. at 201, 2 ELR at 20491.

[151] Appendix B.

[152] 333 F. Supp. at 1368, 1 ELR at 20618.

[153] Appendix B.

[154] 342 F. Supp. at 784, 2 ELR at 20385.

[155] Appendix B.

[156] 325 F. Supp. at 125 note 52, 1 ELR at 20175, note 52.

[157] Appendix B (Gillham Dam), 325 F. Supp. at 744, 1 ELR at 20136.

Judge Eisele compared a proposed Corps dam to the Gila River project.[158] In *SCRAP v. United States*[159] Judge Skelley Wright compared an ICC-approved temporary suspension of rate limits on the rail shipment of recyclable metal scrap to the Cannikin nuclear test, the San Antonio Freeway, and the approval of the abandonment of "a few miles of track."[160]

Because such comparisons have not been attempted in very many cases, for the present it is virtually impossible to isolate the courts' comparative criteria of selection. Thus we have not gone beyond the few trends cited here to attempt to identify a group of common factors which the majority of the courts use to separate "major" and "minor" action, e.g., impact on crowding, wildlife, fauna, land abuse, etc. Such comparative criteria may be isolated in the future, but for the present the most one can say is that an extremely diverse collection of impacts has been considered as the courts have applied NEPA on a case-by-case basis.

Do "Major Federal Action" and "Significant Effects" Constitute One or Two Tests?

A literal reading of the statutory language which defines the federal actions to which NEPA applies may suggest that two statutory standards have to be met. The phrase, "major federal action significantly affecting" the environment, seems to require that actions first have to be found to be "major," and then, after this determination is made, must also be found to have potentially "significant" environmental effects. Courts that have specifically considered this issue are divided but favor a two-test standard. Numerous cases implicitly resolve the question, however, by simply assuming that Congress intended NEPA to cover *all* pending federal actions that may cause significant environmental effects. Thus the use of both "major" and "significantly affecting" appears to have been for emphasis.[161] On the one hand, federal action does not technically become "major" just because it may be accompanied by significant environmental

[158] The Gila River Project was successfully challenged for lack of NEPA compliance in Sierra Club v. Laird, Appendix B.

[159] Appendix B.

[160] 346 F. Supp. at 200, 2 ELR at 20490. He was speaking of the facts in Committee for Nuclear Responsibility v. Seaborg; Named Individual Members of the San Antonio Conservation Society v. Texas Highway Dept.; and City of New York v. United States, all cited in Appendix B.

[161] The legislative history on this point is at best ambiguous. *See* S. Rep. No. 91–296, 91st Cong., 1st Sess. 20 (1969).

effects; on the other, it makes little sense to find a project minor when its effects are significant.

The court in *Citizens for Reid State Park*[162] stated the one-test view well:

> NEPA . . . require[s] all federal agencies to incorporate as an integral part of their planning process consideration of the environmental consequences of *any* proposed action, and wherever such consideration indicates that the action may significantly affect the quality of the human environment to prepare and file a detailed impact statement [emphasis added].[163]

However, as in many opinions which discuss this point, the court never quite comes to grips with the issue. Other language in the opinion indicates that this directive may not be incompatible with the twofold test. Further, the court was assisted in reaching its conclusion by the language in the Department of Defense's own procedural compliance guidelines, which define "major federal action significantly affecting the quality of the human environment" as any decision that would "either affect the environment on a large geographical scale or have a serious environmental effect in a more restricted geographical area."[164]

Many cases implicitly decide that "major federal action" is to be judged by the magnitude of the potential environmental consequences of going ahead. For example, in *Virginians for Dulles v. Volpe*[165] the court concluded:

> The introduction of the "stretch" jets into [Washington National Airport] . . . in 1968 was not a "major action" as that term is used in section 102(2) (C). It bases this conclusion on its findings that the difference between the 727–200 (stretch jet) and 727–100 (the stretch jet's predecessor) is minimal insofar as "affecting the quality of human environment" is concerned.[166]

In *Izaak Walton League of America v. Schlesinger*,[167] the court preliminarily enjoined the issuance of an interim operating license to the 809-megawatt Quad Cities nuclear power station because an impact statement had not been prepared. The court's discussion of whether a "major Federal action" was involved in the licensing

---

[162] Appendix B.

[163] 336 F. Supp. at 788, 2 ELR at 20124.

[164] 336 F. Supp. at 787, note 5, 2 ELR at 20124, note 5. DOD Directive 6050.1, 36 Fed. Reg. 15750 (Aug. 9, 1971), ELR 46066.

[165] Appendix B.

[166] 344 F. Supp. at 578, 2 ELR at 20362.

[167] Appendix B.

focused upon plaintiff's exhibits and affidavits which indicated that the environment might be significantly affected.[168] The court did not also consider the magnitude of the project aside from its environmental impact. Likewise, in a similar case involving the construction of the Astoria nuclear power station on the East River in Queens borough, New York, the court found that the Corps of Engineers had improperly determined that the construction permit which it was about to grant did not involve a "major Federal action significantly affecting the quality of the human environment." In so doing, the court in *Citizens for Clean Air v. Corps of Engineers*[169] discussed potential environmental impacts and did not separately examine the magnitude of the proposed plant.

In *Conservation Society of Southern Vermont v. Volpe*[170] the court focused its attention on "one little hill and one beaver pond," finding that "if there has been any substantial showing of potential serious environmental harm, an impact statement must be filed"[171] for proposed construction on the relevant portion of Route 7 between Bennington and Manchester, Vermont. The court did not discuss the magnitude of the project in deciding whether an impact statement had to be prepared. See also *City of New York v. United States* and *Hiram Clarke Civic Club v. Romney,*[172] where courts again focused on environmental impacts alone. *Pizitz v. Volpe*[173] also deserves mention. The district court in that case appeared to sanction the Federal Highway Administration's funding under a clear one-test standard: "no major adverse impact on the environment."[174] Likewise, environmental effects alone were specifically discussed as determinative of whether a statement had to be prepared in *Maryland–National Capital Parks and Planning Comm'n v. United States Postal Service.*[175]

In *Sierra Club v. Hardin,*[176] the federal actions contemplated were permits for timbering and land patents for pulp mill construction in an Alaskan national forest. Because a statement had been

[168] 337 F. Supp. at 295, 2 ELR at 20042.

[169] 349 F. Supp. 696, 2 ELR 20650 (S.D.N.Y. 1972).

[170] 343 F. Supp. 761, 2 ELR 20270 (D. Vt. 1972).

[171] 343 F. Supp. at 767, 2 ELR at 20272.

[172] Appendix B for both.

[173] —— F. Supp. ——, 2 ELR 20378 (M.D. Ala.), *aff'd,* 467 F.2d 208, 2 ELR 20379 (5th Cir.), *modified on rehearing,* 467 F.2d 208, 2 ELR 20635 (5th Cir. 1972).

[174] —— F. Supp. at ——, 2 ELR at 20379.

[175] Appendix B.

[176] Appendix B.

hastily compiled and circulated, the court assumed without deciding that NEPA applied to the Forest Service's actions. Nevertheless, the court extensively discussed the issue of whether NEPA should apply. Rather than discuss separately whether the several federal permissions were "major" actions with "significant effects," the court went directly to the question of the actual environmental impact of the logging and milling. The court not only implicitly adopted a "substantial effect" test, but it also appeared to say that where it is as yet uncertain what the full environmental consequences are of particular activities, statements must nevertheless be prepared. The court thus drew attention to what will certainly become a major issue in NEPA's application: How can a decision be made whether to require a statement, until it is known just how significant the potential adverse impacts may be? The court very clearly sees that for a large class of federal actions NEPA's language begs the question. Thus, a statement on the actual expected impact of a proposed federal action may be necessary before a decision can be made on whether the statement should have been prepared at all.[177]

Several cases do explicitly state that NEPA imposes a two-part test. Such was the view of the Second Circuit in *Hanly I*[178] when it upheld defendant's contention that "major federal action"

> . . . Refers to the cost of the project, the amount of planning that preceded it, and the time required to complete it, but does not refer to the impact of the project on the environment. We agree with defendants that the two concepts are different and that the responsible federal agency has the authority to make its own threshold determinations as to each in deciding whether an impact statement is necessary.[179]

However, *Hanly I* cites *Citizens for Reid State Park*[180] for this conclusion. As we discussed, *Citizens for Reid State Park* supports the single-test standard. The Second Circuit's view finds somewhat firmer support in the other district court opinion which it cited, *Natural Resources Defense Council v. Grant:*[181]

> An administrative agency may make a decision that a particular project is not major, *or* that it does not significantly affect the quality of the

[177] 325 F. Supp. at 125 note 52, 1 ELR at 20171 note 52. Sierra Club v. Mason, 351 F. Supp. 419, 2 ELR 20694 (D. Conn. 1972), raises the same issue.

[178] Hanly v. Mitchell, Appendix B.

[179] 460 F.2d at 644, 2 ELR at 20218.

[180] Appendix B.

[181] Appendix B.

human environment, and, that, therefore, the agency is not required to file an impact statement [emphasis added].[182]

In *Scherr v. Volpe*[183] federal defendants asked the Seventh Circuit to review a district court ruling which held that the Federal Highway Administration had abused its discretion in finding that an impact statement did not have to be prepared for the conversion to four lanes of Highway 16, a 12-mile, two-lane highway located in the Lake Kettle Moraine area of Wisconsin. Noting that the administrative record was virtually barren of facts supporting the agency's determination, the circuit court affirmed the district court on the basis that the FHWA's regulation regarding impact statement preparation appeared to be drafted specifically to cover highway expansion such as that contemplated for Highway 16. The regulations (PPM 90-1) adopted a two-part test, which the court applied without comment.

Thus by the express terms of the defendants' own regulation, Highway 16 constitutes a major federal action. The second step in the analysis is whether the project is one which will significantly affect the quality of the human environment. [PPM 90-1] . . . sets forth criteria for determining whether a major project "significantly affects the environment."[184]

Additional support for this view appears in *Goose Hollow Foothills League v. Romney.*[185] In the *Goose Hollow* case the Department of Housing and Urban Development had concluded that an impact statement was not required for a loan which HUD had made for the construction of a $3.1 million high-rise apartment building. Since HUD had already asked for a "preliminary environmental worksheet" from the developer, the court concluded that HUD had conceded that the loan was a major federal action. The court thus focused its attention on HUD's contention that the loan would not have a "significant effect" on the environment. Discussing the project's environmental effects in some detail, the court found that a statement had to be prepared, because the project was controversial, would increase traffic, would concentrate population, would change the character of the neighborhood as its first high-rise, and would affect scenic views.[186]

A two-part test was accepted by Judge Wright in *SCRAP v.*

---

[182] 341 F. Supp. at 366, 2 ELR at 20189.
[183] Appendix B.
[184] 466 F.2d at 1033, 2 ELR at 20455.
[185] Appendix B.
[186] 334 F. Supp. at 879, 1 ELR at 20492.

United States.[187] In that case rail shippers obtained from the Interstate Commerce Commission a temporary freight rate increase for rail shipment of recyclable scrap metal. Plaintiffs argued that an impact statement should have been prepared for the increase. In finding for plaintiffs, the court said:

> The Commission stands in violation of NEPA unless it can demonstrate either (a) that the . . . order is not a "major Federal action" or (b) that the . . . order does not "significantly [affect] the quality of the human environment." We conclude that the Commission has failed to prove the truth of either of these propositions.[188]

The court then proceeded to consider both questions in detail. It pointed out that the commission's order appeared to affect "virtually every piece of freight moved by rail in this country" over a period of several months and concluded that such action was certainly as major as "the abandonment of a few miles of track" or "the funding of a highway project through a single locality."[189] On the question of significant environmental effects, the court rejected a conclusory ICC finding that the order would not be subject to NEPA and relied on CEQ Guidelines which require statements if an action will "arguably" affect the environment and if it has generated controversy. The court also cited a CEQ comment on the ICC's draft statement to the effect that an environmental impact from such action was very likely.[190]

Five additional cases deserve mention without extensive discussion. In *Save Our Ten Acres v. Kreger*[191] the court adopted the two-part test, relying upon *Goose Hollow* and, perhaps erroneously, upon *Echo Park.* In *Jicarilla Apache Tribe of Indians v. Morton* and *McLean Garden Residents Ass'n, Inc. v. National Capital Planning Comm'n,*[192] both courts adopted the two-part test without discussion or citation of cases, presumably through a literal reading of the language of §102(2) (C). In *Sierra Club v. Mason*[193] the

---

[187] Appendix B.

[188] 346 F. Supp. at 199, 2 ELR at 20490.

[189] Citing City of New York v. United States, Appendix B, and Named Individual Members of the San Antonio Conservation Society, Appendix B.

[190] 346 F. Supp. at 201, 2 ELR at 20491. This rare instance of a written CEQ comment on a draft agency impact statement illustrates the broader role which CEQ could play in implementing NEPA, were it of a mind to comment publicly and in writing on a larger selection of proposed agency actions. Note the court's reliance on the CEQ analysis.

[191] Appendix B.

[192] Both in Appendix B.

[193] Appendix B.

court implicitly accepted the two-part test by considering whether the Corps was correct in maintaining that a channel dredging project would not have a significant environmental impact, the Corps having already conceded that the project was a major federal action. In *Maddox v. Bradley*,[194] which was decided on plaintiffs' lack of standing and NEPA's nonretroactive application, the court went on nevertheless to imply that a federal proposal to fence 6,800 acres of watershed failed to meet both the tests of major action and significant effects.

Not only is this particular fence project not of major importance to require the Government to make the necessary environmental study in this case, but the evidence here indicates that the creation of the fence will enhance rather than destroy or take away from proper environmental control of the area.[195]

Interestingly, the court in *Maddox* refused to view the entire project—condemnation, reservoir construction, future uses—as the object of impact statement preparation, choosing to focus upon "a relatively small part of the entire project," the fencing of the boundary lines.[196] If plaintiffs had not urged the court to view the fencing as the all-important federal action, perhaps the larger project would have qualified for NEPA treatment, in the absence of the other reasons for dismissal.

Thus the cases are divided on the issue, but appear to favor a two-test standard. However, they do not go so far as to support the illogicality which would follow from construing "major action" without regard to the purposes of Congress in enacting NEPA, which included forestalling repetition of the kinds of environmental errors and disasters that had been brought to its attention. It makes little sense to call a project minor when its environmental effects are significant, because it is just these effects which §102(2) (C) requires to be discussed in the impact statement. To conclude that the prevailing judicial view favors a strict two-test standard would require overlooking the large number of cases which implicitly adopt a one-test standard by evaluating "significant environmental effects" only. Furthermore, no decided case applying a two-test standard has refused to require an impact statement when the environmental effects of a "minor" project were significant.[197] Moreover, in

---

[194] Appendix B.

[195] 345 F. Supp. at 1259, 2 ELR at 20405.

[196] *Id.*

[197] Two cases, however, came close to finding that major projects with insignificant effects do not require statements. *See* Hanly v. Mitchell, Appendix B (the office building), and Save Our Ten Acres v. Kreger, Appendix B.

Indian Lookout Alliance v. Volpe[198] the court almost held that a minor project with significant effects did require statement preparation. The court held that in spite of the "limited size" of a highway project which might cause significant environmental impacts, an impact statement had to be prepared. However, the court also insisted that the statement be coordinated with study of an adjacent segment, thereby redefining the magnitude of the action so that it was no longer as "limited" in size.

The Courts Must Be Able to Decide Whether Actions Are "Major"

Two views emerge from the cases about how responsibility for interpreting NEPA should be divided between agencies and courts. One view would allow agencies wide latitude in determining when the Act should apply and would confine judicial review to the question whether the agency had abused its discretion under the usual standard of arbitrary and capricious decision making. The alternative view would subject the agency decision to close scrutiny under a variety of formulations that seem to amount to *de novo* review.

Generally, the court allows the agency freedom to make specific applications of a broad statutory term, while making legal questions of how the statute is to be construed fully reviewable.[199] Unfortunately, as the Second Circuit has pointed out,[200] the two rules often overlap, and NEPA seems to lie in this border zone. When an agency decides that a statement need not be filed, it is applying the statute to the circumstances of a particular project, but it is also interpreting the words "major action" and "significant effect" as excluding the case at hand. Not surprisingly, therefore, the courts have dealt unevenly with the issue.

The thirteen decided cases on the issue are split, with the majority of opinion paying lip service to wide agency discretion to reverse the threshold decision. The courts' verbal deference to administrative discretion, however, is undercut by what the decisions actually require. In four of the nine cases applying the arbitrary and capricious standard, the courts nevertheless found abuse of discretion. Only three cases relied on that standard to shield the agency deci-

[198] Appendix B.

[199] L. Jaffe, Judicial Control of Administrative Action 546 *ff*. (abridged ed. 1965).

[200] Hanly v. Mitchell (Hanly I), Appendix B, 460 F.2d at 648, 2 ELR at 20220. The court gives currency to the term "threshold determinations," which the agencies must first make regarding §102(2)(C)'s applicability before proceeding.

sion, and even in these cases the courts looked closely at the various project magnitudes and possible environmental impacts before ruling. A circuit decision favored the standard of reasonableness, relying on *Overton Park*,[201] and remanded. Three cases favored *de novo* review.

The leading circuit court decision applying the arbitrary and capricious standard is *Hanly v. Kleindienst*.[202] In its second decision the Second Circuit first confirmed that NEPA commits the threshold determination to the agencies as part of their "screening function." Then the court adopted the arbitrary and capricious standard, stating:

> We see no reason for application of a different approach here since the APA standard permits effective judicial scrutiny of agency action and concomitantly permits the agencies to have some leeway in applying the law to factual contexts in which they possess expertise.[203]

Nevertheless, what the court appeared to give in defining the applicable standard, it took back in interpreting NEPA's requirements as a matter of law. In addition to interpreting the meaning of "significantly," the court spelled out certain requirements for the agency threshold determination. Having earlier held that the determination must be supported by a reviewable environmental record covering a wide variety of factors, the court went on to define obligations for public notice and participation.

> We now go further and hold that before a preliminary or threshold determination of significance is made the responsible agency must give notice to the public of the proposed major federal action and an opportunity to submit relevant facts which might bear upon the agency's threshold decision. We do not suggest that a full-fledged formal hearing must be provided before each such determination is made, although it should be apparent that in many cases such a hearing would be advisable for reasons already indicated. The necessity for a hearing will depend greatly upon the circumstances surrounding the particular proposed action. . . . The precise procedural steps to be adopted are better left to the agency, which should be in a better position than the court to determine whether solution of the problems faced with respect to a specific major federal action can better be achieved through a hearing or by informal acceptance of relevant data.[204]

To reach this result, the court found that §§102(2) (A), (B), and (D) were not limited in scope merely to major actions found by

---

[201] *See* Citizens to Preserve Overton Park v. Volpe, Appendix B (U.S.).
[202] *See* Hanly v. Mitchell (Hanly II), Appendix B.
[203] —— F.2d at ——, 2 ELR at 20720.
[204] —— F.2d at ——, 2 ELR at 20723.

the agency to meet the §102(2) (C) test of threshold significance. The agencies are subject to these three sections whether or not they must prepare impact statements. Specifically, the court found that §102(2) (B) buttressed its holding that the GSA had to provide channels for public notice and participation.

> Since an agency, in making a threshold determination as to the "significance" of an action, is called upon to review in a general fashion the same factors that would be studied in depth for preparation of a detailed environment impact statement, §102(2) (B) requires that some rudimentary procedures be designed to assure a fair and informed preliminary decision. Otherwise the agency, lacking essential information, might frustrate the purpose of NEPA by a threshold determination that an impact statement is unnecessary.[205]

The upshot of the Second Circuit's ruling is that it may be easier to prepare an impact statement than to take the steps which that court thinks necessary to prepare a record supporting a negative threshold determination. The loose standard of review which the court imposed diminishes in importance beside the extensive requirements applicable to the threshold determination.

In *Citizens for Clean Air v. Corps of Engineers*,[206] the court relied upon *Hanly I* to find that the Corps, in connection with its authority to grant permits, had to make a threshold determination regarding the environmental impact of Consolidated Edison's 800-watt fossil-fueled electrical generating plant to be located at Astoria in Queens, New York. As in *Hanly I*, a "reviewable environmental record" had to be compiled. This the Corps had failed to do, the court held, because the record consisted only of outside comments and did not reflect independent Corps efforts to assess the plant's impact. The Corps finding that construction of the plant was not a major federal action with significant potential environmental effects was therefore held to be "arbitrary and capricious action."[207]

The *Goose Hollow* case[208] also held that plaintiffs must show "arbitrary action" before a federal court will overturn agency determinations. There, however, HUD failed to conduct an independent review of the effects of a 221-unit, 16-story high-rise student apartment building, and relied instead on the promoter's own worksheet. This was taken to show that the agency had "ignored cumulative

---

[205] —— F.2d at ——, 2 ELR 20723.

[206] Appendix B.

[207] 349 F. Supp. at 707, 2 ELR at 20654. An identical result was reached in Sierra Club v. Mason, Appendix B, in which the court both distinguished and applied Hanly I. 351 F. Supp. at 426, 2 ELR at 20696.

[208] Goose Hollow Foothills League v. Romney, Appendix B.

effects on the quality of the human environment which are by no means insignificant," and was therefore held to have acted "arbitrarily."[209] It appears that the court introduced its own view of "significance" and hence arrived at the same conclusion that would be reached by strict judicial statutory interpretation, although the case may simply stand for the rule that the agency's determination must be based upon an independent assessment of the facts.

In four additional cases the courts applied the arbitrary and capricious standard without extensive discussion. The court in *Echo Park Residents Committee v. Romney*[210] refused to overturn a HUD determination that no statement would be necessary for a 66-unit apartment project, since the department had not been "arbitrary, capricious, or [abused its] discretion."[211] Similarly, in *Hiram Clarke Civic Club v. Romney*[212] the court found that HUD had not abused its discretion by preparing a "Negative Declaration" of significant environmental impact on a loan it insured for construction of a 272-unit apartment complex.[213] In *Maryland–National Capital Park and Planning Comm'n v. United States Postal Service*,[214] the court applied the arbitrary and capricious standard without discussion and allowed work on a suburban, industrially zoned bulk mail facility to proceed without an impact statement. In *Citizens for Reid State Park v. Laird*,[215] the court said that "the Act plainly commits this preliminary determination [that effects would not be significant] to the agency. . . . The standard of review in such cases is limited."[216] However, the court had before it a full study of the potential effects of the proposed military maneuvers. In each of these cases, however, the court did briefly comment upon the environmental effects which the projects might potentially cause.

The district court in *Pizitz v. Volpe*[217] may have intended to apply the liberal standard when it stated, "this court cannot substitute its judgment for that of the defendants and embark by trial upon an investigation to determine the environmental impact of the project."[218] However, the court may have intended only to find that

---

209 334 F. Supp. at 879, 1 ELR at 20492.
210 Appendix B.
211 —— F. Supp. at ——, 2 ELR at 20337.
212 Appendix B.
213 —— F. Supp. ——, 2 ELR at 20363.
214 Appendix B.
215 Appendix B.
216 336 F. Supp. at 789, 2 ELR at 20125.
217 Appendix B.
218 —— F. Supp. ——, 2 ELR at 20379.

it could not review *de novo* the adequacy of the statement already submitted in that case. The impact statement did contain, however, an agency conclusion that "no major adverse impact on the environment" would flow from the proposed project.[219]

On the side of a larger judicial role in reviewing agency threshold determinations stands the Fifth Circuit decision in *Save Our Ten Acres v. Kreger.*[220] There the court rejected the district court's application of the arbitrary and capricious standard to a GSA decision not to prepare a statement on the proposed construction of a federal office building in downtown Mobile, Alabama.

> This decision should have been court-measured under a more relaxed rule of reasonableness. . . . The spirit of the Act would die aborning if a facile, ex parte decision that the project was minor or did not significantly affect the environment were too well shielded from impartial review. Every such decision pretermits all consideration of that which Congress has directed be considered "to the fullest extent possible." The primary decision to give or bypass the consideration required by the Act must be subject to inspection under a more searching standard. . . .[221] . . . A thorough study of *Overton Park* teaches that a more penetrating inquiry is appropriate for court-testing the entry-way determination of whether all relevant factors [in §102(2) (C)] should ever be considered by the agency [footnote omitted].[222]

In three additional cases courts undertook what amounts to *de novo* review, themselves interpreting and applying the statutory standard to the facts at issue and claiming a leading role for the courts. In *Scherr v. Volpe*[223] Judge Doyle flatly stated that he could "not accept the contention" of the Federal Highway Administration that its decision not to file a statement could only be overturned if arbitrary or capricious. "When [the FHWA's] failure is then challenged," the judge continued, "it is the court which must construe the statutory standards ('major' and 'significantly affecting'), and, having construed them, then apply them to the particular project and decide whether the agency's failure violates the Congressional command."[224] However, he also indicated that this agency determination would in fact be invalid as "arbitrary and capricious" action. In affirming, the Seventh Circuit agreed with this finding and explicitly refused to rule on the question whether the agency could

219 *Id.*
220 Appendix B.
221 —— F.2d at ——, 3 ELR at 20041–42.
222 —— F.2d at ——, 3 ELR at 20042.
223 Appendix B.
224 336 F. Supp. at 888, 2 ELR at 20070.

be reversed on a review of the merits in the absence of arbitrary and capricious action.[225]

Additional support for this line of reasoning is provided by the decision in *Natural Resources Defense Council v. Grant*.[226] The court remarked that "certainly an administrative agency as the Soil Conservation Service may make a decision that a particular project is not major, or that it does not significantly affect the quality of the human environment, and, that, therefore, the agency is not required to file an impact statement."[227] However, the court then quotes the finding of *Scherr*, given above, that upon challenge the construction of the statutory term and the determination whether the agency has properly applied it are functions for the court.

In *Kisner v. Butz*[228] the district court applied the *Scherr* holding and conducted a *de novo* trial on the merits, fully examining plaintiffs' contention that Forest Service officials improperly decided that construction of the final 4.3-mile segment of a gravel-surfaced forest road through black bear and deer habitat was not "major federal action." The court found, however, that the action did not meet the NEPA standard and distinguished *Scherr* on the rather vague ground that in *Kisner* the parties' positions were fully evolved and "mature."[229] Further testimony could add nothing to the record already made, which was apparently not the case in *Scherr*.

The real significance of the dispute over standards of review is that if courts must rely on agency determinations, the agencies themselves can decide when to prepare statements. The very agents of a government whose environmental errors led to the adoption of NEPA would be able to set the threshold below which the Act would not apply. Although in some cases both approaches could lead to the same result, as perhaps they might in *Hanly*, *Citizens for Clean Air*, *Goose Hollow*, and *Scherr*, there could well be cases in which a judge could not find agency action that was clearly arbitrary, although his own view of the facts would favor preparation of a statement.

Statutory exegesis does not seem to be a particularly fruitful way of choosing between the two approaches. Section 102(2) (C) does not plainly commit the determination to the agency involved. Nowhere does the section say that major federal actions, as defined by

---

[225] 466 F.2d at 1032, 2 ELR at 20455.
[226] Appendix B.
[227] 341 F. Supp. at 366, 2 ELR at 20189.
[228] Appendix B.
[229] 350 F. Supp. at 321, 2 ELR at 20715.

the agency, shall be preceded by impact statements. Nor does NEPA's legislative history shed much light on the question. The colloquy between Senator Jackson and Dr. Linton Caldwell at the hearings, which is the only guide to NEPA's action-forcing provision, indicates that OMB was seen as the supervisor of the §102(2) (C) process.[230] The failure of OMB to respond and the entry of the courts into the area of NEPA enforcement suggest that the legislative plan is of doubtful relevance to the way NEPA is or should be interpreted.

Nor does resort to the determinations of an "environmental" agency solve the problem, although the CEQ has evinced some interest in helping decide such questions. In Chairman Russell Train's testimony before a joint session of two Senate committees, the former tax court judge cited language from *Environmental Defense Fund v. Tennessee Valley Authority*[231] to the effect that CEQ's view of the importance of a proposed federal action should be given great weight.

> Such administrative interpretation cannot be ignored except for the strongest reasons, particularly where the interpretation is a construction of a statute by the men designated by the statute to put it into effect.[232]

Two problems exist with this point of view. First, the court was not in fact suggesting that courts should give great weight to CEQ's view in deciding whether the size of a federal project merited an impact statement. Rather, the court was stating that the CEQ's view on *which* ongoing projects merited impact statements was entitled to great weight. While similar, the two determinations are not the same. Second, NEPA clearly does not designate CEQ as the "men charged by the statute to put it into effect." NEPA, interestingly enough, is silent on the CEQ's role, other than to state that it must receive copies of impact statements and review agency programs for compliance with NEPA. NEPA puts the obligation to comply directly

---

[230] See chapter I, text *supra* notes 35 and 36.

[231] 339 F. Supp. 806, 2 ELR 20044 (E.D. Tenn.), *aff'd*, —— F.2d ——, 2 ELR 20726 (6th Cir. 1972).

[232] 339 F. Supp. at 811, 2 ELR at 20046. Cited in *Joint Hearings Before the Senate Comm. on Public Works and the Senate Comm. on Interior and Insular Affairs on the Operation of the National Environmental Policy Act of 1969*, Serial No. 92–H32, 92d Cong., 2d Sess. 21 (March 1, 7, 8, and 9, 1972). The language quoted in Environmental Defense Fund v. Tennessee Valley Authority is taken word-for-word from Judge Eisele's decision in the Gillham Dam case, Environmental Defense Fund v. Corps of Engineers, 325 F. Supp. at 744, 1 ELR at 20136.

upon the agencies proposing federal action. The CEQ's authority to implement NEPA is derived primarily from an executive order[233] and inferentially from NEPA itself. Nevertheless, courts frequently turn to the guidelines for assistance in resolving NEPA issues. A dictum of the Second Circuit, previously quoted, brings out both aspects of the courts' use of the guidelines:

Although the Guidelines are merely advisory and the Council on Environmental Quality has no authority to prescribe regulations governing compliance with NEPA, we would not lightly suggest that the Council . . . has misconstrued NEPA.[234]

The dicta in *Environmental Defense Fund v. Tennessee Valley Authority* on the one hand, and in the *Greene County* case on the other, could provide an interesting starting point for discussion of the CEQ's authority. For instance, in spite of the *Greene County* dictum, it may very well be that a strong case can be made that CEQ in fact does possess the necessary power to require that impact statements be prepared and to rule on their adequacy. Such a power might be fully consistent with active judicial review.

Leaving these possibilities aside, CEQ's existing practices argue against allowing it to assume a major role in determining whether agency proposals should be covered by impact statements. The council does not systematically review the magnitude of projects for which impact statements might be prepared. For those which it does review, it is committed to a policy of informal, *ex parte* consultation with the agency involved. It has not attempted to assert that it has the power to require a recalcitrant agency to prepare a statement. It leaves to the agencies the drafting of procedural compliance guidelines which it will neither endorse nor approve. Thus CEQ review would take place under circumstances extremely difficult to ascertain or challenge. Such determinations should be fully open to public examination and, if necessary, judicial review. These considerations become especially important in light of the council's location within the Executive Office of the President. Political pressures may more immediately affect the council, whose members serve at the pleasure of the President, by exposing it to pressures within the executive branch emanating from the various agencies. Not only does the council lack judicial independence and enforcement power,

---

[233] Exec. Order No. 11514, §3(h), 35 FED. REG. 4247, ELR 45003 (March 5, 1970).

[234] Greene County Planning Board v. Federal Power Comm'n, Appendix B, 455 F.2d at 421, 2 ELR at 20021.

but it may need to seek allies among the very agencies whose procedures are questioned.[235]

The policy and goals of the Act suggest that the approach of *Scherr, Kisner*, and *Natural Resources Defense Council v. Grant* is correct and that *de novo* review of threshold determinations is appropriate. NEPA is unlike the majority of usual regulatory statutes. It neither sets up an agency to supervise private conduct nor supplements existing regulatory authority. Nor does the Act pinpoint a particular ill, for which a precisely focused statute may legislate a cure. Instead, the Act attempts to regulate the way in which all federal agencies make decisions. They are told to consider matters alien to their own limited self-interest, to expend time and money on statement preparation, to delay favorite projects, and to do all this when the benefits of the process do not redound to the agency involved but to the good of the environment. Such considerations suggest the inherent weakness of agency self-policing under NEPA, particularly since §102 duties are not inherently flexible, but demand a strict standard of compliance. Without a judicial check, the temptation would be to short-circuit the process by setting statement thresholds as high as possible within the vague bounds of the arbitrary or capricious standard. The past history of agency "crabbed interpretations" making a "mockery of the Act" leaves little room for confidence.[236]

A second level of analysis looks to the comparative expertise of court and agency. Those cases which allow agencies to apply statutory standards to facts typically rely on agency expertise or specialization in the subject matter at issue, and find that the function should be allocated to the administrators. Such is not the case with NEPA; for example, the Federal Highway Administration's particu-

---

[235] Under the Clean Air Act, §309, discussed in chapter VI at pp. 229 *ff.*, Congress gave the Environmental Protection Agency the responsibility of commenting, in writing, on impact statements within its scope of authority. A finding by the administrator that an impact is "unsatisfactory" requires the administrator to "refer the matter to the Council on Environmental Quality." But authority may not exist for the administrator to determine such questions as whether an agency has to file a statement at all. Some policy considerations argue for such an authority, especially if the CEQ does not undertake such review. In a sense EPA is just as much an environmental watchdog as CEQ. Its wide expertise, larger staff (by a factor of 100), and large regional offices equip it to react to damage threatened by pending federal action. On the other hand, as a mission-oriented agency it definitely has a great deal to accomplish in administering federal environmental protection programs.

[236] Calvert Cliffs' Coordinating Comm. v. Atomic Energy Comm'n, 449 F.2d 1109, 1117, 1 ELR 20346, 20350 (D.C. Cir. 1971), *cert. denied*, 404 U.S. 942 (1972).

lar field of competence is in building roads, not in deciding whether preparation of an environmental statement on a particular project best advances the goals of NEPA. As one court said of the Corps of Engineers, "Engineers are doers, not aesthetics [sic]."[237] This view is echoed in Judge Oakes' dissent in *Scenic Hudson II*: "While judicial deference to administrative expertise is required, not every agency is expert in every aspect of science, technology, aesthetics or human behavior."[238]

Third, NEPA applies to all agencies, cutting across programmatic lines, which suggests that questions of its application should be decided by "generalists." While it may be that courts should defer to those administrators whom the statute designates to implement it, §102(2) (C) does not assign that function to anyone in particular other than the responsible federal official. [The only agency mentioned in §102(2) (C) is the CEQ, and there it appears merely as a recipient of statements and comments, presumably in its role of presidential advisor.] The judiciary has a vantage point which permits it to generalize about the relative effects of different kinds of programs, or similar programs carried out by different agencies, and thus to determine the appropriate level at which a project becomes "major." Where different agencies have overlapping functions, such as the water resources projects of the Corps of Engineers and the Soil Conservation Service, permitting each agency to set NEPA thresholds as it sees fit might mean that different assessments might be made of otherwise identical projects.

Indeed, to take only one example, the action of the court in *Natural Resources Defense Council v. Grant*,[239] which held that the channelization of 66 miles of stream was "major" and that the Soil Conservation Service had erred in refusing to go through the §102 procedure, suggests that the potential for such abuse does exist.

In brief, *de novo* review is appropriate, not only because the circumstances are congenial to a judge-made determination, but because a practical need exists for impartial review. The agencies themselves, the Office of Management and Budget, the Council on Environmental Quality, and the Environmental Protection Agency, all have substantial drawbacks or offer actual resistance to such a role.

---

[237] United States v. 247.37 Acres of Land, Appendix B, —— F. Supp. at ——, 1 ELR at 20517.

[238] Scenic Hudson Preservation Conference v. Federal Power Comm'n (Scenic Hudson II), Appendix B, 453 F.2d at 484, 1 ELR at 20508.

[239] Appendix B.

AGENCIES TO WHICH NEPA APPLIES

The Act's legislative history shows that Congress directed its strongest criticism toward development-oriented agencies for their neglect of environmental values and marked their decision-making processes in particular for reform.[240] But when the question arose whether agency programs dedicated to environmental improvement should also be subjected to NEPA's sweeping requirements, particularly the requirement for impact statements, Congress did not make its wishes as clearly known. A hurried attempt was made at the last minute to clarify the legislative history of §104 so that it would be interpreted to exempt environmentally protective federal activities from NEPA obligations.[241] At least one subsequent court opinion shows just now far short of its goal that attempt fell. "There is no exception [from NEPA], as defendants have argued, carved out for those agencies that may be viewed as environmental improvement agencies."[242]

"All Agencies of the Federal Government"

Virtually every agency of the federal government has by now prepared an impact statement. Most of them have also been subject to NEPA lawsuits which raise issues under §102(2) (C). Yet the courts have decided only one case in which an agency, as an agency, did not have to comply with the impact statement requirement.[243] In this area of implementation, as in others already discussed, the courts have read the widest possible scope into §102(2) (C). Likewise, the courts have found no ambiguity in NEPA's limitation to federal agencies. States, municipalities, and private citizens owe no duties under NEPA.[244]

---

[240] 115 CONG. REC. S. 17458 (daily ed. Dec. 20, 1969) (remarks of Mr. Muskie).

[241] *Id., and see* 115 CONG. REC. S. 12111, 12114 (daily ed. Oct. 8, 1969). *See also* Comment, *Landmark Decision on the National Environmental Policy Act in Calvert Cliffs' Coordinating Comm., Inc. v. Atomic Energy Comm'n,* 1 ELR 10125, especially "Novel statutory interpretation of §104: the Muskie–Jackson Compromise," at 10127 (August 1971).

[242] Kalur v. Resor, Appendix B, 335 F. Supp. at 15, 1 ELR at 20642.

[243] Cohen v. Price Comm'n, 337 F. Supp. 1236, 2 ELR 20178 (S.D.N.Y. 1972).

[244] States and municipalities may, however, be subject to state legislation patterned on NEPA. Such legislation has been enacted in eight states and Puerto Rico; Hawaii achieves the same result by executive order. *See* Introduction, *supra* note 3. Private and state defendants may also be enjoined from action which might prejudice any federal decision that may eventually be

"Environmental" agencies other than the Environmental Protection Agency (EPA) have acquiesced in the preparation of impact statements, although the legislative history contains evidence that Congress possibly intended to excuse them from compliance. For example, Senator Jackson listed the Park Service along with the Federal Water Pollution Control Administration and the National Air Pollution Control Administration as environmental control agencies.[245] While his mention of the air and water agencies can be explained by the conflict of jurisdiction which appeared to be brewing between Senator Jackson's committee—NEPA's sponsor—and Senator Muskie's air and water pollution subcommittee—sponsor of air and water legislation—the inclusion of the Park Service cannot. The phrases otherwise used to describe the agencies in question, however, favor the view that only the pollution programs which EPA was soon to inherit were to be exempted.

The courts have addressed the issue in the litigation over the Refuse Act Permit Program. In *Kalur v. Resor*,[246] the court held both that the Corps of Engineers could not delegate its statutory authority under the Refuse Act to EPA and that it could not escape compliance with NEPA. The Corps defended noncompliance with NEPA on the grounds that the Corps itself was an environmental improvement agency, a view which the court emphatically rejected.[247]

Although Congress did not settle the issue at the time NEPA was enacted, the courts will probably never have to rule on the matter because environmental agencies other than EPA have routinely complied with §102 (2) (C) and because congressional debate has now narrowed to whether EPA alone should be exempted, with

---

necessary in connection with the action. *See* discussion *supra* at pp. 64 *ff.*, especially the analysis of the circuit decisions in Silva v. Romney, Arlington Coalition on Transportation v. Volpe, and Named Individual Members of the San Antonio Conservation Society v. Texas Highway Dept., all in Appendix B. Silva speaks of a "partnership" between a private grantee and HUD; earlier in Lathan v. Volpe, Brooks v. Volpe and Daly v. Volpe (all in Appendix B), state highway departments were held to be virtually federal entities, or quasi-federal ones. *See especially* Daly v. Volpe, 350 F. Supp. 252, 257, 2 ELR 20443, 20444 (W.D. Wash. 1972), the cases cited in the court's footnote 14 and the text above it.

[245] 115 Cong. Rec. S. 17453 (daily ed. Dec. 20, 1969). The reference is in Exhibit 1, "Mayor changes in S. 1075 as Passed by the Senate," in discussion of "Section 102 in General."

[246] Appendix B.

[247] 335 F. Supp. at 15 note 57, 1 ELR at 20652 note 57.

implicit congressional agreement that no other agency may claim the benefit of the earlier, hastily assembled legislative history.

We conclude chapter VIII with a brief discussion of a systematic approach to NEPA compliance and impact statement preparation which might necessitate preparation of statements by subcabinet groups and agencies of the Executive Office of the President such as the Office of Management and Budget, and the Council on Environmental Quality. Impact statements have never been actually prepared by these units of government, and the question arises whether they are included when NEPA speaks of "federal government" [§101(a)] and "all agencies" [§102(2)]. Whether or not they are obligated to prepare statements under appropriate circumstances, and we think that they are, NEPA gives them ample authority if they choose to act.[248] If they are asked to comply in a litigation, the test of their status as an "agency" is set out in *Soucie v. David*.[249] In that case the Office of Science and Technology was held to be an agency because it had been given statutory duties to perform independent of its primary role as presidential advisor.

NEPA's Application to EPA

NEPA contains no exemption from any of its provisions for EPA or any other agency. But, as mentioned above, an attempt took place immediately before the vote on NEPA to exclude "environmental" agencies from NEPA obligations by writing a "clarifying" legislative history for §104. The EPA had not been formed at the time NEPA was enacted,[250] a factor which further blurred Congress' intent. However, the water and air pollution programs soon to be transferred from the Department of the Interior and HEW, respectively, were used as examples for the types of agencies which were intended to be exempted. The other programs subsequently transferred to EPA[251] were never mentioned, which gives credence to the view that the primary concern of Senators Jackson and Muskie in their attempts to write a history for §104 was to protect the air and water programs supervised by Senator Muskie's subcommittee from additional responsibilities and Senate oversight under NEPA.

---

[248] *See* Comment, *NEPA and Federal Policy-Making: NRDC v. Morton, Legislative Impact Statements, and Better NEPA Procedures*, 2 ELR 10038 (April 1972), especially "Impact Statement Preparation by the Executive Office or Inter-Agency Task Forces," at 10042.

[249] 448 F.2d 1067, 1 ELR 20147 (D.C. Cir. 1971). *But see* Environmental Protection Agency v. Mink, —— U.S. ——, 3 ELR 20057 (1973).

[250] Executive Reorganization Plan No. 3 of 1970, ELR 48001 (July 9, 1970).

[251] *Interior:* pesticides; Gulf Breeze Biological Laboratory. *HEW:* solid waste management; radiological health. *AEC:* radiation standards. *Agriculture:* pesticides.

Although the legislative history of §104 was ambiguous, it was interpreted in the CEQ Guidelines as exempting all of EPA's regulatory activities from NEPA.

Because of the Act's legislative history, environmental protective regulatory activities concurred in or taken by the Environmental Protection Agency are not deemed actions which require the preparation of environmental impact statements under Section 102(2) (C) of the Act.[252]

Further complicating the picture, EPA indicated that it would comply to some extent by publishing guidelines which stated that it would prepare impact statements on its approvals of regional water quality management plans (clearly a regulatory action) and on its waste treatment grants.[253] Through December 1972, EPA had in fact prepared twenty-nine statements, primarily on waste treatment grants, but had not filed any statements on regional water quality plans.[254]

For three years the issue languished in perplexing ambiguity. The half-dozen judicial opinions that touched upon the question did not answer it, nor did Congress further clarify its original intent. However, Congress recently took a limited step toward resolving the matter by enacting the Federal Water Pollution Control Amendments of 1972.[255] Section 511(c)(1) exempts[256] EPA from preparing impact statements on all aspects of the water quality program, except for the issuance of discharge permits for new sources and the making of grants for publicly owned waste treatment works. The enactment of §511(c)(1) suggests that Congress has opted for a "go slow," program-by-program approach to the question of EPA's compliance with NEPA.

Litigation which has recently been filed to require EPA to comply with §102(2)(C) in its air and pesticides programs further suggests that unless Congress acts soon, the courts may be the first to resolve

---

252 CEQ Guidelines, §5(d), Appendix C.

253 37 FED. REG. 879 (Jan. 20, 1972).

254 COUNCIL ON ENVIRONMENTAL QUALITY 102 MONITOR, vol. 2, no. 12, pp. 82–84 (January 1973).

255 33 U.S.C. §1151 *et seq., as amended* Oct. 18, 1972, Pub. L. No. 92–500, 86 Stat. 816.

256 Whether §511(c)(1) exempts EPA from NEPA or actually extends NEPA's coverage depends on one's point of view. In Senator Muskie's view, Congress never intended for EPA to have any NEPA responsibilities, thus the two exceptions to §511(c)(1)'s policy that NEPA does not apply to EPA in the water area are in fact *extensions* of NEPA's coverage. And the general rule of §511(c)(1) simply declares the preexisting law, much as §309 of the Clean Air Act can be viewed as declaratory of §102(2)(C)'s requirements. *See* Comment, *Section 309 of the Clean Air Act: EPA's Duty to Comment on Environmental Impacts,* 1 ELR 10146 (September 1971).

the issue for the remaining EPA programs. Moreover, the existing decisions which touch upon the issue do not clearly indicate the direction that the courts will take.

The Water Pollution Program. The new water act exempts most EPA actions under the water quality program from the §102 process. However, §511(c)(1) contains ambiguities which must be resolved before EPA can apply it in the new water quality program.[257] Section 511(c)(1) states:

> Except for the provision of Federal financial assistance for the purpose of assisting the construction of publicly owned treatment works as authorized by section 201 of this Act, and the issuance of a permit under section 402 of this Act for the discharge of any pollutant by a new source as defined in section 306 of this Act, no action of the Administrator taken pursuant to this Act shall be deemed a major Federal action significantly affecting the quality of the human environment within the meaning of the National Environmental Policy Act of 1969 (83 Stat. 852).

At the least, §511(c)(1) exempts important EPA water quality decisions concerning:

The setting of new source performance standards
The establishment of "best practicable treatment" and "best available treatment" standards for categories of waste-discharging facilities
The issuance of guidelines for state control of nonpoint sources
The approval of regional water quality management plans [covered by §102(2) (C) under EPA's old guidelines]
The approval of state discharge permit programs[258]

An important question remains whether §511(c)(1) exempts the water quality program from the §102(2)(C) impact statement requirement *only*, or from this and every other section of NEPA as well. While the issue cannot be litigated as precisely as can issues arising under §102(2)(C), one may nevertheless expect the courts to reach it sooner or later. The requirements of §102(2)(D) and the balancing requirement discussed in the *Calvert Cliffs'* decision will most likely not be foregone without an attempt to require EPA to show how it has complied.

The issue arises primarily because the phrase used in §511(c)(1),

---

[257] The possible effect of §511(c)(1), and of its legislative history as viewed by Senator Muskie, upon *other* EPA programs is treated in the next subsection.
[258] 118 CONG. REC. S. 16884 (daily ed. Oct. 4, 1972) (remarks of Mr. Baker).

"major Federal action significantly affecting the quality of the human environment," was taken from §102(2)(C)—the only NEPA section in which the phrase appears. Section 511(c)(1) by its terms thus provides an exemption only from the impact statement process. However, Senator Muskie and Congressman Jones said on the floors of their respective houses that "the term 'major Federal action' and NEPA are synonymous in the minds of the conferees," and that the exemption was intended to be from all the provisions of NEPA.[259] Their position finds inferential support in the joint statement of the bill's managers: "If the actions of the Administrator were subject to *the requirements of NEPA*, administration of the Act would be greatly impeded" [emphasis added].[260] Senator Muskie elsewhere stressed EPA's blanket exemption.[261] Senator Jackson disagreed, asserting that "there are limitations on how far one can go in interpreting the explicit and unambiguous language of the conference report." He stated his understanding that the §511(c)(1) exemption covered only the §102(2)(C) process.[262] Senators Gaylord Nelson and Philip Hart and Congressman John Dingell stated similar views.[263]

Section 102(2)(D) imposes a requirement wholly distinct and apart from §102(2)(C)(iii) which requires all agencies to "study, develop, and describe appropriate alternatives" to any action which involves unresolved conflicts about natural resource use. Action involving such conflicts will be taken frequently during the implementation of the new water program. Moreover, under the *Calvert Cliffs'* opinion, agencies are required to trade off competing factors in a finely tuned balancing process when it comes time for a final agency decision to be made (see chapter VII). Such tradeoffs involve both the developmental, economic, or social values which the agency may already have to consider, and environmental values newly added to the agency's mandate by NEPA. The availability of alternatives can change the weight which particular factors are assigned.

One can readily see the difference it will make if EPA has been exempted from these requirements as well. The Environmental Protection Agency will become the only federal agency with a mandate

---

[259] *Id.* at S. 16885 (remarks of Mr. Muskie), and at H. 9119 (remarks of Mr. Jones).

[260] 118 CONG. REC. H. 8901 (daily ed. Sept. 28, 1972) (Conference Report, joint explanatory Statement of the Committee of Conference).

[261] 118 CONG. REC. S. 16878 (remarks of Mr. Muskie).

[262] *Id.* at S. 16888.

[263] *Id.* at S. 16889, 16890, H. 9127.

to pursue its statutory mission in water pollution control to the exclusion of all else, including the effect which its decisions may have on other environmental values, such as air pollution and land use. While the shoe will be on the other foot, with developmental interests rather than environmental ones at the disadvantage, there is at least a question whether Congress could have intended such a result.

Before leaving the exemption granted EPA, we should add a brief word of clarification about §511(c)(2), a reworded version of the "Baker amendment," which also affects the obligations of EPA and the other federal agencies under NEPA. Section 511(c)(2) overrules that portion of the *Calvert Cliffs'* opinion which allowed agencies to set higher water quality standards as a condition of obtaining a license or permit than would have been allowed by standards set under the Federal Water Pollution Control Act.[264] However, §511(c)(2) does not require the licensing agency to put water quality considerations aside while it conducts its final balancing judgment. When the agency begins its final balancing judgment

---

[264] 449 F.2d at 1122, 1 ELR at 20353. *See* 118 CONG. REC. S. 16885 (daily ed. Oct. 4, 1972) (colloquy between Mr. Buckley and Mr. Muskie). Judge Wright to the contrary, §104 of the Act does seem somewhat ambiguous:

> Nothing in section 102 or 103 shall in any way affect the specific statutory obligations of any Federal agency (1) to comply with criteria or standards of environmental quality, (2) to coordinate or consult with any other Federal or State agency, or (3) to act, or refrain from acting contingent upon the recommendations or certification of any other Federal or State agency. (Appendix A)

What §104 does not say is whether NEPA is to be interpreted as supplementing the obligations of the federal agency by having it not only "comply," "coordinate," or "act," but also file an impact statement. Given this uncertainty, Senator Jackson's explanation on the Senate floor of the reason this particular language was chosen could very well have carried a good deal of weight, had courts chosen to rely upon it, particularly with reference to the questions of obligations of the AEC and Army Corps of Engineers raised in Calvert Cliffs' and Kalur:

> The language of this section [§103 of the Senate Bill and §104 of the Act] is designed to insure that the provisions of section 16, and particularly section 16(C) of S. 7 [the Water Quality Improvement Act of 1970] are consistent with the requirements of section 102 of S. 1075. Section 16(c) of S. 7 would have the effect of exempting the Corps of Engineers, the Atomic Energy Commission, and some other agencies from the requirement in S. 1075 for a detailed statement on the environmental impact of proposed actions involving any discharge into the navigable waters of the United States. 115 CONG. REC. S. 12114 (daily ed. Oct. 8, 1969).

This language is overlooked in the analysis of legislative intent conducted in Calvert Cliffs'.

under NEPA in deciding whether to grant the permit, it is free to consider water quality effects as one element in the balancing equation. Thus §511(c)(2) does not affect EPA's or any other agency's obligation to prepare an impact statement. It does, however, affect the contents of the section of licensing agencies' statements concerning water quality; and, as indicated, it limits what a licensing agency may do in the balancing process. The factor of water pollution is "frozen" at the level specified by EPA or the state agency. The factor may be taken into account, but may not be varied, so that the net effect of §511(c)(2) is to forbid the imposition of the "benefit" of higher water quality in order to offset some other form of "cost," such as land misuse. If other costs are unavoidable and cannot be reduced by conditioning the license or permit, §511(c)(2) may have the effect of unavoidably tipping the benefit-cost tradeoff against allowing the permit or license at all.

At best, an applicant will design his project to surpass Water Act standards, although the licensing agency cannot formally require such action, in order to make sure. Environmentalists will no longer be allowed to intervene to request tighter water quality conditions on licenses, but they will be permitted to present evidence on water quality effects as those effects bear on the agency's final decision. However, as is more likely, it will usually be very difficult to show that sufficient degradation of water quality will occur, despite compliance with Water Act standards, to require license denial. Licensing agencies' balancing processes are likely to be so imprecise that obtaining such proof will be much more difficult than showing that a tighter water quality requirement would yield benefits exceeding incremental costs.

Other Programs. While Congress has resolved the question of NEPA's applicability to EPA's water quality program, the courts may have the first opportunity to resolve it for EPA's air pollution and pesticide control programs. Whether it will be the courts or Congress which resolves NEPA's applicability to the EPA solid waste, radiation, noise, and ocean dumping authority is less certain. As of this writing, approximately twenty-six air pollution suits have been filed against EPA alleging noncompliance with §102(2)(C) of NEPA.[265] Almost all of the suits have been brought by industry.

---

[265] Comment, *Litigation under the Clean Air Act*, 3 ELR 10007 (March 1973) (table of cases).

Industry has also brought suit alleging that the administrator's recent decision to cancel all DDT registrations should also have been preceded by a full §102(2)(C) review.[266]

The five already existing judicial opinions of conceivable relevance to NEPA's application to EPA afford little guidance to the courts which must decide the pending lawsuits. In *Kalur v. Resor*[267] and *Sierra Club v. Sargent*,[268] involving the now-defunct Refuse Act program jointly administered by EPA and the Corps of Engineers, both district courts indicated that EPA was subject to §102(2)(C). However, in a recent industry challenge to an EPA compliance order under the Clean Air Act, the Third Circuit disapproved those decisions and indicated a contrary view.[269] Getty Oil had argued that EPA was required to prepare an impact statement for a compliance order issued under the Clean Air Act. The Third Circuit held that the issue should have been raised in a §307 proceeding under the Clean Air Act, but went on to say that the cases cited to it were not persuasive that EPA was bound by NEPA[270] and that other requirements of the Clean Air Act sufficiently provided for the achievement of NEPA's goals. The Third Circuit's view has been distinguished as dictum by the district court in *Anaconda Co. v. Ruckelshaus*,[271] which required the preparation of a statement on the strategy which EPA had proposed for controlling sulfur oxide air pollution produced by nonferrous smelters in Montana. Finally, the district court in *Gibson v. Ruckelshaus*[272] enjoined an EPA sewerage treatment grant for failure to prepare an impact statement.

The government's appeal in the *Kalur* case will probably be dropped, because it involves only the future administration of the Refuse Act Permit Program, now superseded after passage of the Federal Water Pollution Control Amendments of 1972. In *Sargent*, the appeal may go on to decision since the case involves the need to prepare an impact statement on a discharge permit already granted

---

[266] Coahoma Chemical Co. v. Ruckelshaus, Civil No. DC–72–73–S (N.D. Miss. filed Oct. 17, 1972), *further proceedings enjoined* Environmental Defense Fund v. Environmental Protection Agency, Nos. 72–1548, 72–2142 (D.C. Cir. Jan. 31, 1973).

[267] Appendix B.

[268] —— F. Supp. ——, 2 ELR 20131 (W.D. Wash. 1972).

[269] Getty Oil Co. v. Ruckelshaus, 342 F. Supp. 1006, 2 ELR 20393 (D. Del.), *rev'd*, 467 F.2d 349, 2 ELR 20683 (3rd Cir. 1972), *cert. denied*, 41 U.S.L.W. 3389 (Jan. 16, 1973).

[270] 467 F.2d at 359 note 17, 2 ELR at 20687 note 16.

[271] —— F. Supp. ——, ——, 3 ELR 20024, 20030 (D. Colo. 1972).

[272] —— F. Supp. ——, 1 ELR 20337 (E.D. Tex.), *rev'd sub nom.* City of Lufkin v. Gibson, —— F.2d ——, 1 ELR 20438 (5th Cir. 1971).

for an existing facility. But again, Congress has radically changed the law regarding water quality permits in the recently enacted legislation. In *Gibson,* the Fifth Circuit dissolved the injunction and remanded for a determination whether the district court could have apppropriately taken jurisdiction, but here, too, the new water act has changed the law by explicitly requiring statements in EPA's sewerage grant program. In *Getty,* plaintiff was turned out of court on the grounds that relief should have been sought through §307 proceedings under the Clean Air Act and that NEPA does not require an impact statement at the enforcement stage. The court devoted only one sentence to the NEPA question.[273] The lengthy opinion in *Anaconda* would appear to offer more precedential value, but it is subject to attack on the same jurisdictional grounds that the court relies upon to distinguish *Getty.* It would appear that the issues in *Anaconda* should be raised only in §307 proceedings under the Clean Air Act.

Thus the courts have said little about EPA's NEPA obligations outside the water area. Because they have yet to rule, the courts may be influenced by Congress' recent action in enacting the Water Pollution Control Amendments of 1972. Just how much they may be influenced depends upon how willing they are to accept Congress' attempt to fill in NEPA's legislative history almost three years after the fact and upon which version of that history they choose to accept.[274]

During consideration of the conference report on the 1972 water bill, the floor managers in the Senate and House stated their view that NEPA as originally enacted did not apply at all to EPA. In their view the provision in §511(c)(1) applying NEPA to two types of EPA actions was an *extension* of NEPA. Senator Muskie stated, "If there have been any doubts as to whether the various provisions of NEPA apply to the activities of EPA, section 511(c)(1) is expressly designed to still such doubts with finality."[275]

While saying that NEPA's legislative history was "ambiguous" on the question, Senator Jackson did not really disagree. He expressed the opinion that the decisions holding NEPA applicable to EPA were "contrary in some respects to NEPA's legislative history."[276]

---

[273] 467 F.2d at 359, 2 ELR at 20687.

[274] The court in Anaconda Co. v. Ruckelshaus, Appendix B, emphatically rejected EPA's argument that §511(c)(1) of the new Water Pollution Control Amendments of 1972 exempted EPA from statement preparation under the Clean Air Act. —— F. Supp. at ——, 3 ELR at 20031.

[275] 118 Cong. Rec. S. 16877 (daily ed. Oct. 4, 1972).

[276] *Id.* at S. 16886.

On the other hand, the bare language of §511(c)(1) tends to support the conclusion that Congress regarded NEPA as applying to EPA except where there is an express exemption. The declaration that "no action [under the Federal Water Pollution Control Amendments] shall be deemed a major Federal action . . ." indicates that EPA's actions would otherwise be subject to §102(2)(C) whenever they fall within its terms. The *Calvert Cliffs'* opinion contains strong language to the effect that NEPA's statutory language is sufficiently clear and all-encompassing to make resort to legislative history unnecessary. The court stated:

> As the Supreme Court often has said, the legislative history of a statute . . . cannot radically affect its interpretation if the language of the statute is clear. . . . It is, after all, the plain language of the statute which all the members of both houses of Congress must approve or disapprove. The courts should not allow that language to be significantly undercut.[277]

If other courts join *Anaconda* in the same approach to §511(c)(1), they may conclude that it allows NEPA to be applied to EPA's programs other than water quality. In this view, §511(c)(1) creates an exception, but only for the water program, to the previously all-embracing scope of NEPA.

Should NEPA Apply to EPA's Programs? The reasons advanced for applying NEPA to EPA's programs are largely the same as they are for any other agency: improved planning and coordination; an increased likelihood that decisions to further one environmental goal will be taken with awareness of the possible impacts on other environmental concerns; fuller use of available expertise through the comment process; the substantial benefits of public participation; and finally, careful decision making through a balanced weighing of costs and benefits. Having NEPA apply to EPA would ensure consistent application of NEPA to all agencies, with the advantages of equal treatment which a uniform government-wide policy would bring. Finally, the decisions which EPA makes do not invariably bring benefits of unalloyed environmental protection; by law the agency must frequently trade environmental quality off against economic benefit. NEPA would help ensure that in making these decisions EPA did not become a "captive agency."

We think that these arguments should prevail. The reasons advanced for exempting EPA from NEPA, although they have considerable merit, do not stand up very well under closer examination. Much of the difficulty in understanding the need for EPA's com-

[277] 449 F.2d at 1126, 1 ELR at 20355.

pliance stems from the impression that Congress had only the reform of development-oriented agencies in mind when it passed the Act. According to this view NEPA was intended to whip recalcitrant developmental agencies into line with the action-forcing procedures of §102, while enlarging their statutory powers to consider environmental factors through the national environmental policy enacted by §101. The Environmental Protection Agency, it is thought, does not need to be forced to act and already possesses a wide range of specific environmental mandates. Why require more than is really needed?

The largest part of the answer lies in EPA's dual nature. While it certainly possesses a plethora of environmental mandates, and has been tough on pollution in its short history, its authority is in fact a two-edged sword. In the air quality program, for example, the setting of primary national air quality standards "requisite to protect the public health"[278] requires the administrator to balance risks and benefits so that some risk of damage to life and health is in fact borne by the public. As unpleasant as it is to interpret legislation in this manner, federal regulation of health and environmental quality may be viewed as a mandate not so much to save lives and protect the environment as to authorize the taking of some life and the sacrificing of some of the environment.[279] The EPA is hardly alone in this matter. Most regulatory agencies must allow costs as well as protect benefits, which is precisely the point: EPA and old-line regulatory agencies alike all make tradeoffs and all need information and consultation to improve the decision-making process. NEPA does this and can provide sauce for the goose as well as for the gander.

While EPA's authority would not be directly affected, and while it would retain wide discretion to set regulatory standards and make enforcement decisions, NEPA would add important new dimensions

[278] Defined in §109(b)(1), 42 U.S.C. §1857c–4(b)(1), of the Clean Air Act, 42 U.S.C. §1857 *et seq., esp. as amended* by Clean Air Amendments of 1970, 84 Stat. 1676, Pub. L. No. 91–604 (Dec. 31, 1970).

[279] The actions which EPA is now empowered to take with respect to pesticides and radiation are tailor-made for the application of section 102(2)(C). These actions involve tradeoffs between the risks of exposure to pesticides and radiation and the benefits to be gained from their continued release to the environment. Thus, leaving aside water and air programs, the application of the requirements of section 102(2)(C) to pesticide and radiation tradeoffs would give much-needed specificity to EPA's articulation of the risks and costs inherent in its decisions. *Joint Hearings, supra* note 232, at 429 (testimony of Frederick R. Anderson).

See also COMMITTEE ON PUBLIC ENGINEERING POLICY, NATIONAL ACADEMY OF ENGINEERING, PERSPECTIVES ON BENEFIT-RISK DECISION MAKING (1972) 157 pp.

to EPA's decision making. If NEPA were applied to all of its programs, EPA would have to prepare an impact statement setting out the consequences to the rest of the environment of each of its major environmentally protective actions, including an analysis of alternative means of accomplishing the same primary objective but with less ancillary environmental damage. The Environmental Protection Agency would have actually to weigh the proposed action, alternatives to it, and other factors mandated by the regulatory statute before making a final decision.

If excused from these duties, EPA would be the only agency of the federal government which could ignore the environmental side effects of its actions. On the other hand, definite benefits would follow from compliance. For example, if EPA has proposed a particular air pollution control plan which could create a solid waste disposal, land use, or water pollution problem, or change consumption patterns of natural resources (fuels) for the worse, or increase emissions of other air pollutants, EPA might reasonably be expected to examine its decision to see whether on balance the overall environmental effect of its proposed action would be beneficial. In so doing EPA might profit from the comments of other agencies that possess expertise in the affected areas. NEPA prohibits EPA's sister agencies from acting without taking the comprehensive look mandated by NEPA. That the "major actions" taken by EPA are in large measure environmentally protective ones is not a sufficiently distinguishing factor. Specialization and "tunnel vision" within the environmental field may be as harmful as they are in other areas.

Other objections to NEPA's applicability to EPA have been advanced. The legislative history of the new water pollution amendments contains explicit remarks by Senator Muskie which establish beyond any doubt that in his view Congress did all that was "humanly possible" to spell out EPA's mission under the new act in such detail that NEPA could not possibly apply in any way to the water pollution control program.[280] The same argument can be made regarding EPA's obligations under the air pollution and other EPA regulatory programs, although they lack either specific exempting legislation such as §511(c)(1) or a favorable legislative history.

The argument against NEPA compliance on this ground proceeds on two different footings. The first is that Congress has specified the factors to be considered by EPA in making decisions and has assigned the relative weights each factor is to have. No room has been left for discretionary balancing under NEPA. But a harder look at

---

[280] 118 CONG. REC. S. 16878 (daily ed. Oct. 4, 1972).

practically all of the statutes under which EPA acts shows that they actually mandate tradeoffs which are part of the same judgmental decision making that all agencies have to perform.[281] The second argument against compliance is that the time frame of compliance with regulatory statutes is inconsistent with the rather lengthy time requirements of adequate §102(2)(C) compliance. Yet in the event incompatibility appears to exist, either the more specific regulatory legislation in fact governs, guidelines specifying incompatible objectives may be rewritten to accommodate compliance with both acts, or EPA may be asked to make a special effort to move the administrative process more rapidly than it might like in order to comply. But new, strict statutory deadlines should not automatically excuse the agency from compliance with preexisting authority.

Even if the administrator has no discretion to "balance" under NEPA, or if he must act quickly and disregard NEPA's requirements, impact statement preparation can still serve the useful purpose of informing the public, the Executive branch, and Congress of environmental impacts so that future remedies may be discussed.[282] Without NEPA, EPA might well be reluctant to spell out the unavoidable risks and negative consequences of a proposed action. Moreover, *Natural Resources Defense Council v. Morton*[283] appears to require discussion of alternatives which the administrator is currently helpless to implement, thereby allowing him to stimulate informed debate on future legislative needs.

Another line of argument concedes the value of infusing NEPA's objectives into EPA's programs but goes on to assert that EPA already meets these goals under the requirements of its existing regulatory authority or under evolving doctrines of judicial review. Compliance with NEPA, the argument goes, is superfluous.[284]

---

[281] *See, e.g.*, 42 U.S.C. §1857c–5(a)(2) (Clean Air Act).

[282] Preservation of NEPA's "integrity" would accomplish this goal. Hanly v. Mitchell (Hanly I), Appendix B, 460 F.2d at 648, 2 ELR at 20220; City of New York v. United States, Appendix B, 337 F. Supp. at 160, 2 ELR at 20277.

[283] Appendix B.

[284] The Administrator is given the responsibility of making policy reviews . . ., annual comprehensive economic cost studies . . ., and periodic reports to Congress. . . . It is apparent that the Clean Air Act itself contains sufficient provisions for the achievement of those goals sought to be attained by NEPA. Getty Oil Co. v. Ruckelshaus, Appendix B, 467 F.2d at 359, 2 ELR at 20687.

See also footnote 130, International Harvester Co. v. Ruckelshaus, —— F.2d ——, ——, 3 ELR 20133, 20149 (D.C. Cir. 1973): "To require a 'statement,' in addition to a decision setting forth the same considerations, would be a legalism carried to the extreme."

Various parts of EPA's authority do require it to hold public hearings, prepare written quasi-judicial records of decision, and specifically document various aspects of agency action. To varying degrees these requirements achieve NEPA-like goals. Evolving standards of judicial review have also buttressed the agencies' duty to explain themselves and to consider alternatives, again in the manner NEPA requires (see chapter II). However, all these requirements together still fall short of the requirements of the §102 process. Circulation of a draft impact statement to federal, state, and local agencies and to the public would reveal EPA's preliminary views on a matter in much greater detail to a wider audience sooner than under any existing legislation. Nor do existing requirements ensure that the five specific points spelled out in §102(2)(C) will be routinely discussed each time the agency acts. Finally, existing requirements are not systematic in the same way §102(2)(C) is. That section provides a set, predictable, fair procedure for review each time the agency acts on a major proposal.

The pesticide regulation program is both the hardest case and an excellent example. To cancel the registration of a pesticide, EPA has to conduct a full-scale hearing at which practically every environmental impact and alternative action imaginable may be examined in a quasi-judicial atmosphere.[285] Pesticide manufacturers, users, citizens' action groups, EPA itself, other federal agencies, and state and local units of government may present their oral and written views and are subject to cross-examination. The written record amassed may run to thousands of pages. The hearing examiner must first prepare a lengthy recommended decision. The administrator must then consider the evidence and make a final decision based upon the substantial evidence in the record.[286]

NEPA, however, can still make a significant additional contribution to this review. Without NEPA, EPA is not required to document its preliminary position on the ancillary environmental impacts of its impending decision or to solicit comment on that position in a systematic fashion. Nor must the five factors detailed in §102(2)(C) be addressed systematically. Alternative means of accomplishing the same ends which lie beyond the administrator's power to implement

[285] Federal Insecticide, Fungicide, and Rodenticide Act, 7 U.S.C. §§135–135k, *as amended* by the Federal Environmental Pesticide Control Act of 1972, 7 U.S.C. §135 *ff.*, Pub. L. No. 92–516, 86 Stat. 973 (Oct. 21, 1972).

[286] *Id. See* §§6, 16(b).

are not required to be studied and set out for public discussion and consideration by other agencies or by Congress. Thus even the quasi-judicial decision-making process that EPA must utilize in order to cancel a pesticide registration does not fulfill several important requirements which NEPA explicitly imposes.

To the extent that NEPA and other regulatory statutes pursue similar goals, it should not be too difficult to integrate compliance with both so that duplication and conflict are controlled. The amount of coordination may vary with the nature of the requirements of the regulatory statute and of the Administrative Procedure Act, with the most work to be done in melding NEPA with APA adjudicatory proceedings. Taking the pesticide program as an example, the current requirement of CEQ Guidelines that a draft statement be prepared might be imposed at the prehearing stage. The burden of preparing the draft would rest upon the EPA staff. The final statement conceivably could be incorporated in the hearing examiner's recommended decision, which would accompany his "proposal through the existing agency review processes."[287] While "agency review processes" commence somewhat before the final statement has been completed under this plan, strict technical compliance may have to be forgone in order to effect a reasonable integration of the APA-style hearing and decision with the requirements of NEPA. Influence upon the final decison can thus be exerted only through reactions to the draft, which are included in the record and perhaps in the examiner's recommended decision. This is as it should be, because the administrator is bound by law to make his decision on the record as developed at the hearing.

The additional work and delay which NEPA may very well cause are not made any less formidable by dismissing EPA's concerns as typical agency reluctance to change its ways and increase its responsibilities. The sheer administrative burden of compliance is only part of the picture. Industry and environmental groups alike are watching the new agency very closely. Never have so many private attorneys general—nor such unlikely ones as some industry plaintiffs—prowled the corridors of the administrative agencies looking for a breach of duty. The Environmental Protection Agency is in real danger of being caught between the strict deadlines set by newer environmental control legislation and the challenges mounted on every side under both NEPA and the basic regulatory framework. If

[287] NEPA §102(2)(C), Appendix A.

EPA moves quickly to comply fully with NEPA, it should be able to win the pending lawsuits.[288]

In the end, the most important component of the additional administrative burden that NEPA would impose is the preparation of impact statements for each of the licenses, permits, registrations, approvals, etc., which EPA grants. For pesticides alone, registrations now number around 40,000. A principal reason that the §511(c)(1) exemption was included in the new water act was that it relieved EPA of separately preparing a statement for each one of the thousands of water quality permits which it might have to process for existing discharges.

The answer to the dilemma may lie in the suggestion, set out briefly in chapter VII (pages 290 *ff.*), that agencies should prepare impact statements in "tiers," beginning with the most general and comprehensive statements on legislative proposals, following with statements on implementing policies, programmatic initiatives, and rule making, and ending finally with statements on separate regulatory decisions at the licensing or permit level. Statements nearer the end of the process which involve thousands of separate and frequently minor decisions might be quite brief and still accomplish two important purposes. First, such statements could refer back to earlier statements in the chain, documenting the least-cost, gradually circumscribed winnowing of alternatives that took place. Second, the statements should detail any special consequences likely to evolve from decisions on the specific action, so that NEPA does in fact apply effectively to the lowest levels of agency decision making, without imposing impossible demands upon the administrative process' capacity for returning to first principles each time a routine "major federal action" is taken. In this fashion, EPA can both comply with NEPA and avoid unnecessary delay in implementing its basic programs.

A Postscript on Sovereign Immunity

While there is no doubt that the federal agencies must bear the burden of NEPA compliance, there is lingering doubt, arising from the doctrine of sovereign immunity, whether the agencies can be

---

[288] See chapter I, "Industry's standing to sue," and 4 NATIONAL JOURNAL 1871 (Dec. 9, 1972); *see also, e.g.*, Anaconda Co. v. Ruckelshaus, Appendix B. EPA's chances of avoiding such litigation entirely by prevailing on motions to dismiss for lack of standing to prosecute a NEPA claim were discussed in

sued in court to compel compliance. The doctrine of sovereign immunity rests upon an axiom of Anglo-American jurisprudence which states that government may not be sued without its consent. Early jurists reasoned that the courts dispensed the king's justice, and that the king could not grant justice against himself. Under modern conditions, the theory is anachronistic and poses little obstacle to NEPA suits against administrative agencies, since courts allow it to be readily circumvented through a legal fiction. Plaintiff need only bring suit against a named official who, it is claimed, is acting outside his authority. As the Gillham Dam case explained:

As stated in *Dugan v. Rank*, 372 U.S. 609, 621, 622 (1963), there are recognized exceptions to the *Larson* rule: "Those exceptions are (1) action by officers beyond their statutory authority and (2) even though within the scope of their authority, the powers themselves or the manner in which they are exercised are constitutionally void."[289]

As this case indicates, the doctrine does not ordinarily stand in the way of NEPA suits against federal officials, where plaintiffs allege that the action taken fails to comply with the requirements of the statute. It is surprising that it has been so often put forth by federal defendants; it was rejected by the courts in *Environmental Defense Fund v. Corps of Engineers* (Cross-Florida Barge Canal), *National Helium,* and *Kalur,*[290] as well as in the Gillham Dam case. The Fifth Circuit went so far as to say that raising the defense in the NEPA suit of *Ragland v. Mueller* was "frivolous."[291]

A more complicated problem exists when state agencies or officials are also named as NEPA defendants. The *Bartlett* district court indicated:

Even though Bartlett is sued individually, relief can only realistically be granted against the State itself. In McMorran's case, the Court found that the Eleventh Amendment of the United States Constitution immunized him from suit since the State had not consented to suit or waived its sovereignty. For the reasons so well expressed by the Court in the Hudson Valley decision, I hold that plaintiffs are precluded from maintaining this action against Secretary Bartlett since it is in reality a

---

chapter II (pp. 39 *ff.*). The preferable alternative, however, is to argue estoppel, laches, primary jurisdiction, etc., as discussed at the end of chapter II (pp. 44 *ff.*).

[289] Environmental Defense Fund v. Corps of Engineers, Appendix B, 325 F. Supp. at 733, 1 ELR at 20132.

[290] All in Appendix B.

[291] 460 F.2d 1196, 1197, 2 ELR 20320 (5th Cir. 1972).

suit against the Commonwealth of Pennsylvania to which it has not consented and which immunity it has not waived.[292]

However, the Third Circuit in dictum questioned the wisdom of this rule by saying that "a state may, by engaging in activities of paramount federal interest, waive its immunity under the eleventh amendment."[293]

At least four cases dealing with the state role in highway enterprises have ruled that NEPA must be followed by state defendants. While not characterizing the issue as one of immunity, *Named Individual Members of the San Antonio Conservation Society v. Texas Highway Dept.*[294] prohibited a state, once committed to a federal project, from avoiding federal requirements by withdrawing.

> [As a federal project] the North Expressway is subject to the laws of Congress, and the State as a partner in the construction of the project is bound by those laws. . . . No one forced the State to seek federal funding, to accept federal participation or to commence construction of a federal aid highway. The State, by entering into this venture, voluntarily submitted itself to federal law.[295]

In a similar case the Fourth Circuit in *Arlington Coalition on Transportation v. Volpe*[296] viewed the key issue as pendent jurisdiction.

> Action of a state highway department, challenged because furthering a project that under federal law allegedly must be reconsidered, is a matter in controversy arising under the laws of the United States. Federal jurisdiction over such state action is essential to preserve federal question jurisdiction in the application of federal statutes.[297]

Arlington thus seems to undercut that part of an earlier Fourth Circuit decision, *Ely v. Velde*,[298] which found that NEPA applied to actions of the Law Enforcement Assistance Administration as a funding agency, but not to the basic project decisions which the state had made. The court said that the National Historic Preservation Act "and NEPA, by their very language, impose no duties on the states and operate only on federal agencies."[299] This view is

---

[292] Pennsylvania Environmental Council v. Bartlett, Appendix B, 315 F. Supp. at 246–47, 2 ELR at 20755.
[293] *Id.*, 454 F.2d at 625, 1 ELR at 20627.
[294] Appendix B.
[295] 446 F.2d at 1027, 1 ELR at 20388.
[296] Appendix B.
[297] 458 F.2d at 1329, 2 ELR at 20163.
[298] Appendix B.
[299] 451 F.2d at 1139, 1 ELR at 20615.

questionable, since the point of the cases cited earlier is that state participation in a federal project may result in just such duties. The difference in outcomes may result from the difference in Federal Highway Administration and LEAA procedures; the former made the states responsible for statement preparation, while the latter took the responsibility on itself.

The court in *Ward v. Ackroyd*[300] dealt at length with the issue and found that the state had waived its sovereign immunity by accepting federal funds. The court placed controlling emphasis upon the distinction between suits for damages and those for injunctive relief.

In the latter case I believe that the defense of immunity should yield more readily where the issue sought to be litigated is the compliance or noncompliance by the state with the preconditions set forth in a federal statute or regulation for the enjoyment by the state of federal aid for the specific project for which the state has applied for the aid. In seeking the federal aid the state has necessarily agreed to be bound by the terms of the federal statutes and regulations which govern the availability and disposition of the federal aid.[301]

The court went on to say that although the issue of sovereign immunity was not raised in *Arlington Coalition*, the rationale of the Fourth Circuit's view of pendent jurisdiction was essentially the same as that which the *Ward* court applied to sovereign immunity.[302]

Finally, the court in *La Raza Unida v. Volpe*[303] expressly dealt with the question of sovereign immunity and concluded that state participation in the work caused the state to forfeit such protection.

In the present case, however, there is a question as to (1) the constitutionality of the state action, and (2) whether the officials involved have complied with state and federal statutes. Under *Ex Parte Young* 209 U.S. 123 (1908), these two situations are generally considered exceptions to the sovereign immunity doctrine. A state official who exceeds his authority or who violates the Constitution is not covered by the protective mantle of sovereign immunity.[304]

STATEMENTS ON LEGISLATIVE PROPOSALS

NEPA requires that impact statements be submitted with the legislative proposals that federal agencies submit to Congress. This

---

300 344 F. Supp. 1202, 2 ELR 20405 (D. Md. 1972).
301 *Id.*, 344 F. Supp. at 1215, 2 ELR at 20410.
302 *Id.*, 344 F. Supp. at 1216, 2 ELR at 20410.
303 Appendix B.
304 337 F. Supp. at 225, 1 ELR at 20643.

little-discussed requirement appears in the text of the statute before the wording upon which the courts have relied in applying NEPA to "other major federal actions," making the latter appear to be a draftsman's afterthought.

> . . . All agencies of the Federal government shall . . . include in every recommendation or report on proposals for legislation and other major Federal action significantly affecting the quality of the human environment a detailed statement.[305]

Very few legislative impact statements have been filed in NEPA's three years of existence. Of the 118 draft and 45 final statements filed by November 30, 1972, a large number are on environmentally protective legislation, a category for which statements are much less useful than for development-oriented proposals.[306] The House Committee on Merchant Marine and Fisheries has estimated that, if legislative impact statements were being prepared as they should be, 800 statements would be submitted during each session of Congress.[307]

In spite of the potential usefulness of the §102 process in enabling Congress to survey a wide array of alternatives well in advance of potential environmental harm, the implementation of the legislative impact statement requirement has languished. For reasons that are not entirely clear, Congress, OMB, CEQ, and the agencies have shown little interest in policing the requirement. Nor have many lawsuits been filed to compel preparation, probably because appropriate relief is difficult to formulate.

The four cases which discuss legislative impact statements do not shed much light upon the scope of the requirement or upon how it is likely to be interpreted by the courts in the future. In *Environmental Defense Fund v. Tennessee Valley Authority*,[308] plaintiffs alleged that TVA had to comply with NEPA in constructing the Tellico

---

[305] NEPA, §102(2)(C) Appendix A.

[306] Computer printout of legislative impact statements furnished the author by the Council on Environmental Quality, Mr. Bryan P. Jenny, analyst ("Search Date: 12-06-72"). Before computerized impact statement records began to be kept on July 1, 1972, 88 drafts and 29 finals had been filed, according to Mr. Jenny.

[307] *House Comm. on Merchant Marine and Fisheries, Administration of the National Environmental Policy Act of 1969 (P.L. 91–190),* H.R. Rep. No. 92–316, 92d Cong., 1st Sess. 15, 17, 23, 26 (1971). *See also Hearings on the Administration of the National Environmental Policy Act Before the Subcomm. on Fisheries and Wildlife Conservation of the House Comm. on Merchant Marine and Fisheries*, Serial 91–41, Part 1, 91st Cong., 2d Sess. 60 (1970).

[308] 339 F. Supp. 806, 2 ELR 20044 (E.D. Tenn.), *aff'd*, —— F.2d ——, 2 ELR 20726 (6th Cir. 1972).

Dam and Reservoir Project which, besides the dam and reservoir, included a new town for 50,000 inhabitants. The project had received its initial appropriation in 1966, and $29 million of the total project budget of $69 million had already been spent. The court held that an impact statement was required for the project and for appropriation requests submitted after NEPA's enactment.

> Since the Tellico project is funded by annual Congressional appropriations, it would appear that a request for such an appropriation would be a "proposal for legislation" within the meaning of §102(2) (C). Consequently, each appropriation request after January 1, 1970 would be required to be accompanied by a detailed environmental impact statement.[309]

The Sixth Circuit affirmed, relying upon congressional intent and CEQ and TVA guidelines explicitly covering appropriations. In affirming, the court rejected arguments (1) that Congress intended a distinction between "routine authorizations to disburse funds" and "changes in existing law," with statements to be prepared only on the latter, (2) that impact statement preparation would impose an intolerable administrative burden, and (3) that passage of the Tellico appropriation implicitly exempted the project from NEPA.[310]

The courts did not discuss the type of relief appropriate for enforcing the legislative impact statement requirement alone, because they apparently had concluded that one all-purpose statement could be prepared immediately which might also be submitted later with the annual appropriation request to Congress. Assuming that this interpretation is correct, an injunction against the continuing project was an appropriate form of relief for the failure to file a legislative impact statement. For legislative proposals which do not involve particular projects that are already under way, appropriate relief against the agency may be harder to fashion.

In similar circumstances in *Environmental Defense Fund v. Froehlke*,[311] the court held as a conclusion of law that NEPA was applicable to requests for subsequent appropriations for the Truman Dam. The court did not discuss the basis for its holding but, viewing the case as a whole, its finding appears to buttress the court's overall emphasis on Congress' role in using the impact statement to decide the ultimate fate of the project.

---

[309] 339 F. Supp. at 811, 2 ELR at 20046.
[310] —— F.2d ——, 2 ELR 20726 (6th Cir. 1972).
[311] 348 F. Supp. 338, 2 ELR 20620 (W.D. Mo. 1972).

In *Environmental Defense Fund v. Volpe*,[312] EDF and two congressmen sought to compel the secretary of transportation to file a statement that would accompany several transportation needs reports which the secretary was required by statute to submit to Congress. Although the reports did not contain the actual language of legislative proposals, they did contain the secretary's recommendations to Congress concerning federal funding for states that had completed their part of the interstate network, his recommendations for the functional realignment of the federal-aid highway system, his estimate of future highway needs, and his recommendations for the federal highway program from 1976 to 1990. Ruling from the bench, the court held that these recommendations were for long-term study only and were not the kind of "proposals for legislation" contemplated by NEPA.[313]

The lawsuit was filed quite late, close to the statutory deadline for submission of the transportation reports. Nevertheless, the court appears to err in focusing upon the long-term effects of the secretary's reports and in not adopting the view of the CEQ Guidelines that legislative "reports" are covered by NEPA as completely as are distinct legislative proposals.[314]

A fourth district court also focused upon the speculative, long-term impact of a federal program in dismissing another lawsuit requesting that a legislative impact statement be prepared. In *Scientists' Institute for Public Information (SIPI) v. Atomic Energy Comm'n*,[315] plaintiffs challenged the commission's failure to prepare a comprehensive impact statement on the Liquid-Metal Fast Breeder Reactor Program (LMFBR) which the commission launched in the late 1960s. In arguing that an impact statement had to be prepared for the program's annual appropriations request, plaintiffs pointed out that the program is a line item in the commission's annual authorization and appropriation requests and that extensive testimony had been taken by congressional committees on various aspects of the program. Ruling from the bench, the district court held that the commission was not required to prepare a statement for the overall LMFBR program and that NEPA was satisfied if the commission had prepared impact statements for individual

---

[312] 312 Civil No. 151–72, ELR Dig. [224] (D.D.C. Feb. 15, 1972).

[313] *See* Comment, 2 ELR 10038, 10041–42 (April 1972).

[314] CEQ Guidelines, §5(a)(i), Appendix C.

[315] —— F. Supp. ——, 2 ELR 20642 (D.D.C. 1972). *See also* ELR Dig. [182] and Comment, *Challenging the Environmental Impact of a Major Long-range Agency Program: SIPI v. AEC*, 1 ELR 10090 (June 1971).

facilities that are part of the program, such as the first demonstration plant. The case has been appealed and argued before the District of Columbia Circuit and is awaiting decision.

Although the assumptions of the Sixth Circuit are not clear, it appears that its holding in *Environmental Defense Fund v. Volpe* is in direct conflict with the D.C. court's holding in *SIPI.* In the Tellico case, the court ordered an impact statement to accompany annual appropriations requests for the project; in *SIPI* the court acquiesced in the preparation of fragmented statements on project components, passing up the opportunity to apply NEPA comprehensively at the appropriations request stage on grounds similar to those relied upon by the court in *Environmental Defense Fund v. Volpe.*

Both *Environmental Defense Fund v. Volpe* and *SIPI* appear to be wrongly decided for the same reasons. For NEPA to have its maximum effect, the Act must be applied as early as possible in agency decision-making processes, and as comprehensively as possible to agency initiatives. Furthermore, both cases are also in conflict with *Natural Resources Defense Council v. Morton,*[316] which held that "when the proposed action is an integral part of a coordinated plan to deal with a broad problem, the range of alternatives that must be evaluated is broadened."[317]

> *NRDC v. Morton* puts agencies on notice that if actions which they propose to carry out have not been sufficiently examined in light of the national environmental policy, the proposed action will be held up until a full review of basic, underlying policies, resource commitments and alternatives takes place. The district courts in *SIPI v. AEC* and in *EDF v. Volpe,* both ruling from the bench, would allow agencies to delay impact statement preparation until opportunity for a prospective, wide-ranging examination of them had passed. These district court decisions effectively deny that NEPA has a policy-oriented primary thrust and make impossible full compliance with NEPA by the federal agencies. Clearly *SIPI v. AEC* and *EDF v. Volpe* interpret NEPA in precisely opposite ways from *NRDC v. Morton.*[318]

One commentator has said that the main reason why lawsuits challenging agency failures to prepare legislative impact statements have not been filed more often is that appropriate relief is difficult to fashion. The former chairman of the Administrative Conference has indicated that judicial relief may be impossible to obtain, because Congress cannot realistically be enjoined from considering

---

[316] Appendix B.
[317] 458 F.2d at 835, 2 ELR at 20033.
[318] Comment, *supra* note 248, 2 ELR 10038, 10042 (April 1972).

bills which are not accompanied by impact statements.[319] He perhaps overstates the case. A request for mandatory injunctive relief and declaratory relief against the agency might succeed if it were filed either before the proposal is submitted, or after it is submitted but before Congress has begun to consider it. The order might require the agency to comply with its own guidelines on the subject and with the guidelines of CEQ and OMB. If the court finds these guidelines inadequate, it may order the agency to prepare procedures which comply with the Act.[320] For this purpose CEQ and OMB may have to be joined as parties. In the event of noncompliance, congressional committee members might question the propriety of proceeding with hearings or proposals on which the court had ruled. By refusing to consider the legislation Congress itself might participate to some extent in enforcing a judgment against a recalcitrant agency.

Congressional committees, not the courts, are more likely "enforcers" of the legislative impact statement requirement, because they make primary use of the information gathered. Congress, however, has been surprisingly uninterested in the information which it was anxious for agency decision makers to receive. The loss to improved governmental decision making is all the greater because congressional review of impacts which proposed legislation may have can be much more useful than agency attempts to mitigate impacts after an ill-conceived statute has been enacted. Congress has more latitude to set policy through its power to legislate than an administrator has to avoid environmental impacts after a congressional mandate has been delivered.[321]

Congressional interest in legislative impact statements has not been totally lacking, although it appears to be weaker today than immediately after NEPA's enactment. The Subcommittee on Fisheries and Wildlife Conservation of the House Committee on Merchant Marine and Fisheries held hearings in late 1970 which touched briefly on the matter of NEPA's legislative record. The report on these hearings, filed six months later, attacked the agencies' desultory performance in preparing legislative impact statements and laid the

[319] Joint Hearing, *supra* note 232, at 417. (Prepared remarks of Roger Cramton.)

[320] *See* Calvert Cliffs', Greene County and Sierra Club v. Froehlke (Trinity River–Wallisville Dam), all in Appendix B.

[321] Natural Resources Defense Council v. Morton, Appendix B, 458 F.2d at 833, 2 ELR at 20032, quoting from Defendants' Memorandum (Dec. 8, 1971), at 11.

blame at the CEQ's door.[322] On the other hand, a General Accounting Office study of selected agencies' compliance with NEPA tended to take the OMB to task for not enforcing the requirement.[323] Neither the subcommittee report nor the GAO study blames Congress for failing to require the submission of statements. Such a requirement in fact is imposed only by the Senate Committee on Public Works, which has amended its rules of procedure to require that statements be submitted.[324]

After some early indications that the CEQ and OMB would share the policing of the legislative impact statement requirement, systematic, publicly visible[325] executive branch enforcement has virtually disappeared, unless we are to count the promising but as yet unrealized potential of EPA review under §309 of the Clean Air Act (see pages 229 *ff.*). The OMB has even taken the recent step of relaxing its formal guidelines regarding agency compliance except for water resources projects.

As indicated in chapter I (pages 11 *ff.*), there is some suggestion in NEPA's legislative history that Congress viewed OMB as the appropriate agency to ensure overall compliance with NEPA. The basic decision not to make OMB the guardian of the §102 process was apparently made at the presidential level soon after passage of the Act. Executive Order 11514 gave CEQ primary responsibility to

---

[322] *See* note 307, *supra.*

[323] Comptroller General of the United States, REPORT TO THE SUBCOMM. ON FISHERIES AND WILDLIFE CONSERVATION OF THE HOUSE COMM. ON MERCHANT MARINE AND FISHERIES, IMPROVEMENTS NEEDED IN FEDERAL EFFORTS TO IMPLEMENT THE NATIONAL ENVIRONMENTAL POLICY ACT OF 1969 51, 55 (May 18, 1972).

[324] The Committee's Rule 13 states, "No project shall be approved or other action taken unless the Committee has received an Environmental Impact Statement relative to it."

[325] CEQ and OMB have "agreed" off the record to do a good deal more than their respective guidelines reveal, as the following letter, dated September 7, 1972, from Chairman Train to former OMB Director Weinberger shows:

Dear Mr. Weinberger:

> Our staffs have held useful follow-up conversations to discuss recommendations of the Comptroller General in his Report "Improvements Needed in Federal Efforts to Implement the National Environmental Policy Act of 1969" (May 1972). These conversations have focussed on the recommendations contained in that Report for improving the implementation of NEPA through use of the legislative clearance process.
>
> In summary, we have agreed on the following:
>
> 1. Both OMB and CEQ will continue to cooperate in reminding agencies of their responsibility to prepare impact statements on proposed legislation of environmental significance.
>
> 2. OMB will continue to refer legislative items thought to have environmental significance to CEQ for comment. In connection with such

"issue guidelines to Federal agencies for the preparation of detailed statements on proposals for legislation."[326] The guidelines prepared by CEQ gave an expansive interpretation to the legislative impact statement requirement. The guidelines require statements on "agency recommendations on their own proposals for legislation and . . . agency reports on legislation initiated elsewhere," except that in the latter case only the agency which has primary responsibility for the subject matter must prepare a statement. But, at the same time that CEQ defined the scope of the requirement, it went on to say that the Office of Management and Budget would supplement these general guidelines with specific instructions relating to §102(2)(C) and OMB's legislative clearance process.[327]

Rather than accepting responsibility for enforcement of the legislative impact statement process, OMB has actually loosened its formal requirements and has attempted to refer the matter back to the CEQ. While OMB initially required that statements be submitted with legislative proposals,[328] its current position is that "information copies of required 102 statements should be submitted to OMB if available at the time clearance is requested."[329] Water resources projects still require statements. For other unanswered questions

---

referrals both OMB and CEQ staff will remain alert to the need for a 102 statement in cases where one has not been prepared and will request the initiating agency to prepare such a statement promptly.

3. The CEQ and OMB staffs will cooperate in giving further guidance to particular agencies in identifying types of repetitive legislation requiring environmental impact statements (such as certain bills affecting Federal lands or transportation policy or annual construction authorizations or appropriations). The objective of such guidance will be to assure that such statements will regularly be prepared in draft for consideration by the initiating agency *prior* to submission of such legislative proposals to OMB and be circulated within the Executive Branch at the same time as the legislative proposal.

Through these practices we believe OMB can assist both the Council and the agencies in maintaining a high standard of implementation of NEPA while at the same time giving appropriate recognition to the purposes and needs of the legislative clearance process.

Sincerely yours,
(sgd) Russell E. Train
Russell E. Train
Chairman

[326] *Supra* note 233, §3(h).
[327] *See* Appendix C, §5(a)(i).
[328] OMB Bulletin 72–3.
[329] OMB Bulletin 72–6 (Sept. 14, 1971), but especially OMB's more recent Circular No. A–19 (Revised) (July 31, 1972). *See also* chapter I, *supra* notes 39, 41, 42 and accompanying text.

OMB states that the impact statement requirement "is dealt with generally in guidelines issued by the Council on Environmental Quality." As the General Accounting Office concluded, "under current OMB procedures the agencies have little incentive to prepare such statements in order to receive OMB's legislative clearance."[330] The principal reasons for such inaction, according to OMB staff interviewed by the GAO, are that the environmental impact statement would probably contain more detailed information than OMB needed, and that the commenting process could disrupt congressional requirements.[331]

The reasons for OMB's not requiring compliance confuse the effect of compliance on Congress, which might judge for itself, with the effect of compliance on OMB. Further, OMB apparently did not take into account that there might be value to the public and to the sponsoring agency, in having OMB enforce the requirement, even if the statements tell OMB more than it wants to know. The weight of the point is increased by the absence of any other authority, with a mechanism comparable to OMB's legislative clearance process, which could police the impact statement requirement. The GAO study concludes its discussion of OMB compliance with a recommendation which states very well the advantages of a more active OMB role:

> The objective of section 102 is to build into agencies' decision-making processes an appropriate and careful consideration of the environmental aspects of proposed actions. The environmental impact statement provides visible evidence of the factors known by the agency and of the way the factors were considered. Therefore both the sponsoring agency and the commenting agencies benefit from the information presented in the statement. Although OMB feels that the statement probably would contain more detailed information than they need, its value to the sponsoring agency should not be disregarded.[332]

POSSIBLE LIMITS ON NEPA'S APPLICABILITY

When read literally, NEPA says that no federal agency may propose any action without considering its environmental impact, and that any such proposed action which is "major" must be accompanied by an impact statement. For an Act with such broad scope, exceptions have been surprisingly few. If we leave aside for the

[330] Comptroller General's Report, *supra* note 323, at 55.
[331] *Id.* at 53–54.
[332] *Id.* at 54.

moment the kinds of exceptions that are common to all legislation (e.g., nonretroactive application), the courts have partially limited NEPA's application to "actions" and "agencies" only in the areas of national security and temporary or emergency agency actions. Additional possible exceptions may exist by guideline, such as actions by sister agencies of "lead" agencies. In these areas, which are discussed below, the exceptions are sometimes more apparent than real and have not been uniformly allowed.

National Security and Military Installations

Courts have traditionally been reluctant to interfere with decisions of the Executive which involve questions of national security. The case law under NEPA, however, far from holding uniformly that a "major federal action" involving national security and military installations is exempt, indicates that whatever exemption exists is only partial and that each case will be judged on its facts.

Of the courts which have decided NEPA cases, the Tenth Circuit seemed the most unwilling to apply NEPA to the military, although the action at issue was not of immediate military importance. In *McQueary v. Laird*[333] plaintiffs sought to compel the army to prepare a statement on the effects of continued storage of chemical and bacterial warfare agents at the Rocky Mountain Arsenal in Colorado. The court held:

> Public disclosure relating to military defense facilities creates serious problems involving national security. We hold that NEPA does not create substantive rights in the plaintiffs-appellants here to raise the environmental challenge in regard to the Rocky Mountain Arsenal. In its proprietary military capacity, the Federal Government has traditionally exercised unfettered control with respect to internal management and operation of federal military establishments.[334]

Unfortunately, the opinion considers neither the extent to which NEPA could conceivably limit this unfettered discretion, at least in regard to the environmental hazards posed by military facilities, nor the effect of the applicable army regulations for implementing the Act.

The court in *Nielson v. Seaborg*[335] appeared to follow *McQueary* in refusing to apply NEPA to the AEC's Nevada nuclear tests; however, its ruling on defendants' motion for summary judgment rests

[333] 459 F.2d 608, 1 ELR 20607 (10th Cir. 1971).
[334] 449 F.2d at 612, 1 ELR at 20609.
[335] 348 F. Supp. 1369, 2 ELR 20765 (D. Utah 1972).

on narrow grounds involving deficiencies in plaintiffs' complaint and briefs. Plaintiffs alleged that the tests were "in contravention of the policies and purposes and intended effect" of NEPA. The court observed that plaintiffs had not thereby clearly charged the AEC with abuse of discretion but, even if they had, the tests were not reviewable, being wholly committed to agency discretion. With respect to plaintiffs' argument that NEPA and Reorganization Plan Number 3[336] required all radiation safety regulatory functions to be transferred to the Environmental Protection Agency, the court replied that plaintiffs' unsupported allegations could not overcome defendants' uncontested affidavit stating that test safety for nuclear tests was handled by a different office than that specifically transferred to EPA. Finally, plantiffs failed to allege the insufficiency of the AEC's impact statements in their complaint. To do so only in the briefs was insufficient to withstand the government's motion for summary judgment. On balance, the court follows the unfettered discretion approach of *McQueary*, and the criticisms applicable to *McQueary* also apply to this case.

Two additional cases, while decided on other grounds, seem to evince a similar reluctance to interfere with military discretion. *Citizens for Reid State Park v. Laird*[337] upheld a determination that a training exercise was sufficiently minor not to require a statement, possibly because the court doubted the wisdom of enjoining the exercise. *State Committee to Stop Sanguine v. Laird*[338] involved a challenge to a navy plan for an ultra-low frequency communications system, to be created by burying cable over a thousand-square-mile grid in northern Wisconsin. The court, strangely enough, confined its discussion of NEPA to §102(2)(E), which refers to the need to recognize the worldwide nature of environmental problems. The court relied on plaintiffs' failure to allege that the worldwide character of environmental problems was not recognized by the defendant and on the view that the amount in controversy, measured by the Seventh Circuit's "plaintiff's benefit" rule, did not meet the $10,000 figure required for federal jurisdiction.[339] The court's failure to consider the issue as a federal question arising under the Administrative Procedure Act and NEPA suggests that it was searching for some way to duck the possible national security issues

[336] *Supra* note 250.
[337] Appendix B.
[338] 317 F. Supp. 664, 2 ELR 20100 (W.D. Wisc. 1970).
[339] 317 F. Supp. at 667, 2 ELR at 20101.

that would have been involved had the case been decided on the merits.

A recent district court opinion from the district of Hawaii stands in stark contrast to the decisions which apparently are reluctant to apply NEPA to the military. In *People of Enewetak v. Laird,*[340] the court held that the §102(2)(C) requirement applied to the Pacific Cratering Experiments conducted on Enewetok, a Pacific atoll used extensively in the past for test detonations. The court's brief order did not comment upon the effect on national security of NEPA compliance in this instance.

An additional, well-known NEPA challenge involving a nuclear test is widely thought to be the leading litigation on NEPA's application to national security. The courts involved had no difficulty with NEPA's application to security-sensitive underground nuclear testing. However, the several decisions resulting from Project Cannikin in *Committee for Nuclear Responsibility v. Schlesinger*[341] present a much more complicated picture and deal peripherally with national security, as the background of the case shows.

Project Cannikin involved testing a 5-megaton nuclear warhead by detonating it deep beneath Amchitka, an island in the Aleutian chain off Alaska. The Atomic Energy Commission assumed that NEPA applied to the test and prepared both a draft and a final impact statement. Plaintiffs contested the adequacy of the commission's final statement, primarily relying upon a series of secret studies prepared by the government and not included in the NEPA review process. The studies, some of which were ordered to be made public late in the litigation, allegedly cast greater doubt on the safety and environmental soundness of the test than did the commission's "official" impact statements.

The challenge to Project Cannikin was heard in the district and circuit courts three times each before an appeal requesting an injunction in aid of jurisdiction was denied by the Supreme Court. Yet, much less new law was made than might be expected from such a prolific litigation.

The seven rulings—three in the District Court, three in the Court of Appeals and one in the Supreme Court—certainly offered the prospect of much new judge-made law on NEPA. Surprisingly, however, the precise holdings in the litigation add little of substance to NEPA's judicial history. The Supreme Court's ruling, after the first oral argu-

[340] —— F. Supp. ——, 2 ELR 20739 (D. Hawaii 1972), —— F. Supp. ——, 3 ELR 20190 (D. Hawaii 1973).

[341] Appendix B.

ment to be presented to it under the act, leaves much in doubt. The majority of four (Brennan, Douglas and Marshall, dissenting) merely denied the relief requested without giving reasons or delivering an opinion. The Circuit Court refused to rule either on the correctness of the district court's holding that the AEC had complied with NEPA or on the court's refusal to release certain memoranda. Perhaps the most important result in the cases was a holding in the Circuit Court that responsible scientific opinion which disagreed with the agency assessments of environmental impact must be included in the agency's impact statement, but even this holding is open to criticism that it is restrictive and confusing and that the holding ten months earlier on exactly the same point in an Arkansas district court more clearly and fully interpreted the intent of the act. See *EDF v. Corps of Engineers*, 1 ELR 20130 (E.D. Ark. 1971). Finally, the district court in its three rulings never prepared a written opinion.[342]

Furthermore, the failure of the plaintiff environmental groups to obtain the relief requested does not indicate that the government was successful in pressing its several defenses, including the claim that essential considerations of national security exempted it from doing any more than it already had to comply with NEPA. In fact, the litigation has been judged to be a NEPA victory. Had the litigation been pursued after the blast, as the circuit court indicated that it could, perhaps it would have formally vindicated the view of the circuit court and Justices Brennan and Marshall that a substantial question existed as to the adequacy of the commission's compliance with §102(2)(C).

The eventual loss before the Supreme Court, [and] the subsequent uneventful detonation of the device . . . should not be allowed to overshadow the importance for NEPA of the Cannikin litigation. It is significant that the plaintiff groups were allowed to appear at all before a special Saturday session of the court (requested by the Chief Justice just hours before the blast) to challenge a major nuclear test of allegedly vital importance to national security, which Congress had debated and the President had personally ordered to go ahead, on the ground that the act required the Commission to use the 102 process for informing Congress, executive decision-makers and the public of the full range of environmental risks which the project posed. The litigation, it must be stressed, was not a loss for NEPA. Rather, even though they narrowly failed to obtain the relief requested, the environmental groups conducted a litigation the net result of which, whatever the merit of specific holdings, is a warning to federal agencies that NEPA may not be cast aside, even in the circumstances of Project Cannikin, nor may agencies alone determine what their impact statements must contain.[343]

---

[342] Comment, *Project Cannikin and the National Environmental Policy Act*, 1 ELR 10161, 10162 (October 1971).

[343] *Id.*

With respect to the effect of national security upon the outcome of the litigation, the circuit court made it very plain that military secrets were not involved. The government has the right to protect military and diplomatic secrets, the court said, but plaintiffs "indicated clearly" that they did not seek to have such secrets included in the impact statement.[344] However, citing "overriding requirements of national security," the court refused on two occasions to grant an injunction.

Plaintiffs' application for a stay *pendente lite* confronts the Court with an inherent limitation on its scope and information. The AEC release yesterday, reflecting the President's approval of the test, states that the Executive Branch has considered problems of environmental damage and has given precedence to "overriding requirements of national security." The Court is concerned solely with the question of legality of the AEC action under NEPA and its obligation to determine that question. It is in no position to consider or appraise the national security aspects of the test underlying the President's determination. It is the responsibility of the Executive to take into account both the considerations of national security and the serious issues of legality, identified by the opinions of this Court, relating to the claim that the AEC has failed to comply with NEPA and thus to permit the informed appraisal, by the Executive, Congress and the public, contemplated by that statute. The Court limits its actions in this litigation to matters within the judicial province; it is in no position at this juncture to enter a stay order that would interject the Court into national security matters that lie outside its province.[345]

And in a second opinion a few days later:

While the Government's assertion of monetary damage from an injunction is not minimal, it does not weigh as heavily with us as its assertions of potential harm to national security and foreign policy—assertions which we obviously can not appraise—and given the meager state of the record before us, we are constrained to refuse an injunction. Cf. Reynolds v. Sims, 377 U.S. 533, 585 (1964); Note, *Developments in the Law—Injunctions*, 78 Harv. L. Rev. 994, 1005–08 (1965). While we deny preventive relief, it should be clear that plaintiffs may yet prevail in their claim that the AEC failed to comply with NEPA in approving the Cannikin test.[346]

To the extent that the Cannikin litigation made new law at all, it concerned whether injunctive relief is ever available under any statute in view of an Executive determination that national security

[344] 463 F.2d at 792, 1 ELR at 20530.
[345] 463 F.2d at 795, 1 ELR at 20532.
[346] 463 F.2d at 798, 1 ELR at 20534.

is at stake. While it would perhaps have been desirable for the litigation to have been carried through to a formal vindication of plaintiffs' assertion that the commission's statement was inadequate, nevertheless the courts strongly implied that the alleged inadequacy did exist and could in time be fully reviewed. Furthermore, by agreeing to hear the case under extraordinary demands of time and somewhat exceptional circumstances, the courts confirmed that they will closely police attempted agency end runs around NEPA and will admit exceptions to its applicability only under the most compelling circumstances.

Temporary or Emergency Actions

The problem in this area stems from the discord between the time required for NEPA consideration of issues and the need for quick action associated with programs such as wage and price stabilization. The cases do not yet reveal a clear trend.

In *Cohen v. Price Commission,*[347] Judge Weinfeld ruled that the "long range" aims of NEPA and its "prolonged decision-making process" were incompatible with the nature of the Price Commission (which was to exist only until March 1973), and its need to "act upon matters within its authority with dispatch." Where the grave financial condition of the New York Transit Authority meant that the legality of a fare increase would have to be speedily decided, the court found that there was "substantial doubt" that Congress intended to have the commission operate in accordance with the Act, and denied plaintiff's plea for a preliminary injunction.

It should be noted that this holding is not dispositive, particularly in cases where more time would be allowed for a decision on a price increase. In particular, it could be argued that NEPA would apply to commission decisions made after the delay inherent in public hearings.

The case of *Port of New York Authority v. Interstate Commerce Comm'n*[348] took a position similar to *Cohen* on a challenge to an interim increase in lighterage rates in New York Harbor. The court reasoned that environmental issues could be raised and a statement prepared in the ongoing proceeding to determine *final* rates. (Should the rates finally granted be lower than those collected in the interim,

[347] Appendix B.
[348] 451 F.2d 783, 2 ELR 20105 (2d Cir. 1971).

the difference would be refunded to the shippers.) The court said of the interim proceeding:

> The detailed evaluation of benefits and costs required by NEPA is not possible at the stage of review of the lawfulness of the proposed tariffs under focus here. . . . Suspension proceedings are informal. . . . They are simply not amenable to the careful balancing analysis mandated by NEPA.[349]

These two cases may be contrasted with the decision of a three-judge panel in *SCRAP v. Interstate Commerce Comm'n.*[350] Judge Wright dealt with the commission's argument that the rate increase granted there required speedy action by saying:

> The Commission's position appears to rest on the *non sequitur* that because an action is taken quickly it is therefore unimportant. Yet it hardly requires argument to demonstrate that some of the most important federal actions in our history have been taken with great alacrity. To the extent that the need for speed is relevant at all, it goes not to the importance of the federal action but to the provision in NEPA which requires compliance only "to the fullest extent possible."[351]

The court concluded that under the facts of the case, which included protracted negotiations over the increase, the ICC had not carried its burden under the "fullest extent possible" standard. *Cohen* was distinguished as dealing with a temporary *agency*, which the ICC certainly was not.[352]

It is difficult to say that these cases show a definite trend at all. To the extent that they do, however, they indicate that a case-by-case approach will evolve.

We now turn from the two main categories of possible exemption from NEPA's requirements, and take a look at a third possibility, which exists for an agency that does not have the predominant or lead role in proposing and carrying out an action that involves more than one agency. Under CEQ Guidelines, with the apparent approval of several courts, the "lead agency" must coordinate and prepare a comprehensive statement. (For a fuller discussion of the lead agency concept, see pages 196 *ff.*)

Although the exempted agency escapes preparing an impact statement on a proposal for which it otherwise would definitely have to prepare one, the "lead agency" concept does not create a true ex-

---

[349] 451 F.2d at 789–90, 2 ELR at 20108.
[350] Appendix B.
[351] 346 F. Supp. at 199, 2 ELR at 20490.
[352] 346 F. Supp. at 199 note 13, 2 ELR at 20490 note 13.

emption from NEPA. First, a full statement must still be prepared and circulated in advance of the first federal step toward approval of the action. In theory, a statement equivalent to the one that would be submitted by the exempted agency must be prepared. Second, the exempted agency must fully assist the lead agency so that in theory the writing of the statement is not so much a delegated task as a cooperative one. The "lead agency" concept is shaky and is criticized on several grounds in chapter VI (see pages 197 *ff.*).

V

Problems of Transition: Projects and Programs in Progress When NEPA Was Enacted

THE implementation of NEPA has been marked by controversy over its applicability to actions begun before its enactment but still incomplete on January 1, 1970, when the Act took effect. Since many major undertakings take years to proceed from initiation to completion, it is not surprising that the issue of whether to apply the Act to actions still under way has figured in over fifty of the cases so far decided.

To be sure, §102(1) commands that agencies follow NEPA procedures "to the fullest extent possible." But opinions may differ as to when work is sufficiently well advanced to make change impossible, or, as now rephrased by the CEQ Guidelines, "impracticable." This formulation presumably covers those situations where an impact statement, though theoretically possible, would be submitted too late to make changes in the project feasible.[1]

Unfortunately, neither the statute nor the legislative history provides specific guidance when views of practicality or possibility differ. While Senate Report 91-296 does contain a tantalizing reference to applying the Act to the "ongoing activities" of the federal government,[2] in context this may simply mean "day-to-day operations" rather than those projects only partly completed as of enactment. Here, as in many other areas, the courts have had to evolve rules to flesh out the statute.

The judicial response has been characterized by concern over the "retroactive" application of the Act. This term ordinarily refers to

---

[1] It might be argued that preservation of the integrity of the §102 process might still require statement preparation, if the information which it provided would be of future use to decision makers confronted with similar projects. See chapter IV, *supra* note 282 and accompanying text.

[2] SENATE COMM. ON INTERIOR AND INSULAR AFFAIRS, NATIONAL ENVIRONMENTAL POLICY ACT OF 1969, S. REP. NO. 91–296, 91st Cong., 1st Sess. 19 (July 9, 1969).

legislative attempts to alter the legal status of a deed already done; for example, deciding after the fact that an act should be made a crime and the actor punished. The principal objection is that this makes it impossible for an individual to assess in advance the legality of his plans, putting him instead at the mercy of later legislation. The legal doctrine which has grown up to prevent this inequity, the retroactivity doctrine, is of uncertain relevance in NEPA cases. Certainly it would apply to actions completed in every respect by January 1, 1970. But for actions with "spillover" beyond this date, much depends on the judicial definition of the action involved.

The controversy thus centers on actions straddling the effective date of NEPA, which are only partly completed at that time. Different courts have taken the word "action" in either of two senses: (1) the formal legal event which commits the federal government to a project, such as signing a contract; (2) the substantial physical event, such as the work necessary to build a dam or road. In the first view, if the formal event—the "critical action"—took place after January 1, 1970, the Act applies. Those courts adopting the second view hold that NEPA is applicable when "substantial action" or substantial tasks remain to be performed before the project is complete. In effect, the work remaining is itself viewed as a post-1969 major federal action. In a related formulation, the practicality of applying the §102 procedure depends on whether the costs of altering or abandoning the work thus far completed clearly outweigh the possible benefit to the environment of doing so.[3]

This chapter will discuss three classes of cases. We will first discuss those situations where a critical action was identified, where the federal role was simply to approve the plans of a *nonfederal* actor (as in a permit or contract), and where no subsequent federal responsibility or involvement lingered. When the government has thus washed its hands of a matter before NEPA's enactment, courts tend to conclude that the decision has become final, that rights have been vested, and that reconsideration of the federal approval would be objectionably retroactive.

The second set of cases focuses on the progress of the action or project itself. If the federal government is carrying out the work (e.g., building a dam or a federal facility), the government is not

[3] Arlington Coalition on Transportation v. Volpe, 332 F. Supp. 1218, 1 ELR 20486 (E.D. Va. 1971). The test stated here was enunciated by the Fourth Circuit when it reversed, 458 F.2d 1323, 1332, 2 ELR 20162, 20164, *cert. denied sub nom.* Fugate v. Arlington Coalition on Transportation, 41 U.S.L.W. 3249 (Nov. 7, 1972).

irrevocably committed to allowing the action to continue. Work may be halted without prejudicing the rights of others. Alternatively, the remaining work may itself be viewed as a new major action.

By far the greatest difficulties come in the third set of cases, where argument can be made for either the "critical action" or "substantial action remaining" approach. Typically, these are cases where formal federal action occurred before 1969 (the commitment may be fairly tenuous), but the federal presence continued to permeate the project thereafter. The paradigm is the federal-aid highway program, where procedural confusion creates a great degree of ambiguity as to the real significance of the formal actions. We will argue that the better approach in such cases is for a court to consider the actual progress of the work and to use discretion in applying NEPA. Given such continuing federal involvement, "practicability" should be tested by the actual progress of the ongoing action, which may be veiled by the formalities of administrative action.

THE "CRITICAL ACTION" APPROACH

The courts have found that the "critical action" approach to the problem of ongoing action is most relevant when it is possible to isolate a single federal decision on whether or not to permit a non-federal actor to proceed with a project, and when the federal government thereafter bears no responsibility for the project. Should these conditions be met, the cases indicate that NEPA is to be applied only if the federal decision became final after January 1, 1970. In *Pennsylvania Environmental Council v. Bartlett*,[4] for example, plaintiff sought an impact statement on a federal grant of funds for relocation of a secondary federal-aid road. After examining the relevant statute,[5] the circuit court concluded:

> . . . The only duty remaining on the Secretary after November 20, 1969 was to make a final inspection of the project. . . . For all practical purposes, therefore, final federal action on the project took place prior to January 1, 1970, the effective date of NEPA.[6]

Two cases apply a similar rule, but reach a different conclusion on the facts. In *Sierra Club v. Hardin*,[7] the court approved the *Bartlett*

[4] 315 F. Supp. 238, 2 ELR 20752 (M.D. Pa. 1970), *aff'd*, 454 F.2d 613, 1 ELR 20622 (3rd Cir. 1971).

[5] 23 U.S.C. §117(a).

[6] 454 F.2d at 624, 1 ELR at 20627.

[7] 325 F. Supp. 99, 1 ELR 20161 (D. Alas. 1971).

rule. "The rule that NEPA should not be given retroactive effect to frustrate activities to which the Government had committed itself prior to the passage of the Act is, in principle, sound."[8] However, the court went on:

> The difficult question involves pinpointing that moment in time when the federal government can be said to have committed itself. . . . While it is true that U.S.P. had expended considerable sums of money investigating the Echo Cove mill site and had definitely decided to use that location prior to the effective date of NEPA, the Act speaks to the federal government, not individuals doing business with the government. Prior to April 24, 1970, when the first use permit was issued, neither party was bound to construct the mill at Echo Cove. It was the granting of the permit in that location which constituted a major federal action, and as the permit was granted after January 1, 1970, NEPA was applicable.[9]

In implicit accord is *National Helium Corp. v. Morton*,[10] where the district and circuit courts applied NEPA to a decision by the secretary of the interior to exercise the termination clause of a preexisting contract. The district court noted that the procedures for implementation of the Act adopted by the department had been "made applicable to continuing major federal actions having a significant effect on the environment, even though arising from projects initiated prior to enactment of that Act."[11] Since the decision to terminate the contract constituted a major agency action, and since it had been made in 1971, the court concluded that the failure to prepare the statement justified injunctive relief pending trial on the merits.

A second set of "critical action" cases appears at first to involve a retroactive application of the Act. In *Zabel v. Tabb*[12] the plaintiff Zabel had applied for a dredge-and-fill permit from the Corps of Engineers in 1966, had seen it denied the following year, had instituted suit, and had received a favorable ruling from the district court in 1969. The question before the Fifth Circuit in 1970 was whether the Corps could consider environmental factors when acting on such applications, or whether it was limited to assessing effects on navigation. Judge Brown concluded that the Corps was not so limited, and gave as one reason for this opinion the indicated

---

[8] 325 F. Supp. at 126 note 52, 1 ELR at 20171 note 52.

[9] *Id.*

[10] 326 F. Supp. 151, 1 ELR 20157 (D. Kan. 1971), *aff'd*, 455 F.2d 650, 1 ELR 20478 (10th Cir. 1971).

[11] 326 F. Supp. at 156, 1 ELR at 20159.

[12] 430 F.2d 199, 1 ELR 20023 (5th Cir. 1970), *cert. denied*, 401 U.S. 910 (1971).

policy behind NEPA. "This Act essentially states that every federal agency shall consider ecological factors when dealing with activities which may have an impact on man's environment."[13] The court went on to apply the Act to plaintiff's application, since "although the Congressional command was not in existence at the time the permit in question was denied, the correctness of that decision must be determined by the applicable standards of today."[14] While the court might be viewed as applying the Act retroactively, it should be remembered that the action sought by plaintiff—an order compelling the Corps to award the permit—would have taken place after January 1970. The court may have felt that remanding the case to the Corps would be futile, since the ecological factors which led to the 1967 denial could again be considered, and since there was an independent reason to believe that the Corps already could look to such factors under the Fish and Wildlife Coordination Act.[15] For another case directly in point, see *Bankers Life and Casualty Co. v. The Village of North Palm Beach, Florida*[16] (application for reissuance of a permit that lapsed in 1963 was deferred until 1969; case was decided in 1972). See also *Delaware v. Penn Central Transportation Co.*,[17] where similar issues are present, although not reached in the opinion.

A different conclusion was reached on the facts in *Petterson v. Froehlke*,[18] where NEPA was held not to apply to the issuance of a dredging permit on August 5, 1969, since no action by the Corps was subsequently necessary.

THE "SUBSTANTIAL ACTION REMAINING" APPROACH

When a project is being carried out under direct federal sponsorship, courts approach the question of ongoing action somewhat differently. In such cases, most of which involve reservoir or channel work under the aegis of the Corps of Engineers, impact statements are generally required when the work remaining to be performed after 1970 would in and of itself qualify as "major." These projects

---

[13] 430 F.2d at 211, 1 ELR at 20030.

[14] 430 F.2d at 213, 1 ELR at 20030.

[15] 16 U.S.C. §§ 661–666.

[16] —— F.2d ——, 2 ELR 20528 (5th Cir. 1972).

[17] 323 F. Supp. 487, 1 ELR 20105 (D. Del. 1971).

[18] Petterson v. Resor, 331 F. Supp. 1302, 2 ELR 20013 (D. Ore. 1971), *sub nom.* Petterson v. Froehlke, —— F. Supp. ——, 2 ELR 20747 (D. Ore. 1972).

usually take years or decades from start to finish, and involve a number of stages, such as planning, land purchases, and the actual construction of the various parts. Should work on a particular stage not have started before the passage of the Act, the trend is to require an impact statement on the decision to proceed.

Section 11 of the CEQ Guidelines is the bench mark used by courts in making the case-by-case decision on the necessity for a statement. That section provides:

> 11. *Application of section 102(2)(C) procedure to existing projects and programs.* To the maximum extent practicable the section 102(2)(C) procedure should be applied to further major Federal actions having a significant effect on the environment even though they arise from projects or programs initiated prior to enactment of the Act on January 1, 1970. Where it is not practicable to reassess the basic course of action, it is still important that further incremental major actions be shaped so as to minimize adverse environmental consequences. It is also important in further action that account be taken of environmental consequences not fully evaluated at the outset of the project or program.[19]

The key terms in section 11 are "practicable" and "further incremental major actions." Similar concepts were expressed by the district court in the Gillham Dam case[20] which found that a statement should be submitted when:

> In terms of the estimated total cost, approximately two thirds of the project has been completed. However, no work has yet been done on the dam itself. . . . The work on the spillway and the outlet works is complete or substantially complete.[21]

The work remaining, the court ruled, was itself a "further major federal act" within the meaning of the guidelines. Although the court did not discuss practicality, since the interim guidelines were phrased in terms of "the fullest extent possible,"[22] the court did find a strong congressional policy making environmental protection a continuing responsibility. Moreover, defendants had put themselves in a poor position to question practicality, since they had already prepared a statement which was found inadequate.

Another suit against the Corps by the Environmental Defense

---

[19] CEQ Guidelines, §11, Appendix C.

[20] Environmental Defense Fund v. Corps of Engineers, 325 F. Supp. 728, 1 ELR 20130 (E.D. Ark. 1970–71), —— F. Supp. ——, 2 ELR 20260 (E.D. Ark.), 342 F. Supp. 1211, 2 ELR 20353 (E.D. Ark.), *aff'd*, 470 F.2d 289, 2 ELR 20740 (8th Cir. 1972).

[21] 325 F. Supp. at 744, 1 ELR at 20136.

[22] 325 F. Supp. at 743, 1 ELR at 20136.

Fund also halted work in progress, this time on the Cross-Florida Barge Canal.[23] Work on the project had been congressionally authorized in 1942, and was one-third complete at the time of the suit. Nevertheless, the court found that NEPA placed a "continuing" responsibility on the federal government, and also found that the balance of equities favored plaintiff. Not only might further construction "irreparably damage marine and plant life and a primary source of drinking water for the State of Florida,"[24] but in addition, the delay resulting from a temporary injunction would be slight compared with time required for the canal to be completed.

In a third case between the same parties, over the Tennessee–Tombigbee waterway,[25] the court found even more reason to apply the Act. Although the project was authorized in 1946 and plans were approved in 1962, actual construction was not scheduled to start until October 1971.

In three other cases dealing with long-term water resource projects, courts have also found the "substantial action remaining" test to be superior to the "critical action" approach. In *United States v. 247.37 Acres of Land*[26] the controversy concerned a dam authorized in 1938, but for which land was not condemned until 1969 and construction not begun by the time of NEPA's enactment. The court began the discussion by speaking of authorization, condemnation, and the like as "rungs" on the "ladder" which led to completion—a "critical action" viewpoint. However, the "rung" involved here was uncertain and not clear; accordingly, it was held that the Act "constitutes a direction to proceed thereafter only with consideration given to the various policies and requirements of that Act."[27]

Environmental Defense Fund v. Tennessee Valley Authority (Tellico)[28] was brought against a project at a stage of completion similar to that of the Gillham Dam. The concrete portion of the dam had been completed in 1969, and two-thirds of the necessary land had been acquired. In all, some $29 million out of a total cost of $69 million had been expended. Relocation of roads, however, was only one-sixth complete as of the trial, and work had barely begun on the

---

[23] Environmental Defense Fund v. Corps of Engineers (Cross-Florida Barge Canal), 324 F. Supp. 878, 1 ELR 20079 (D.D.C. 1971).

[24] 324 F. Supp. at 880, 1 ELR at 20079.

[25] 331 F. Supp. 925, 1 ELR 20466 (D.D.C. 1971), 348 F. Supp. 916, 2 ELR 20536 (N.D. Miss. 1972).

[26] —— F. Supp. ——, 1 ELR 20513 (S.D. Ohio 1971).

[27] —— F. Supp. at ——, 1 ELR at 20517.

[28] 339 F. Supp. 806, 2 ELR 20044 (E.D. Tenn.), *aff'd*, —— F.2d ——, 2 ELR 20726 (6th Cir. 1972).

earthfill portion of the dam. The court held that a statement must be prepared on the ongoing project and that work was to be halted in the interim. The court found support in the CEQ Interim Guidelines, in the absence of any clause in the Act which excluded ongoing works, and in the Act's stress on inclusive applicability for its conclusion that "NEPA is not retroactively applied to an ongoing project."[29] Without distinguishing the "critical action" rule as applying to different situations, the opinion criticized four cases which ruled against applying the Act,[30] since "none of these opinions analyze NEPA to determine whether the theory has any merit beyond noting the absence of language requiring retrospective application."[31]

On appeal the Sixth Circuit affirmed, taking what was if anything an even stronger position in favor of applying NEPA to uncompleted actions. Again, the court did not distinguish the situations where the critical action rule might apply. Instead, it issued a sweeping holding:

> An agency must file an impact statement whenever the agency intends to take steps that will result in a significant environmental impact, whether or not these steps were planned before January 1, 1970, and whether or not the proposed steps represent simply the last phase of an integrated operation most of which was completed before that date [footnote 9]. Although this formulation might compel the preparation of impact statements for projects that are so nearly complete that there is no reasonable prospect that the decision to proceed as planned would be reversed, that is no reason to adopt a lesser standard and thereby encourage bureaucratic evasion of responsibility. [Citing *City of New York v. United States*, 337 F. Supp. 150, 160 (E.D.N.Y., 1972).][32]

In *Natural Resources Defense Council v. Grant*[33] the district court found that NEPA encompassed a Soil Conservation Service watershed project which had been approved by Congress in 1966, and where, "much planning and preparation has occurred prior to January 1, 1970. However, a construction contract remains to be let and construction upon the installation of the project has yet to

[29] 339 F. Supp. at 811–12, 2 ELR at 20046.

[30] Pennsylvania Environmental Council v. Bartlett; Investment Syndicates v. Richmond; Brooks v. Volpe (before reversal by 9th Cir.); and Elliot v. Volpe, all in Appendix B.

[31] 339 F. Supp. at 811–12, 2 ELR at 20046.

[32] —— F.2d at ——, 2 ELR at 20731.

[33] 341 F. Supp. 356, 2 ELR 20185 (E.D. N. Car. 1972), *remanded*, —— F.2d ——, 2 ELR 20555 (4th Cir.), —— F. Supp. ——, 2 ELR 20647 (E.D. N. Car.), —— F. Supp. ——, 2 ELR 20648 (E.D. N. Car. 1972), —— F. Supp. ——, 3 ELR 20176 (E.D. N. Car. 1973).

begin."[34] Interestingly enough, the discussion of the retroactivity question in *Grant* pointed out the aspect distinguishing "critical action" from "substantial action remaining." "In so holding [critical action] these courts have focused upon a specific federal action or action which occurred on a particular date or dates."[35] However, the *Grant* opinion did not seem to recognize that this approach might be valid under different circumstances, since it referred to the "proper approach" as being the *Morningside–Lenox*[36] rule. As a result of the semantic confusion which plagues discussions of "retroactivity," the court dismissed precedents rather than distinguishing them.

The principle behind the "substantial action remaining" rule is so well established that in seven cases of continuing federal involvement, the applicability of the Act was conceded by defendants.[37]

There are three cases which illustrate what stage a project must have reached before NEPA became law in order to be exempted from the impact statement requirement, cases where the action remaining was evidently not thought substantial. *Investment Syndicates v. Richmond*[38] was an early case where plaintiff property owners challenged the action of the Bonneville Power Administration in adding a new transmission line. Not only had congressional approval been granted in 1967, but twenty-two of the necessary fifty easements had been acquired before 1970. The court found in such circumstances:

> The letting of the contract, clearing of the right-of-way and construction of the line itself, although occurring after January 1, 1970, are merely a small portion of the work required to complete the project. I cannot believe that Congress intended that the NEPA apply to "major Federal actions" which had reached this stage of completion as of the date of enactment.[39]

The principle may be correct, but its application to the facts would seem contrary to the result reached in the later Gillham Dam and Tellico cases. One wonders how "actual construction" could

[34] —— F. Supp. at ——, 2 ELR at 20188.

[35] *Id.*

[36] Morningside–Lenox Park Ass'n v. Volpe, 334 F. Supp. 132, 1 ELR 20629 (N.D. Ga. 1971).

[37] Sierra Club v. Froehlke (Kickapoo River); Conservation Council of North Carolina v. Froehlke; Allison v. Froehlke; Akers v. Resor; Environmental Defense Fund v. Armstrong; Save Our Ten Acres v. Kreger; Zlotnick v. D.C. Redevelopment Land Agency, all in Appendix B.

[38] 318 F. Supp. 1038, 1 ELR 20044 (D. Ore. 1970).

[39] 318 F. Supp. at 1039, 1 ELR at 20045.

be a small part of the work required on the project. Only if it is assumed that the purchase of rights-of-way accounted for the bulk of the cost of the project does this conclusion seem at all justified. The result might be better reached on the alternate grounds given by the court, that of balancing equities in order to arrive at some idea of the practicality of applying NEPA. The court indicated that it was more probable than not that the decision would be upheld upon further study, particularly since plaintiffs' interest was not in environmental protection but in the value of property it had bought *after* the right-of-way had been laid out. "Investment Syndicates presented several alternative plans but none appeared to offer any great improvement and some would either worsen the situation esthetically or be entirely impractical."[40]

The second case, *Virginians for Dulles v. Volpe*,[41] seems to be the counterpart of *Sierra Club v. Mason*.[42] The court found that operations at an airport built before 1970 could continue in the absence of an impact statement:

> The airport has certainly reached the stage of completion that the "costs already incurred" in adopting and using it as a commercial jet airport "so outweigh the benefits of altering or abandoning" it as such that "no feasible and prudent alternative to the use" would exist. [Citing *Arlington Coalition on Transportation v. Volpe*.] No case has yet held that NEPA's requirements apply to such an ongoing project . . .[43]

The key difference between this case and *Mason* is that here there was no action with a life of its own; plaintiffs claimed that the introduction of stretch jets in 1968 was such, but the court found the effect of the change on the environment to be minimal.

The final case, *Maddox v. Bradley*,[44] offered an example of insubstantial remaining action. In this case, where the entire project involved the construction of a reservoir, and it had been planned before 1970 to fence it, building the fence itself was found to be a "relatively small part of [the] entire project," and hence not subject to NEPA. The court relied on *Ragland v. Mueller*,[45] where the Fifth Circuit found that completion of 4 miles of highway, where 16 miles of the segment had already been built, was not a substantial remaining action. The project was so close to completion, reasoned

[40] 318 F. Supp. at 1040, 1 ELR at 20045.
[41] 344 F. Supp. 573, 2 ELR 20360 (E.D. Va. 1972).
[42] 351 F. Supp. 419, 2 ELR 20694 (D. Conn. 1972).
[43] Virginians for Dulles v. Volpe, 344 F. Supp. at 577–78, 2 ELR at 20362.
[44] 345 F. Supp. 1255, 2 ELR 20404 (N.D. Tex. 1972).
[45] 460 F.2d 1196, 2 ELR 20320 (5th Cir. 1972).

the court, that there was no point in requiring an impact statement (see pages 174 *ff.*).

HYBRID CASES

Some cases have been interpreted with the aid of either the "critical action" or the "substantial action remaining" rule, depending on the judicial view of the situation. The key factor in these rulings is that there was some formal federal action taken before 1970, but federal involvement persisted to some degree after NEPA's enactment. Challenges to highways built under the federal-aid program raise particular problems here, owing to the fragmented process whereby federal approval is granted for such undertakings. Intermediate cases generally have led to a wide variety of results, which courts are often strained to reconcile. The use of the concept of retroactivity tends to gloss over significant differences in the underlying situations.

Federal Action Before 1970 with Subsequent Involvement

The first decision handed down under the Act, *Texas Committee on Natural Resources v. United States*,[46] presented a problem of action essentially complete by 1970 but with some federal involvement thereafter. Plaintiffs sought to compel the FHA to prepare a statement on the grant of a loan for construction of a park and golf course. While the loan had been formally approved before NEPA's enactment, the court did not feel that this approval disposed of the question.

> The plaintiff, on the other hand, would have us look at the whole picture. What is involved here is not merely a loan, but an extensive federal project, the consequences of which will have substantial and detrimental environmental and ecologic consequences. Not one federal dollar has been expended towards this project, not one step of actual construction has been undertaken. The only thing that has occurred has been the processing of papers.[47]

Although the court did not name the "critical action" and "substantial action remaining" tests as such, it did seem aware of the difference involved. As the opinion put it, the applicability of NEPA "depends on one's point of view." The search for a critical action

---

[46] —— F. Supp. ——, 2 ELR 20574 (W.D. Tex. 1970), *vacated,* 430 F.2d 1315 (5th Cir. 1970).

[47] —— F. Supp. at ——, 2 ELR at 20575.

would have focused on the legal significance of processing papers, and would have probably concluded that the Act was inapplicable. The *Texas Committee* court, however, was intent on assessing the equities of the situation and accordingly looked at the underlying project. Since no physical action had been taken, substantial action still remained to which the Act could be applied.

Texas Committee may be contrasted with *Pennsylvania Environmental Council v. Bartlett*,[48] where both the district court and the Third Circuit seem to have chosen to view the action in a formalistic way only. The opinion of the circuit court in particular focused on the federal decision of November 20, 1969, to grant assistance under the federal-aid program for secondary roads. The court does not consider whether continued funding after enactment might have "federalized" the ongoing work, particularly since construction did not begin until February 1970.

The *Bartlett* district court attempted to distinguish *Texas Committee* on the question of degree of federal involvement. While admitting that in neither case had federal money been paid out before NEPA's enactment, the court contended that construction had already begun in *Bartlett*, but not in the Texas case. However, a close reading of the case indicates that the construction in *Bartlett* began *after* the Act; all that had occurred before passage was the awarding of contracts.[49] The real difference was that in *Bartlett* plaintiffs had not filed suit until construction was under way. Thus, the result might be better rationalized on the grounds of laches, rather than those of retroactivity. In any event, the district court appears to have erred in believing that the language "to the fullest extent possible" indicated a "moderate, flexible, and pragmatic approach to the immediate application of the Act" (see chapter III).

The next pair of cases, *Boston Waterfront Residents Ass'n v. Romney*[50] and *San Francisco Tomorrow v. Romney*,[51] again illustrates the dichotomous judicial approach to ongoing projects. The facts are strikingly similar; in both, plaintiffs sought to halt urban renewal projects funded by the Department of Housing and Urban Development. The Boston grant of $30 million was approved in 1965, and $14 million was spent before 1970; in San Francisco $31

---

[48] Appendix B.

[49] 315 F. Supp. at 246, 2 ELR at 20754: "February 2, 1970 [was] the first day of construction. . . ."

[50] 343 F. Supp. 89, 2 ELR 20359.

[51] 342 F. Supp. 77, 2 ELR 20273 (N.D. Cal. 1972), *aff'd in part, rev'd in part*, —— F.2d ——, 3 ELR 20125 (9th Cir. 1973).

million had been approved in 1966 and $17 million in "amendatory grants" had been approved by HUD after January 1970. Yet it was the *Boston Waterfront Residents* court which ruled that NEPA had to be observed before the historic buildings involved there could be demolished, while the *San Francisco Tomorrow* court ruled that despite formal project changes after 1970, no statement was necessary.

The reasoning of the two opinions illustrates this contrast. The Boston court looked to the project itself and found:

> The Waterfront project is a major one which will extend into the future and which involves continuing federal involvement on a large scale. In the light of the strong policy declarations of NEPA the court cannot find that such a project should be exempted from the statute, when particular parts of the project are still planned for the future and where there is still time and opportunity for the consideration of alternatives. [Citing *Arlington Coalition on Transportation v. Volpe*.][52]

In contrast, the San Francisco court found that the projects were in an advanced stage, "at least as far as Federal approvals or other Federal action was concerned." The court purported to draw the following rule from other cases on ongoing actions:

> That if, subsequent to January 1, 1970, there is any significant departure from the original design having ecological significance or if, subsequent thereto, a design feature of ecological significance left open in the original design is resolved or one previously provided for is significantly changed, an "impact statement" must be prepared; absent one of these circumstances, NEPA is not applicable.[53]

The problem, of course, is that such a rule is not consistent with the bulk of the cases discussed above. One would like to know how the court arrived at this strange conclusion, but this is impossible, since the opinion states that "it would serve no purpose at this juncture to review each of these decisions in detail." Having nicely covered its tracks, the court went on to conclude:

> In the present case all relevant design and planning phases of the two projects had been determined prior to January 1, 1970, and it does not appear that any changes thereafter made or contemplated require impact statements under the foregoing test.[54]

This was true even though "little actual construction has begun on the project"; the court did not attempt to weigh the practicality of applying NEPA, but saw the completion of formal approval as an

[52] Boston Waterfront, 343 F. Supp. at 91, 2 ELR at 20360.
[53] 342 F. Supp. at 82, 2 ELR at 20275.
[54] *Id.*

absolute barrier. The contrast between this analysis and that of *Boston Waterfront Residents* is made more striking when it is noted the two opinions were handed down only six days apart.

The weakness of the critical action formula when there is continuing federal involvement is also demonstrated in the district court opinion in *Jicarilla Apache Tribe of Indians v. Morton.*[55] In *Jicarilla,* plaintiffs sought to compel the secretary of the interior to issue a statement on an array of federal actions involving the "Four Corners" complex of six power plants. The court arrived at a conclusion of law, without supporting authority, which was that "NEPA cannot be enforced retroactively as to proposals involving major Federal action where the Agency has already initiated its review processes of the proposal prior to January 1, 1970."[56] Again, this holding contradicts a number of cases cited earlier. In the Gillham Dam case, for example, project plans had not only been reviewed by the Corps of Engineers and Congress, but had been partially implemented. Why the formal start of a review process should exempt an agency from NEPA is not clear from the *Jicarilla* opinion, especially since a statement may be vital to an informed review.

The *Jicarilla* court's subsequent attempt to interpret the notion of "practicability," as expressed in the CEQ Guidelines, suffered from a further flaw. The opinion set forth as a conclusion of law that the basic course of action involved in the building of the power plants could not be practicably reassessed. From other language in the opinion,[57] it appears that the court assumed that the demand for additional generating capacity required the work to go forward, but this conclusion was arrived at without consideration of the possible environmental losses. The circularity involved should be evident; it was not "practicable" to reassess a plant in the light of environmental effects because such effects were a priori ruled out as a reason for changing the course of action. The court in the Cross-Florida Canal case[58] seems to have taken a more correct view of the Act. There the threat to the state's water supply justified reconsideration of a highly expensive ongoing project. At the very least, in close cases the NEPA procedure is to be preferred, because the impact statement might itself indicate whether the threatened dangers

[55] —— F. Supp. ——, 2 ELR 20287 (D. Ariz. 1972), *aff'd,* —— F.2d ——, 3 ELR 20045 (9th Cir. 1973).

[56] —— F. Supp. at ——, 2 ELR at 20295.

[57] —— F. Supp. at ——, 2 ELR 20293–94 (Findings of Fact Nos. 54, 66, 72).

[58] Environmental Defense Fund v. Corps of Engineers, Appendix B.

were sufficiently great to warrant reopening the basic decision to proceed.

Jicarilla raised a third question when it indicated that the NEPA procedure was not mandatory in shaping future action, when the basic decision could not be reopened. This result was produced by the inherent ambiguity of the CEQ Guidelines, §11, which does not indicate what procedure is to be followed in "taking account" of newly discovered consequences, or in "shaping" future action. On one reading §11 could be read to permit the agency to follow whatever procedure it wished. However, given the basic thrust of NEPA itself, and the duty of courts to fill in the gaps in the guideline's interpretation of the statute, a very different result is possible. If the quantum of discretionary action remaining were itself large enough to qualify as major (as in *Mason*), and if it appeared that alternatives could be overlooked on merely informal review, the situation would be tailor-made for applying the Act. This latter approach better accords with the concern expressed by Congress over the effects of incremental and uncoordinated decision making, a concern to which the *Jicarilla* district court seemed blind, and to which the Ninth Circuit did not adequately respond in its affirmance. The Four Corners development might well undergo the type of probing analysis brought to bear in *Sierra Club v. Froehlke* (Trinity River–Wallisville Dam).[59]

Ongoing Regulatory Action

A second type of hybrid case concerns multistage federal regulation, when one stage preceded NEPA and the other followed enactment. *Calvert Cliffs'* established the undisputed general rule that NEPA applies to those stages of the decision which postdate the Act.[60]

The problem in *Calvert Cliffs'* was not with the rule that NEPA applies to the stages of regulatory decision making which postdate the Act, but rather with the Atomic Energy Commission's attempt to put off the Act's effective date by giving itself a generous amount of time to prepare for NEPA compliance. The court held that the commission's delay in modifying its procedures to incorporate NEPA's requirements was unacceptable.

> No doubt the process of formulating procedural rules to implement NEPA takes some time. Congress cannot have expected that federal

[59] Appendix B.

[60] Calvert Cliffs' Coordinating Comm. v. Atomic Energy Comm'n, 449 F.2d 1109, 1120 note 25, 1 ELR 20346, 20351 note 25 (D.C. Cir. 1971).

agencies would immediately begin considering environmental issues on January 1, 1970. But the effective date of the Act does set a time for agencies to begin adopting rules and it demands that they strive, "to the fullest extent possible," to be prompt in the process. The Atomic Energy Commission has failed in this regard. . . . The Commission cannot justify its 11-month delay.[61]

On this ground, the court held that the AEC's action in not considering environmental issues for license applications before March 4, 1971 was invalid.

The delayed compliance date . . . cannot be justified by the Commission's long, drawn-out rule-making process. . . . And, in any event, the obvious sense of urgency on the part of Congress should make clear that a transition, however "orderly," must proceed at a pace faster than a funeral procession. . . . [The spectre of a national power crisis] . . . must not be used to create a blackout of environmental considerations in the agency review process. . . . Perhaps there may be cases in which the need for rapid licensing of a particular facility would justify a strict time limit on a hearing board's review of environmental issues; but a blanket banning of such issues until March 4, 1971 is impermissible under NEPA.[62]

The court also considered whether the Act would apply to later operating license applications where the construction license had been granted before NEPA was enacted. In the face of AEC procedural compliance guidelines limiting NEPA's applicability, the court suggested that it would be undesirable to wait until the later proceedings before considering alternatives which might reduce environmental damage, and that ongoing construction should be reviewed to see if "backfitting" the facilities was feasible. "Although the Act's effective date may not require instant compliance, it must at least require that NEPA procedures, once established, be applied to consider prompt alterations in the plans or operations of facilities approved without compliance."[63]

The problem of the "footdragging" agency also came up in *City of New York v. United States*,[64] where the Interstate Commerce Commission was slow in reacting to NEPA's mandate.

And although it is evident that the Commission has been slow in reacting to the directive of the CEQ . . . and of NEPA itself, that each federal agency establish formal procedures to guide the preparation of §102(2)

[61] 449 F.2d at 1120, 1 ELR at 20352.
[62] 449 F.2d at 1121–22, 1 ELR at 20352–53.
[63] 449 F.2d at 1121, 1 ELR at 20352.
[64] 337 F. Supp. 150, 2 ELR 20275 (E.D.N.Y.), 344 F. Supp. 929, 2 ELR 20688 (E.D.N.Y. 1972).

(C) environmental impact statements, this cannot excuse the Commission's failure to consider adequately the provisions of NEPA once the Act had become effective.[65]

Although not exclusively concerned with regulatory programs, other courts have been more tolerant of the absence of procedures and have allowed agencies to avoid compliance where there were as yet no guidelines spelling out standards for compliance. Thus the Third Circuit in *Transcontinental Gas Pipeline Corp. v. Federal Power Comm'n*[66] held that in the absence of CEQ and agency guidelines defining the scope of §102(2)(C) a reasonable man would not think NEPA applicable to the action at hand, and excused the commission from compliance.

A more reasonable, if not tidy, approach to the problem of agency delay was employed by the Second Circuit in *Greene County Planning Board v. Federal Power Comm'n.*[67] There the court applied the impact statement requirement to only one of three power transmission lines servicing the same pumped storage project, although all three had received commission approval after NEPA's passage. The court's ruling appeared to be based in part on petitioners' failure to make timely objection to the two lines not enjoined, but more importantly, the court appeared to rely heavily upon its findings that there was no showing of deliberate delay by the commission in complying with the new Act. The court said, "It would be unreasonable to expect instant compliance with all of the Act's procedural requirements . . . and there is no indication (as there is with respect to the Gilboa-Leeds line) of obstinate refusal to comply with NEPA."[68]

Federal Action Under the Federal-Aid Highway Program

The peculiar nature of the Federal-Aid Highway Program compounds the already difficult task of those courts which must relate the Act to ongoing action. Highway undertakings are planned at the state level and built by private contractors, but are funded in large part through matching grants from the federal highway trust fund (approximately 90 percent of the cost of interstate highways and 50 percent of the cost of other highways are paid in this manner).[69]

---

[65] 337 F. Supp. at 158–59, 2 ELR at 20276.

[66] 464 F.2d 1358, 2 ELR 20495 (3rd Cir. 1972).

[67] 455 F.2d 412, 2 ELR 20017 (2d Cir. 1972), *cert. denied* 41 U.S.L.W. 3184 (Oct. 10, 1972).

[68] 455 F.2d at 425, 2 ELR at 20023.

[69] 23 U.S.C. §§104 *passim.*

The fund is disbursed without the need for congressional appropriation, in accordance with a preestablished formula which sets maximums for each state. Each specific undertaking for which federal aid is requested must, however, meet Federal Highway Administration (FHWA) standards, and so must be reviewed and approved at several points in the planning process.

This statutory scheme supplemented by FHWA directives results in a uniquely fragmented system of approvals. In an excellent article, analyzing the administration of the federal-aid highway program,[70] Peterson and Kennan provide a guide to this bewildering maze. They indicate that road building is segmented into "projects" that may be lengths of highway or items of work on a particular stretch. The reason they adduce is:

Apportionments of federal-aid highway funds under section 104 are calculated in lump sums for each of the categories of federal-aid highways within a state. At the time of apportionment, the funds are not designated for use in relation to any particular highway. . . . Whenever the state receives the necessary program approval of a project under section 105(a), FHWA transfers the federal share of that project's cost from the state's "unprogrammed balance" account to an "approved program balance" account, and these funds are considered to be available for obligation by the state only on that project. . . . Because states do not wish to restrict their use of federal-aid highway funds in this way, they tend to submit many small projects for FHWA approval.[71]

The procedures for approval compound the confusion:

The principal source of the confusion is easily identified. The FHWA's published procedures are incomplete, outdated, and virtually inaccessible. . . . Most of the FHWA's operating procedures, on the other hand, are published in several categories of voluminous, obscure directives. . . . Even after one obtains copies, the operating procedures are difficult to interpret. Many terms are not defined, and some are used with obviously different meanings in different contexts. Some operating procedures still apparently in force have not been revised or updated for fifteen years.[72]

The authors distinguish three types of federal approvals which are required under the statutes: program approval; approval of plans, specifications, and estimates; and authorization to proceed with work.[73] In theory the federal government becomes contractually

---

[70] R. Peterson and R. Kennan, *The Federal-Aid Highway Program: Administrative Procedures and Judicial Interpretation* 2 ELR 50001 (April 1972).

[71] *Id.* at 50004–05.

[72] *Id.* at 50001–02.

[73] *Id.* at 50006 *ff.*

liable for the costs of the work after the second approval stage (known as "PS&E approval"). In fact, as the authors note:

> FHWA officials have also indicated to us that there is often no single PS&E approval. Rather, PS&E approval frequently consists of a series of approvals which may not even refer explicitly to section 106(a). An obvious difficulty with this practice is that it is impossible in some cases to fix one date after which the federal government is obligated to pay its share of a project's costs.[74]

The discussion of the federal role in cases arising under NEPA tends, however, to be in terms of two other types of approvals, which were established by FHWA directive PPM 20-8.[75] These are "location approval" and "design approval," each of which must be preceded by a public hearing.[76] PPM 20-8 has resulted in a further complication which crops up in the cases. When deciding whether a project had federal approval before the date of that directive, courts must look for some ill-defined equivalents. As Peterson and Kennan state:

> A second reason for the difficulty in implementing PPM 20-8 has been the novelty of its requirements. No single date after which the requirements of PPM 20-8 become applicable is specified. Rather, PPM 20-8 is drafted so as to be applicable only if the requests for "location approval" or "design approval" were made after its effective date. Because there was no counterpart to either location approval or design approval before PPM 20-8 was issued, it has been difficult to determine how to apply the new requirements to highway building underway at that time.[77]

PPM 20-8 is closely connected with PPM 90-1, which is the FHWA's directive implementing NEPA. The latter applied the Act to all highway "sections" (an FHWA term) which received design approval after February 1, 1971, and set up special procedures for undertakings receiving design approval before that date.[78]

Judicial bewilderment with the Alice-in-Wonderland nature of the FHWA's operations pervades the reported cases. Opinions frequently misstate the process; for example, the Ninth Circuit in *Lathan v. Volpe*[79] adopted the description of the process which the FHWA had offered.

---

[74] *Id.* at 50009.

[75] FHWA Policy and Procedure Memorandum (PPM) 20–8 (Jan. 14, 1969), ELR 46505, 23 C.F.R. ch. I, pt. 1 App. A (1972).

[76] Peterson and Kennan, *supra* note 70, at 2 ELR 50012.

[77] *Id.* at 50013.

[78] FHWA PPM 90–1 (Aug. 24, 1971), ELR 46106.

[79] 455 F.2d 1111, 1 ELR 20602 (9th Cir. 1971), *modified on rehearing*,

Defining "any section of federally-funded roadway" as a "project," the court's description outlines five "successive stages" in the construction of a federally financed highway: (1) "program," (2) "routing," (3) engineering design," (4) "right-of-way acquisition," and (5) "actual construction."[80]

Peterson and Kennan succinctly observe that "This description is inconsistent in every respect with the procedures we have described."[81]

When such rulings are used for precedential guidance (and despite this flaw, *Lathan* is by no means a bad opinion), the muddle is complete; the real issues are often veiled by the elusive complexity of FHWA procedure. When asked to decide whether NEPA applies to ongoing road building, many courts try to distinguish the various stages of formal federal approval and to pinpoint one as the "critical action." Peterson and Kennan suggest that the stronger set of cases consists of those which look to the underlying project and which thus come nearest to the "substantial action remaining" point of view.

The courts that have considered these problems have adopted commonsense solutions, for the most part avoiding entrapment in the semantic confusion that characterizes many legal arguments about the federal-aid highway program.[82]

Even those "critical action" cases which form the next subject of discussion frequently turn to other factors of substance, such as the existence of contractual obligations or the actual state of the work, in order to support the conclusions reached on more formalistic grounds.

FHWA Cases—The "Critical Action" Approach. At least six cases have found that NEPA does not apply to ongoing work because all discretionary federal decision making preceded enactment. The first of these is *Bartlett* (discussed on page 144), which differs from the others in that it involved a special type of grant—that for the relocation of a secondary road—for which only one federal approval was required. In the second case, *Elliot v. Volpe*,[83] the court found as a matter of fact, but from conflicting testimony, that

455 F.2d 1122, 2 ELR 20090 (9th Cir.), 350 F. Supp. 262, 2 ELR 20545 (W.D. Wash. 1972).

[80] Peterson and Kennan, *supra* note 70, at 2 ELR 50017, quoting Lathan v. Volpe, Appendix B, 455 F.2d at 1115–16, 1 ELR 20602–03.

[81] Kennan and Peterson, *supra* note 70, 2 ELR at 50017–18.

[82] *Id.*, 2 ELR at 50019.

[83] 328 F. Supp. 831, 1 ELR 20243 (D. Mass. 1971).

the planning of I-93 through Somerville, Massachusetts, had been completed before the effective date of the Act. The court's discussion centered on the meaning of "design approval" as used in PPM 20-8. although that directive had not been in existence on the date when in the court's view such approval had been given. The opinion mentioned in addition that right-of-way acquisition and site preparation had taken place before January 1, 1970. Also, one construction contract had already been let and another one advertised for bids.

The *Elliot* court saw the matter for decision as one of the "retroactivity" of the Act, and inferred that Congress had not intended the Act to so apply, at least when contract rights had vested or planning had been completed.

> It must be presumed that Congress was aware that there were unfinished and incomplete federally aided highway projects in various stages of development when the Act was passed and made effective January 1, 1970. If Congress had intended to authorize federal officers to require changes in the design and construction plans of highway projects after construction contracts had been let or construction bids invited, it could easily have chosen language to express such intention clearly.[84]

This conclusion is seriously weakened, however, by the way in which the court treated the language of NEPA §102(1), "to the fullest extent possible." Without undertaking an independent examination of the legislative history of the phrase, the *Elliot* court relied on the approach of the *Bartlett* district court, that these words were

---

[84] 328 F. Supp. at 835, 1 ELR at 20245.

Bartlett and *Elliot* both view the existence of construction contracts before January 1, 1970, as a reason against applying NEPA. The argument seems to be that the contract is an irrevocable commitment, since if it is subsequently broken by the government through an order to halt work or change plans in order to comply with §102, the government is then liable for the breach. At least two cases, however, hold that such liability is not an absolute barrier to applying NEPA, but merely one cost factor to consider. The court in Harrisburg Coalition Against Ruining the Environment v. Volpe, 330 F. Supp. 918, 1 ELR 20237 (M.D. Pa. 1971) observed that:

> As the testimony indicates, remedies against the Commonwealth are available to the contractor for his monetary loss. . . . And in light of the two-week shutdown already experienced by the contractor, these remedies will more than likely be invoked in any case. Finally, I must take cognizance of the fact that the Supreme Court in *Citizens to Preserve Overton Park v. Volpe* [Appendix B] considered the equities of a somewhat similar situation and did not hesitate to halt construction work by the contractor at the time of oral argument of that case. 330 F. Supp. at 925, 1 ELR at 20239.

Similarly, the court in Morningside–Lenox Park Association, Inc. v. Volpe, Appendix B, found that the existence of other remedies meant that there was no barrier to applying NEPA:

> The existence of executed contracts and the performance of prior contracts

inserted only to provide flexibility of application. As indicated in chapter III, this interpretation is incorrect.

Only one circuit court has come out in clear support of the formalistic approach. The Third Circuit, which earlier affirmed *Bartlett*, followed this approach in *Concerned Citizens of Marlboro v. Volpe.*[85] It found that NEPA did not apply when federal approval of funding for the purchase of rights-of-way took place in 1967, and where no federal funding of construction had been contemplated.

> Under the circumstances here present federal involvement, for purposes of determining the applicability of the FAHA [Federal-Aid Highway Act], ended with government approval and the commitment of government funds. The same reasoning controls our disposition of plaintiffs' contention concerning the NEPA. . . . The effective date of the NEPA, however, is January 1, 1970, and like §128 it does not apply retroactively. Federal approval for the Route 18 project having occurred before 1970, the NEPA was not here applicable. [Citing *Bartlett*[86] and *Wildlife Preserves, Inc. v. Volpe.*[87]][88]

The circuit court apparently equated the federal approval here with "final design approval." Interestingly enough, the opinion gives no indication of the actual state of the project as of 1970, although it seems from the relief requested that the federal right-of-way funds were still being expended.

In the fourth case, *Citizens to Preserve Overton Park v. Volpe*[89]

---

do not grant the project immunity from a requirement of thorough compliance with Section 102 of NEPA. Those who do business with a regulated federal agency, directly or indirectly, cannot necessarily object to what the agency is required to do or not do by virtue of subsequent statutory regulation [Citing Federal Housing Administration v. The Darlington, Inc. 358 U.S. 84 (1958).] 334 F. Supp. at 145–46, 1 ELR at 20634.

The decisions in Environmental Law Fund v. Volpe, 340 F. Supp. 1328, 2 ELR 20225 (N.D. Cal. 1972), and in Arlington Coalition on Transportation v. Volpe, Appendix B, suggest that the costs of such remedies would be weighed in deciding whether a project was so far advanced as to be beyond the reach of NEPA. Such costs should not be automatically equated with the entire sum due under the contract as this could lead to undeserved windfalls. Rather, the contractors would likely be limited to the sum expended in reliance on the contract.

[85] 459 F.2d 332, 2 ELR 20207 (3rd Cir. 1972).

[86] Pennsylvania Environmental Council v. Bartlett, Appendix B.

[87] —— F.2d ——, 1 ELR 20316 (3rd Cir. 1971). This non-NEPA case involved the public hearing provisions of 23 U.S.C. §128(a).

[88] Concerned Citizens of Marlboro, Inc. v. Volpe, Appendix B. —— F.2d at ——, 2 ELR at 20209.

[89] 309 F. Supp. 1189 (W.D. Tenn.), *aff'd*, 432 F.2d 1307, 1 ELR 20053 (6th Cir. 1970), *rev'd*, 401 U.S. 402, 1 ELR 20110 (U.S. 1971), 335 F. Supp. 873, 1 ELR 20447 (W.D. Tenn. 1972), —— F. Supp. ——, 2 ELR 20061 (W.D. Tenn. 1972).

(on remand from the Supreme Court's determination of issues under 23 U.S.C. §138 relating to use of parklands), plaintiffs sought to amend their pleadings to include a NEPA claim. The district court denied the motion, cryptically remarking:

> NEPA is not applicable here because of undisputed reliance on the approval of the park route prior to Secretary Volpe's approval in November, 1969 and because of reliance on his approval prior to the effective date of NEPA.[90]

It could be asked, though, to what extent defendants were entitled to "rely" on an administrative action that was being litigated at the time, and which was later held to have been erroneously given. It might be that the court reached its conclusion because it found that design approval, as opposed to approval of the use of parklands, had been correctly given before 1970. Still, since the court ordered that the case be remanded to the secretary of transportation for a new parklands determination, and since alternatives could still have been introduced and selected as a result of that determination, the §102 procedure could have been found "practicable," especially since construction in the park had not begun. (In contrast, see *Sierra Club v. Morton,*[91] where, on remand from a Supreme Court determination of the issue of standing, plaintiffs were permitted to add a NEPA claim of action to a suit initiated before 1970.)

The most recent cases which rule against plaintiffs on a theory of "critical action" attempt also to apply some test of the equities of strictly following NEPA. *Environmental Law Fund v. Volpe*[92] made NEPA applicable automatically to those projects receiving design approval after January 1, 1970, and to some projects approved before then, depending on the presence or absence of four factors. The court began with the observation that it was "most difficult to place a practical meaning" on the Act's effective date, since "most highway projects, for example, take six to eight years from the time that location approval is sought until construction is completed." Next to be considered were the CEQ Guidelines, which

---

[90] 335 F. Supp. at 884, 2 ELR at 20061.

[91] Sierra Club v. Hickel (Mineral King), —— F. Supp. ——, 1 ELR 20010 (N.D. Cal. 1969), *rev'd*, 433 F.2d 24, 1 ELR 20015 (9th Cir. 1970), *aff'd sub nom.* Sierra Club v. Morton, 405 U.S. 345, 2 ELR 20192, *motion for leave to amend complaint granted,* —— F. Supp. ——, 2 ELR 20469 (N.D. Cal.), 348 F. Supp. 219, 2 ELR 20576 (N.D. Cal. 1972).

[92] Appendix B.

Make a clear distinction between projects initiated *before* January 1, 1970 and projects initiated *after* January 1, 1970. For the latter an environmental impact statement is required "to the fullest extent possible" but for the former a statement is required only "to the maximum extent *practicable*." This distinction is in concert with the legislative history and the general principles against retroactive application of statutes.[93]

Judge Peckham then decided that the date of design approval marked the end of the planning stage, and was thus the appropriate date to use in deciding whether to apply the Act.

Thus, even if a project were initiated prior to January 1, 1970, if the planning phase of the project did not take place until after January 1, 1970, a NEPA statement is required. No balancing of factors can be permitted in such a case; the state highway department *must* file a statement in compliance with Section 102(2) (C). However, if all the planning for a project took place *prior* to January 1, 1970—that is, if design approval preceded the passage of NEPA—a Section 102(2) (C) statement is required only if "practicable."[94]

The court went on to do what had not been done in *Bartlett* or *Investment Syndicates*, which was to state a four-factor test for practicability, to be used when design approval (or better, the "planning stage") preceded the Act:

The participation of the local community in the planning of a project
The extent to which the state department involved has attempted to take environmental factors into account in regard to a particular project
The likely harm to the environment if the project is constructed as planned
The cost to the state of halting construction while it compiles an environmental impact statement[95]

Under the circumstances of the case, the court held, the balance of factors favored defendant. It was emphasized, though, that:

This is not to say that in most highway cases where design approval was received before January 1, 1970 an environmental impact statement would not be required. Indeed, the court can envision many situations where a 102(2) (C) statement would be required even though design approval took place prior to January 1, 1970.[96]

---

[93] 340 F. Supp. at 1331, 2 ELR at 20226.
[94] 340 F. Supp. at 1331–32, 2 ELR 20226.
[95] 340 F. Supp. at 1333, 2 ELR at 20226–27.
[96] 340 F. Supp. at 1334–35, 2 ELR 20227–28.

Further language suggests that the four tests are to be considered in the alternative.[97] Were a court to feel unsatisfied about any one factor, an injunction should issue.

The modification of the critical action approach used in *Environmental Law Fund v. Volpe* won subsequent approval in *Conservation Society of Southern Vermont v. Volpe*[98] and was cited with approval in *Keith v. Volpe*[99] and in *Ward v. Ackroyd*[100] (although the court in *Ward* nevertheless felt bound by the Fourth Circuit's rule in *Arlington Coalition on Transportation v. Volpe*, see page 170). However, the particular factors chosen in *Environmental Law Fund v. Volpe* are open to serious question. "Community participation," in and of itself, might provide little assurance that *environmental* factors were considered in the decision. Too, in the absence of an impact statement it is unlikely that the court will have any reviewable record of the weight given the environment in the state's decision. Indeed, the *Environmental Law Fund* court said in a footnote that it was unable to tell what "criteria and factors" were used by the local federal division engineer who found that a statement would not be practicable.[101] With regard to the third factor, likely harm to the environment, the *Environmental Law Fund* test would place the burden of proof with plaintiff, even though defendant had the greater ability to gather such information. Finally, the fourth factor is equated with the state's loss of federal funds, yet this might be a bookkeeping loss only, as federal outlays would be correspondingly reduced.

A more general deficiency inherent in the *Environmental Law Fund* test was noted by the court in *Committee to Stop Route 7 v. Volpe*:[102]

> Defendants assert as equities the extent of environmental consideration already given by the state officials, and the possible added expense due to inflation and the possible safety hazards that may ensue if there is a delay in construction.
>
> This court declines to accord much weight to these factors because whatever validity they might have rests on a totally erroneous conception of one of NEPA's essential purposes. NEPA is designed to ensure not merely that a major federal action will be taken with minimum damage

---

[97] 340 F. Supp. at 1337, 2 ELR at 20228.

[98] 343 F. Supp. 761, 2 ELR 20270 (D. Vt. 1972).

[99] —— F. Supp. ——, 2 ELR 20425 (C.D. Cal. 1972).

[100] 344 F. Supp. 1202, 2 ELR 20405 (D. Md. 1972).

[101] 340 F. Supp. at 1332 note 7, 2 ELR at 20226 note 7.

[102] 346 F. Supp. 731, 2 ELR 20446 (D. Conn.), *motions to amend the judgment denied*, —— F. Supp. ——, 2 ELR 20612 (D. Conn. 1972).

to the environment. It also requires an agency decision . . . as to whether or not a major federal action should be taken at all. . . .

When defendants ask me to weigh the possibility of increased cost of construction if these projects are delayed, they are assuming that the projects will be built, and that the preparation of an impact statement will have no effect whatever on the decision whether or not to build. I cannot accept that assumption. . . .

Similarly, the weight to be given the state's commendable prior consideration of environmental effects is diminished because that consideration was made on the assumption that the expressway would be built.[103]

Since it would be difficult indeed to find a state highway agency which asked itself if other forms of transportation should be preferred alternatives to a proposed road, it is unlikely that the scope of inquiry under the second factor could ever be as broad as that produced under §102(2)(C). The *Environmental Law Fund* test remains controversial.

FHWA Cases—"Critical Action" Consistent with "Substantial Action Remaining." A number of cases rely ostensibly on a critical action view, but would reach the same result were the alternative test used. These cases are of the type distinguished in *ELF v. Volpe*; they hold the Act applicable on the grounds that federal approval came after 1970. In *Harrisburg Coalition Against Ruining the Environment v. Volpe*[104] a district court within the Third Circuit found that the Act applied to two highway sections where the "critical determinations" under 23 U.S.C. §138 (Parklands) which allowed the project to proceed were not made until May 1970. The court distinquished *Bartlett* on this ground and ordered the projects halted, although a construction contract had already been let on one of them. Similarly, *Nolop v. Volpe*[105] ordered a statement on a project where design approval was not received until July 1970, and no contract was let until the following October. Likewise, where approval of federal funding did not become final until August 1970, and where there had been a continuing controversy over location, the Fifth Circuit found NEPA applicable (*Named Individual Members of the San Antonio Conservation Society v. Texas Highway Department*).[106] [See also *Scherr v. Volpe* (design approval in January 1971);[107] *Conservation Society of Southern Vermont v.*

[103] 346 F. Supp. at 738, 2 ELR at 20448.

[104] Appendix B.

[105] 333 F. Supp. 1364, 1 ELR 20617 (D.S. Dak. 1971).

[106] Appendix B.

[107] 336 F. Supp. 882, 2 ELR 20068 (W.D. Wisc.), 336 F. Supp. 886, 2 ELR 20068 (W.D. Wisc. 1971), 466 F.2d 1027, 2 ELR 20453 (7th Cir. 1972).

Volpe (design approval in January 1971);[108] *Keith v. Volpe* (five out of eight segments received design approval after January 1970);[109] and *Committee to Stop Route 7 v. Volpe* (design approval in March 1970).[110]]

Other cases in this group have raised the related question of whether statements should be prepared on projects which obtained location approval before 1970, but where no design approval had been sought at the time of the suit. In *Lathan v. Volpe*,[111] the Ninth Circuit answered in the affirmative. It ruled that location approval sufficiently federalized a project, and that it was important to complete the §102 procedure as soon as possible thereafter, since flexibility would be increasingly lost as planning continued. In the court's words:

> I-90 already qualifies as a major federal action, even though final federal approval of design or construction plans has not yet been given. Federal highway officials approved the proposed location of I-90 in 1963 and have continued to authorize acquisitions of property for right-of-way since that time. Given the purpose of NEPA to insure that actions by federal agencies be taken with due consideration of environmental effects and with a minimum of such adverse effects, it is especially important with regard to federal-aid highway projects that the §102(2) (C) statement be prepared early. If defendants' contention were accepted—that no environmental impact statement is required until the final approval stage—then it could well be too late to adjust the formulated plans so as to minimize adverse environmental effects.[112]

La Raza Unida v. Volpe[113] presented the same question in more subtle form. Federal location approval was granted for the highway there in question in 1966, and land was being acquired by the state, but no request had been made for the later stages of federal approval which were necessary before federal funds would support the project.

> The State has not requested, nor has it obtained, any federal funds for this project. There is a dispute as to whether the State will eventually

---

[108] Appendix B.

[109] Appendix B.

[110] Appendix B.

[111] Appendix B. Citing Calvert Cliffs' and CEQ Interim Guidelines, 35 FED. REG. 7390, 1 ELR 46001 (April 30, 1970).

[112] 455 F.2d at 1120–21, 1 ELR at 20605. In accord is Brooks v. Volpe (circuit decision), Appendix B. *See also* Willamette Heights Neighborhood Ass'n v. Volpe; and Fayetteville Area Chamber of Commerce v. Volpe, all in Appendix B.

[113] 337 F. Supp. 221, 1 ELR 20642 (N.D. Cal. 1971), —— F. Supp. ——, 2 ELR 20691 (N.D. Cal. 1972).

request federal funds, plaintiffs saying "definitely" and defendants saying "probably not."[114]

The court indicated that the strong congressional policy in favor of environmental protection resolved the uncertainty in favor of NEPA compliance. The very existence of the option to later seek federal funding was held to demand that NEPA be observed.

> The court believes that for the purpose of applying the various federal statutes and regulations a federal-aid highway is any project for which the state has obtained location approval. The state should not have the considerable benefits that accompany an option to obtain federal funds without also assuming the attendant obligations. Any project that seeks even the possible protection and assistance of the federal government must fall within the statutes and regulations.[115]

As the court saw the case for immediate compliance:

> . . . Common sense dictates that the federal protective devices apply before federal funds are sought. . . . All the protections that Congress sought to establish would be futile gestures were a state able to ignore the spirit (and letter) of the various acts and regulations until it actually receives federal funds. Given the realities of actual highway displacement and construction, the statutes and regulations must apply immediately or their purpose will be frustrated.[116]

FHWA Cases—The "Substantial Action Remaining" Approach. The first case to enunciate clearly the "substantial action remaining" criterion for the use of §102 was *Morningside–Lenox Park Association v. Volpe.*[117] There the equivalent of location approval was made final in 1965 and that of design approval in 1967 or 1969. The final authorization of funds, however, did not take place until 1971, since the secretary's approval of the use of parklands was not in hand until February of that year. Most of the right-of-way had been purchased, but actual construction was yet to begin.

The court began by rejecting plaintiff's contention that some critical "approval event" had taken place after 1970:

> In this regard, it is not unreasonable to suggest, as does defendant Volpe, that the "authorization of funds (for construction) is a formalization of the federal government's commitment rather than an agency

---

[114] 337 F. Supp. at 224, 1 ELR at 20643.

[115] 337 F. Supp. at 227, 1 ELR at 20644.

[116] 337 F. Supp. at 231, 1 ELR at 20645. *See also* Civic Improvement Committee v. Volpe, —— F. Supp. ——, 2 ELR 20170 (W.D. N.Car.), *motion for partial preliminary injunction granted,* —— F.2d ——, 2 ELR 20249 (4th Cir.) (Craven, J.), *partial preliminary injunction vacated,* 459 F.2d 957, 2 ELR 20249 (4th Cir. 1972).

[117] Appendix B.

decision of substantive nature, as would be approval of the location and design." Indeed, the Court deems this location or design "approval event" to be clearly the most appropriate time for consideration of these values. . . . The court declines to concur with plaintiff's emphasis of the subsequent federal action of authorizing construction funding as a further critical "approval event" and as the determinative element in this case, for to do so, the Court believes, would, at least tacitly, give the Court's imprimatur to the "piecemeal approval" approach declared unlawful in *San Antonio*.[118]

Defendant's satisfaction must have been short-lived, for the court proceeded to state a rule which favored plaintiff even more. Critical actions were not dispositive; instead:

As did the Courts in *Defense Fund* [*EDF v. Corps* (Gillham)] and *Calvert Cliffs'*, this Court relies upon the clear legislative mandate that Section 102 of the NEPA be implemented "to the fullest extent possible." Also, the interim guidelines of the Council on Environmental Quality, which strongly suggest the application of the Section 102 procedures to ongoing projects, are entitled to considerable weight. While much work has already been done, the Court is not dealing with a *fait accompli.* In short, the Court holds that compliance with Section 102 of the NEPA is required as to an ongoing federal project on which substantial actions are yet to be taken, regardless of the date of "critical" federal approval of the project.[119]

NEPA's application to such ongoing projects, the court indicated, would not be retroactive, although the court also saw *Zabel v. Tabb*[120] as a true example of retroactive application.[121]

The strongest support to date for using a substantial action test in highway cases comes from the decision of the Fourth Circuit in *Arlington Coalition on Transportation v. Volpe*.[122] The section of I-66 there challenged had had its corridor fixed in 1959. The county had relied on the location of the route for its planning, and acquisition of rights-of-way had been approved in 1966. The project overall was 80 percent complete. However, actual construction had scarcely begun, and PS&E approval was yet to be given.

---

[118] 334 F. Supp. at 144, 1 ELR at 20633.

[119] *Id.*

[120] 430 F.2d 199, 1 ELR 20023 (5th Cir. 1970), *cert. denied*, 401 U.S. 910 (1971).

[121] The court also cites Note, *Retroactive Application of the National Environmental Policy Act*, 69 MICH. L. REV. 732 (1971). This excellent note takes the position that there is no barrier to applying NEPA retroactively. *See also* Note, *Retroactive Application of the National Environmental Policy Act*, 39 TENN. L. REV. 735 (1972).

[122] Appendix B.

Significantly enough, the Fourth Circuit refused to accord much weight to the fact that design approval had been given on January 21, 1971. Judge Craven's response to this argument was:

> Manifestly the date of design approval alone does not accurately measure whether Arlington I-66 has reached the crucial stage, and determining the applicability of Section 102(C) by this standard alone would be arbitrary and capricious agency action and an abuse of administrative discretion.[123]

This concept of "crucial stage" refers primarily to the amount of work remaining on the project itself; formal administrative approvals are secondary. This concept lies at the heart of the *Arlington* court's proposed test.

> Doubtless Congress did not intend that all projects ongoing at the effective date of the Act be subject to the requirements of Section 102. At some stage of progress, the costs of altering or abandoning the project could so definitely outweigh whatever benefits that might accrue therefrom that it might no longer be "possible" to change the project in accordance with Section 102. At some stage, federal action may be so "complete" that applying the Act could be considered a "retroactive" application not intended by the Congress. The congressional command that the Act be complied with "to the fullest extent possible" means, we believe, that an ongoing project was intended to be subject to Section 102 until it has reached that stage of completion, and that doubt about whether the critical stage has been reached must be resolved in favor of applicability.[124]

The court then considered the facts of the case under this standard and concluded that application would certainly not be retroactive.

Arlington thus would impose a substantive "balancing" approach. This approach is open to the criticism that "how one balances interests depends on how heavy one thinks they are." The use of the phrases "at some stage," "might," and "so definitely outweigh" seems to cloud rather than clarify the issues. Some doubt as to the possibility of favorable alteration would always exist prior to the preparation of a statement. The test also overlooks the cost of preparing the statement itself, as well as the expense incurred by delay while a statement is being readied.

One case shows how the balance may be varied according to the

---

[123] 458 F.2d at 1332, 2 ELR at 20164–65.

[124] 458 F.2d at 1331, 2 ELR at 20164. Note that the court makes no reference to the CEQ Guidelines.

weight given the factors. *Civic Improvement Committee v. Volpe*[125] was decided the month before *Arlington* was handed down, but took a similar balancing approach. There, no statement was required on a segment of I-81 through Charlotte, North Carolina, because:

> . . . The location of the road has been established and the construction contract has been let, and the beginnings of earth moving are taking place. Unless super highways are to be outlawed in Charlotte, it is unlikely that the route or the elevation of the road . . . will be changed by environmental studies.[126]

The key ambiguity is contained in the word "unlikely," which might or might not mean the same as *Arlington*'s "so definitely outweigh." The court in *Civic Improvement Committee* mitigated its decision, however, by retaining equity jurisdiction and asking for specific recommendations for improvements.

The *Arlington* rule appears to differ from that of *Environmental Law Fund v. Volpe* in a number of ways. Where *Environmental Law Fund* turns to four stated factors, *Arlington* suggests that a court could choose any factors that seemed relevant in a particular case. Too, *Arlington* would place the burden of proof on defendants, as doubts are to be resolved in favor of §102. In *Environmental Law Fund*, plaintiff had to show that environmental factors had not been adequately considered. Perhaps most importantly, *Arlington* places far less weight on the formal event of design approval.

The *Arlington* approach has been adopted by other district and circuit courts. In *Ward*,[127] the Maryland district court indicated that it preferred the *Environmental Law Fund* approach, but felt compelled to follow the lead of its governing circuit court. *Conservation Council of Southern Vermont v. Volpe*[128] occupied the middle ground between the two holdings, using *Arlington* to reject defendant's "narrow view" and *Environmental Law Fund* to reject plaintiff's "broad view." Similarly the court in *Keith v. Volpe*[129] felt that *Arlington* went further than *Environmental Law Fund*, but did not feel compelled to state a preference. The Seventh Circuit specifically endorsed *Arlington* in deciding *Scherr v. Volpe*,[130] but that case is also consistent with *Environmental Law Fund*, since

---

[125] Appendix B.
[126] —— F. Supp. at ——, 2 ELR at 20171.
[127] Ward v. Ackroyd, Appendix B.
[128] Appendix B.
[129] Appendix B. In accord is Sierra Club v. Volpe, 351 F. Supp. 1002, 2 ELR 20760 (N.D. Cal. 1972).
[130] Appendix B.

design approval did not come until after 1970. Courts in Wisconsin, Iowa, and Virginia have used *Arlington* in situations where the *Environmental Law Fund* rule might have led to a different result: *Northside Tenants' Rights Coalition v. Volpe*,[131] *Indian Lookout Alliance v. Volpe*,[132] and *Thompson v. Fugate*.[133]

In *Environmental Defense Fund v. Tennessee Valley Authority* (Tellico),[134] the Sixth Circuit applied *Arlington* to construction of a TVA navigation project. It saw clearly the deficiencies inherent in analyzing a project by "critical states."

> The separate stage analysis would not only attach an unwarranted importance to the accounting or construction methods utilized by particular agencies (and might encourage the adoption of methods intended to circumvent the requirements of the NEPA), but would also, we believe, be difficult to apply.[135]

The court's conclusion was that statements would be required until the construction which caused the environmental effects was completed. The degree to which such effects had already occurred was found to be related to the decision whether to classify the undertaking as involving a "proposal for action" subject to NEPA.[136]

If not read carefully, *Arlington* could complicate any discussion of retroactivity in two ways. First, it might be read as requiring statements on projects which received federal approval before January 1970, even when there was no continuing federal involvement. This would be true if the underlying private project had not progressed to the stage where the balance was turned against applying NEPA. However desirable such a result might be, it would mean reopening a final, not ongoing, federal decision, and thus raise grave questions of retroactivity. The only way to avoid this result is to distinguish clearly the federal action involved from the underlying private activity. In *Arlington* the court apparently saw federal involvement as amounting to more than formal approval; there, the highway undertaking itself was conceded by defendants to be a

[131] 346 F. Supp. 244, 2 ELR 20553 (E.D. Wisc. 1972), —— F. Supp. ——, 3 ELR 20154 (E.D. Wisc. 1973).

[132] 345 F. Supp. 1167, 3 ELR 20051 (S.D. Ia. 1972).

[133] —— F. Supp. ——, 1 ELR 20369 (E.D. Va.), *injunction expanded*, —— F.2d ——, 1 ELR 20370 (4th Cir.), *injunction further expanded, aff'd in part, rev'd in part*, 452 F.2d 57, 1 ELR 20599 (4th Cir. 1971), 347 F. Supp. 120, 2 ELR 20612 (E.D. Va. 1972).

[134] Appendix B.

[135] —— F.2d at ——, 2 ELR at 20733.

[136] *Id.*

major federal action. This would not be true in all cases of permits for private activity.

The second problem stems from *Arlington*'s downplaying of the significance of design approval. It is conceivable that in some areas of activity—e.g., reactor licensing or a *Jicarilla*-type[137] situation—some federal approval remained to be given, but the cost of abandonment would weigh against applying NEPA. The inclusion of the word "altering" must be stressed in applying *Arlington*, since even where the basic balance favors going ahead, NEPA may suggest how best to proceed. If this is kept in mind, the problem raised earlier in the discussion of *Jicarilla* (page 155)—should §102 be followed when the question is how, not whether, to proceed, or how to treat newly discovered effects—is resolved in favor of the NEPA process. This conclusion was indicated in *Northside Tenants' Rights Coalition*.[138]

Even assuming that location and design approval together with right-of-way acquisition so freezes Park Freeway-West that it is "not practicable to reassess the basic course of action," there would still appear to be room to shape specific construction projects "so as to minimize adverse environmental consequences" that might be discovered after a NEPA impact study.[139]

Similarly, *Keith v. Volpe*[140] criticized defendants' failure to examine the effects of a freeway on air pollution.

In planning the Century Freeway, moreover, defendants made virtually no attempt to evaluate the effect of the highway on air pollution in the Los Angeles basin. Paragraph 11 of the Council's guidelines provides that even when the application of NEPA to an ongoing project is not practicable, "It is . . . important in further action that account be taken of environmental consequences *not fully evaluated at the onset of the project* or program." . . . The failure to closely examine the effect of the proposed freeway on air pollution was an egregious omission.[141]

FHWA Cases Finding Actions Completed Beyond the NEPA Stage. How complete must a project then be for NEPA not to apply? Three cases which looked to the substantive state of the work have allowed it to proceed in the absence of an impact statement. In *Ragland v. Mueller*[142] the Fifth Circuit disposed of a challenge to a 20-mile section of I-295 in Florida.

---

137 Jicarilla Apache Tribe of Indians v. Morton, Appendix B.
138 Appendix B.
139 346 F. Supp. at 248, 2 ELR at 20554.
140 Appendix B.
141 —— F. Supp. ——, 2 ELR 20425 (C.D. Cal. 1972).
142 460 F.2d 1196, 2 ELR 20320 (5th Cir. 1972).

While this court and others have held that in certain situations, the National Environment Act of 1969 may be applied retroactively [citing *Named Members of San Antonio Conservation Society* and *Arlington*], this is clearly not such a case. Analysis of the facts reveals that when NEPA became effective January 1, 1970, sixteen of the twenty miles of the disputed highway had already been fully *completed* and the right of way for the remaining four miles had been acquired. It is simply unreasonable to assume that Congress intended that at this point in time, construction should halt.[143]

Pizitz v. Volpe[144] presented the Fifth Circuit with an even clearer situation of fact. The road there had been planned with exit ramps and had been completed but for those ramps. Their construction, contemplated in the original design, was held to be beyond the scope of the Act. For a similar holding, see *Willamette Heights Neighborhood Association v. Volpe.*[145]

Summary of FHWA Cases. Hybrid cases, particularly frequent in the highway field, have been decided according to the rule which the court chooses to apply. The Third Circuit chose a "pure" critical action test (*Bartlett*, *Marlboro*), and the Fourth and Seventh decided to balance equitably the extent of the action that remains incomplete. A third, competing formulation is the compromise put forth in *Environmental Law Fund v. Volpe.* There is a clear conflict among the Circuits and only the Supreme Court can establish uniformity.

From the discussion it appears that the best solution is that proposed in *Arlington,* perhaps modified in two ways to make it more consonant with the *Environmental Law Fund* approach. First, no balancing would be required when design approval postdates the Act. Second, courts could arrive at a list of factors which frequently occurred (thus correcting *Arlington*'s lack of specificity), although the weight accorded each could change from case to case. Peterson and Kennan also see the substantive test as preferable to that of critical action when dealing with the confusion which surrounds the formal system of federal approval of highway grants. Of *Morningside–Lenox* and *Arlington* they write:

These courts wisely looked beyond the calendar to determine whether the application of NEPA or other similar federal laws is meaningful in a

[143] 460 F.2d at 1197, 2 ELR at 20320.

[144] —— F. Supp. ——, 2 ELR 20378 (M.D. Ala.), 467 F.2d 208, 2 ELR 20379 (5th Cir.), *modified on rehearing,* 467 F.2d 208, 2 ELR 20635 (5th Cir. 1972).

[145] Appendix B.

particular case. No particular FHWA approval should necessarily immunize federal-aid highways from new federal statutory requirements. As we have seen, neither location nor design approval obligates the federal government to reimburse a state for project costs. These approvals are administrative contrivances . . . [and in some cases] are given years before changes in ownership or physical alterations of the land occur. . . . If these or other particular FHWA approvals were considered unchangeable, the authority of Congress to establish new environmental or social policies with respect to the federal-aid highway program would be severely curtailed.[146]

This view expressed by the Third Circuit is thus untenable and should be rejected, for it sees only darkly or not at all through the semantic glass. It represents an unwarranted extension of the "critical action" approach, which is properly used only where there was a single federal approval of a nonfederal undertaking before January 1, 1970, and no subsequent federal involvement in implementation.

CONTINUING PROJECTS AND PROGRAMS

So far in this chapter we have been concerned only with agency actions which were initiated before NEPA's enactment and usually had a definite completion date falling after enactment. The lawsuits discussed above offer such examples as a federally funded highway project, a federal license or permit, and a federal resource development project. However, we have not dealt with continuing projects or programs which have no foreseeable completion dates.

Although there has been almost no litigation on such projects and programs, they are nevertheless important enough to be discussed briefly here. As the court in *Lee v. Resor*[147] said:

The reason for distinguishing between continuing and ongoing projects is that an environmental impact statement might be ill-advised for an ongoing project which was near completion when NEPA was enacted, while an impact statement might be advisable and within the scope and intention of NEPA requirements for a continuing project, even though it was begun before NEPA was enacted [footnote omitted].[148]

It would be ironic if projects which threatened endlessly repetitious environmental injury could escape NEPA's reach more easily than projects which had only one chance.

In *Lee v. Resor* the court held that NEPA applied to a 20-year-

---

[146] Peterson and Kennan, *supra* note 70, at 2 ELR 50020.
[147] 348 F. Supp. 389, 2 ELR 20665 (M.D. Fla. 1972).
[148] 348 F. Supp. at 394, 2 ELR at 20666.

old Corps program which used herbicides to control water hyacinths growing in the St. Johns River in Florida. Although preliminary injunctive relief was denied, the court held that it had jurisdiction to order the statement prepared while the spraying continued. The court's interpretation of NEPA's legislative history and purposes convinced it that the Corps' guidelines requiring that impact statements be prepared on all of its continuing programs within three years correctly interpreted the law. The court concluded its interpretation of NEPA's legislative history by stating:

> It would be ironic if Congress did not intend to affect those projects and agency decisions that provided the impetus for the Act. Congress doubtless intended that NEPA have some application to the type of situation presented here [footnote omitted].[149]

Although the court in *Sierra Club v. Mason*[150] found that the harbor dredging at issue in that case was a new major federal action with "a life of its own" which began after NEPA's enactment, in dictum the court affirmed that NEPA was intended to apply to such a continuing agency program. The court apparently was not content to accept the Corps designation of the project as ongoing maintenance, which would have required the Corps by its own guidelines to prepare a statement within three years of NEPA's enactment. The court viewed the guidelines as creating a temporary exception to NEPA's blanket applicability and refused to allow the project at issue to fall within it.

A final case is *New York v. Department of the Army*,[151] in which a long-standing Corps program to dump sewage sludge and dredged spoil in the New York harbor area was challenged for lack of an impact statement. The Corps usually gathered the material to be dumped, but issued permits to private barge owners to dump it in Corps-designated areas. During litigation the Corps conceded NEPA's applicability by preparing an impact statement on the program.

These three cases are closely related to the discussion in chapter VIII suggesting a "multiple-tiered" approach to impact statement preparation. While that discussion focuses upon new federal programs, its logic applies to existing programs as well. The impact statement may be a useful device for comprehensively evaluating entire programs, old or new. Such statements need not be limited to

---

[149] 348 F. Supp. at 395, 2 ELR at 20667.
[150] Appendix B.
[151] —— F. Supp. ——, 2 ELR 20507 (S.D.N.Y. 1972).

modest programs such as hyacinth spraying and harbor maintenance; their preparation would be useful for larger programs, such as the stream channelization program of the Soil and Conservation Service, the public lands grazing permit program of the Bureau of Land Management, the federal grant-in-aid highway construction program, and other long-standing federal programs that are due for an overall environmental assessment.

VI

Preparation and Content of Impact Statements

ONCE AN agency settles down to the task of preparing an impact statement, after the preliminary issues regarding NEPA's applicability have been resolved (see chapters IV and V), it must fulfill a number of additional statutory requirements. These concern the timing of statement preparation, the extent to which the agencies may share statement preparation or delegate it to nonfederal parties, the content and adequacy of the statement, and the commenting procedures. All of these phases of actual statement preparation are also conditioned by the requirement of compliance "to the fullest extent possible" (see chapter III). Before action may be taken, an agency *must* take the specific steps and discuss the particular factors mentioned in the Act.

TIMING OF STATEMENT PREPARATION

The courts have interpreted the requirement of strict compliance in light of certain of Congress' purposes to hold that NEPA impact statements must be prepared at the earliest possible time for every distinct stage of agency decision making. In *Calvert Cliffs'*,[1] in considering whether the Atomic Energy Commission could defer compliance with NEPA until after the commission's hearing boards had completed their review, the circuit court said:

> Compliance to the "*fullest*" possible extent would seem to demand that environmental issues be considered at every important stage in the decision-making process concerning a particular action—at every stage where an overall balancing of environmental and nonenvironmental factors is appropriate and where alterations might be made in the proposed action to minimize environmental costs.[2]

---

[1] Calvert Cliffs' Coordinating Comm. v. Atomic Energy Comm'n, 449 F.2d 1109, 1 ELR 20346 (D.C. Cir. 1971), *cert. denied* 404 U.S. 942 (1972).
[2] 449 F.2d at 1118, 1 ELR at 20350.

This view, which the court went on to apply in holding that compliance was required before the AEC conducted hearings, has been cited in numerous opinions as the correct formulation of NEPA's requirements regarding the timing of impact statement preparation. However, the passage does not bring out adequately the importance of *early* NEPA compliance. This aspect of the timing of statement preparation is stressed in *Daly v. Volpe*,[3] which cites other language in *Calvert Cliffs*': "[The] 'detailed statement' should 'be prepared at the earliest practicable point in time,' " the court said, relying upon *Calvert Cliffs*' and Department of Transportation guidelines which reflect the request in the CEQ Guidelines for preparation "as early as possible and in all cases prior to agency decision."[4]

The court in *Daly* also cited *Lathan v. Volpe*[5] which likewise stressed early preparation.

> Given the purpose of NEPA to insure that actions by federal agencies be taken with due consideration of environmental effects and with a minimum of such adverse effects, it is especially important with regard to the federal-aid highway projects that the statement be prepared early.[6]

Finally, the court in *Citizens for Clean Air v. Corps of Engineers*[7] has made perhaps the clearest statement uniting the concept of compliance at every distinct stage with the need for early preparation.

> Where several federal permits or approvals are required for a project, NEPA requires a §102 review at the point where the action is "distinctive and comprehensive." Once a project has reached a coherent stage of development it requires an environmental impact study. The comprehensive review contemplated by the Act can only be efficacious if undertaken as early as possible.[8]

---

[3] 326 F. Supp. 868, 1 ELR 20242 (W.D. Wash. 1971), 350 F. Supp. 252, 2 ELR 20443 (W.D. Wash.), *opinion on rehearing*, 350 F. Supp. 252, 3 ELR 20032 (W.D. Wash. 1972).

[4] 350 F. Supp. at 256, 2 ELR at 20444. *See* CEQ Guidelines, §2, Appendix C.

[5] 455 F.2d 1111, 1 ELR 20602 (9th Cir. 1971), *modified on rehearing*, 455 F.2d 1122, 2 ELR 20090 (9th Cir.), 350 F. Supp. 262, 2 ELR 20545 (W.D. Wash. 1972).

[6] *Cited in* Daly v. Volpe 350 F. Supp. at 256, 2 ELR at 20444.

[7] 349 F. Supp. 696, 2 ELR 20650 (S.D.N.Y. 1972).

[8] 349 F. Supp. at 708, 2 ELR at 20655. Other cases are in accord. Committee for Nuclear Responsibility v. Seaborg; Arlington Coalition on Transportation v. Volpe; Zabel v. Tabb; Izaak Walton League v. Schlesinger; Businessmen Affected Severely by the Yearly Action Plans, Inc. v. District of Columbia

The district court's second decision on remand after the Supreme Court's decision on standing in *Sierra Club v. Morton*[9] serves to illustrate the proper application of the early preparation requirement. Plaintiffs sought to have a new third cause of action added to their complaint alleging noncompliance with §§102(2)(C) and (D). Defendants contended that plaintiffs had failed to state a claim because they did not allege that defendants had begun construction or issued permits for the proposed recreational development. The court refused to dismiss the cause of action, accepting for purposes of the motion plaintiffs' contention that the solicitation of bids and the issuance of preliminary surveying permits were sufficient under *Calvert Cliffs'* and the CEQ Guidelines to require NEPA compliance.[10]

The reasons for the courts' interpretation of the strict compliance standard can be traced to NEPA's legislative history regarding coordinated advance planning and to the obvious inference that action taken before NEPA review may result in commitments of money and prestige which a tardy NEPA review will be powerless to alter. The circuit court in *Natural Resources Defense Council v. Morton*[11] stated very well the congressional purpose in enacting NEPA.

> What NEPA infused into the decision-making process in 1969 was a directive as to environmental impact statements that was meant to implement the Congressional objectives of Government coordination, a comprehensive approach to environmental management, and a determination to face problems of pollution "while they are still of manageable proportions and while alternative solutions are still available" rather than persist in environmental decision-making wherein "policy is established by default and inaction" and environmental decisions "continue to be made in small but steady increments" that perpetuate the mistakes of the past without being dealt with until "they reach crisis proportions." S. Rep. No. 91-296, 91st Cong., 1st Sess. (1969) p. 5.[12]

City Council; Natural Resources Defense Council v. Grant; Sierra Club v. Froehlke (Trinity River–Wallisville Dam), all cited in Appendix B.

[9] Sierra Club v. Hickel (Mineral King), —— F. Supp. ——, 1 ELR 10010 (N.D. Cal. 1969), *rev'd*, 433 F.2d 24, 1 ELR 20015 (9th Cir. 1970), *aff'd sub nom.* Sierra Club v. Morton, 405 U.S. 345, 2 ELR 20192, *motion for leave to amend complaint granted*, —— F. Supp. ——, 2 ELR 20469 (N.D. Cal.), 348 F. Supp. 219, 2 ELR 20576 (N.D. Cal. 1972).

[10] 348 F. Supp. at 2 ELR at 220, 2 ELR at 20576.

[11] 337 F. Supp. 165, 2 ELR 20028 (D.D.C.), 337 F. Supp. 167, 2 ELR 20089 (D.D.C. 1971), *motion for summary reversal denied*, 458 F.2d 827, 2 ELR 20029 (D.C. Cir.), *dismissed as moot*, 337 F. Supp. 170, 2 ELR 20071 (D.D.C. 1972).

[12] 458 F.2d at 836, 2 ELR at 20033.

The Fourth Circuit in *Arlington Coalition on Transportation v. Volpe*[13] reasoned that NEPA's requirement that the decision maker actually consider altering the remainder of the project in light of facts unearthed in the impact statement mandated enjoining the project. Thus further investment prior to and during study of the statement would not gradually diminish available options until none remained. For this viewpoint the circuit court relied upon language in §101 and in §102(2) and upon the decisions in *Calvert Cliffs'*,[14] *Lathan*,[15] and *National Helium Corp. v. Morton*.[16] The consequences a premature commitment of resources could have for the proper implementation of NEPA were discussed in *Calvert Cliffs'* when the court ordered full interim compliance for facilities that received construction permits before NEPA's passage but which had yet to obtain operating approval. The commission wanted to defer any action under NEPA until the operating license hearing. The court said:

> Once a facility has been completely constructed, the economic cost of any alteration may be very great. In the language of NEPA, there is likely to be an "irreversible and irretrievable commitment of resources," which will inevitably restrict the Commission's options. Either the licensee will have to undergo a major expense in making alterations in a completed facility or the environmental harm will have to be tolerated. It is all too probable that the latter result would come to pass.
>
> By refusing to consider requirement of alterations until construction is completed, the Commission may effectively foreclose the environmental protection desired by Congress. It may also foreclose rigorous consideration of environmental factors at the eventual operating license proceedings. If "irreversible and irretrievable commitment[s] of resources" have already been made, the license hearing (and any public intervention therein) may become a hollow exercise. This hardly amounts to consideration of environmental values "to the fullest extent possible."
>
> A full NEPA consideration of alterations in the original plans of a facility, then, is both important and appropriate well before the operating license proceedings.[17]

---

[13] 332 F. Supp. 1218, 1 ELR 20486 (E.D. Va. 1971), *rev'd*, 458 F.2d 1323, 2 ELR 20162 (4th Cir.), *cert. denied sub nom.* Fugate v. Arlington Coalition on Transportation, 41 U.S.L.W. 3249 (Nov. 7, 1972).

[14] Appendix B.

[15] Appendix B.

[16] 326 F. Supp. 151, 1 ELR 20157 (D. Kan. 1971), *aff'd*, 455 F.2d 650, 1 ELR 20478 (10th Cir. 1971).

[17] 449 F.2d at 1128, 1 ELR at 20356. Other cases adopt the same rationale. Northside Tenants' Rights Coalition v. Volpe, 346 F. Supp. 244, 249, 2 ELR 20553, 20555 (E.D. Wisc. 1972); Daly v. Volpe, 326 F. Supp. 868, 1 ELR 20242 (W.D. Wash. 1971), 350 F. Supp. 252, 258, 2 ELR 20443, 20445

Agencies may not seize upon the concept of "distinctive and comprehensive" stages of decision making to escape impact statement preparation on the grounds that early action is premature or inchoate or involves little environmental impact in itself. The court in *Citizens for Clean Air v. Corps of Engineers*[18] rejected the Corps' attempt to permit Consolidated Edison to continue construction of its Astoria plant during NEPA review, since such construction was certainly a "distinctive and comprehensive" stage of the overall project. Moreover, the *Clean Air* court cited *Named Individual Members of the San Antonio Conservation Society v. Texas Highway Dept.*[19] and *Scherr v. Volpe*[20] as applying the same rationale to highway projects.[21]

Although the cases clearly hold that each major federal action must be preceded by strict compliance with NEPA, several cases suggest that the impact statement requirement may be met after early, and conceivably quite important, planning actions have already been taken. This problem has already been discussed in connection with the timing of federal involvement in highway projects (see chapter IV, pages 64 *ff.*) but deserves further brief discussion here.

Dicta in *Citizens for Clean Air v. Corps of Engineers*[22] and in *Committee to Stop Route 7 v. Volpe*[23] suggest that not every step in the planning or even the early construction of a project may be sufficiently "federal" or "distinctive," or may involve sufficient environmental impact, to merit statement preparation. Yet it hardly needs restating that Congress intended NEPA to affect agency decision making from the very beginning, while federal plans are still but a glimmer in officialdom's eye. Of course, not every action subject to NEPA is specifically subject to the impact statement requirement; compliance with NEPA is required even when the actions involved do not yet have sufficient concreteness and size to

(W.D. Wash.), *opinion on rehearing*, 350 F. Supp. 252, 3 ELR 20032 (W.D. Wash. 1972); Citizens for Clean Air v. Corps of Engineers, Appendix B, 349 F. Supp. at 707, 2 ELR at 20655; Arlington Coalition on Transportation v. Volpe, Appendix B, 458 F.2d at 1333, 2 ELR at 20165.

[18] Appendix B.

[19] Appendix B.

[20] 336 F. Supp. 882, 2 ELR 20068 (W.D. Wisc.), 336 F. Supp. 886, 2 ELR 20068 (W.D. Wisc. 1971), 466 F.2d 1027, 2 ELR 20453 (7th Cir. 1972).

[21] 349 F. Supp. at 708, 2 ELR at 20655.

[22] *Id.*

[23] 346 F. Supp. 731, 735, 2 ELR 20446, 20447 (D. Conn.), *motions to amend the judgment denied*, —— F. Supp. ——, 2 ELR 20612 (D. Conn. 1972).

require an impact statement. But the impact statement is Congress' assurance that an agency has considered environmental factors and has documented their importance well before major federal action is taken. A very persuasive case can be made, therefore, that statements should accompany the agency's earliest initiatives, as evidence that from the very beginning federal plans took environmental impacts into account.

The mischief that a more restrictive view may cause is illustrated by *Jicarilla Apache Tribe of Indians v. Morton.*[24] The district court held as a matter of law that the Kaiparowitz Project did not yet have to be covered by an impact statement because no major federal action regarding the project had been taken since NEPA was enacted.[25] However, the project, which was launched in 1965, involves resource development, not power plant construction, and has already resulted in extensive coal leases from the Bureau of Land Management, a contract to take 102,000 acre-feet of water from Lake Powell annually, and extensive planning for right-of-way and plant sitings.[26] These facts appear to argue that the federal role, even if it has not yet produced a discrete "major federal action" in the period since NEPA's enactment, is sufficiently settled to merit full NEPA compliance. The same criticism can be made of *National Forest Preservation Group v. Butz,*[27] where compliance after the forest supervisor and regional forester had acted, but before the chief forester and the secretary of agriculture had approved the action, was accepted by the court. A similar criticism could have been made of *Upper Pecos Ass'n v. Stans,*[28] had not the government confessed error and prepared a statement on the federal grant necessary for road construction.

The requirement that NEPA must be complied with before a major federal action is initiated obviously precludes compliance at the same time, or after, action goes forward. The court in *La Raza Unida v. Volpe*[29] made the rationale plain.

In addition to the strong policy statements and the wording of the statutes and regulations, common sense dictates that the federal protec-

---

[24] —— F. Supp. ——, 2 ELR 20287 (D. Ariz. 1972), *aff'd,* —— F.2d ——, 3 ELR 20045 (9th Cir. 1973).

[25] —— F. Supp. at ——, 2 ELR at 20296.

[26] —— F .Supp. at ——, 2 ELR 20294.

[27] 343 F. Supp. 696, 2 ELR 20571 (D. Mont. 1972).

[28] 328 F. Supp. 332, 1 ELR 20228 (D. N.M.), *aff'd,* 452 F.2d 1233, 2 ELR 20085 (10th Cir. 1971), *vacated,* 93 S. Ct. 458 (1972).

[29] 337 F. Supp. 221, 1 ELR 20642 (N.D. Cal. 1971), —— F. Supp. ——, 2 ELR 20691 (N.D. Cal. 1972).

tive devices apply before federal funds are sought. It does little good to shut the barn doors after the horses have run away.[30]

Again, in *Lathan v. Volpe*,[31] only this time the horses were stolen:

Defendants suggest that they will conduct additional research on its environmental effects after the highway is constructed. NEPA, however, does not authorize defendants to meet their responsibilities by locking the barn door after the horses are stolen [citing *Calvert Cliffs*'].[32]

A particularly difficult case in this regard is *Environmental Defense Fund v. Froehlke*,[33] the Truman Reservoir case. The case is made difficult by the court's attempt, through a series of pretrial conferences and stipulations, to force the parties into the posture of accepting a detailed NEPA compliance schedule which forestalled some federal actions but allowed others to go forward while the impact statement was being prepared. The court focused upon the minimal environmental harm that would occur before the impact statement had been completed. Total reliance upon this argument is misplaced, however, as *Calvert Cliffs*' shows. Commitments of funds, the selection and condemnation of rights-of-way, the relocation of residents and roads, the letting of supply contracts, all narrow the range of realistic choices once the final statement has been prepared. The court's response to this argument was that the influence exerted by these commitments on Congress' choice of alternatives is "not a proper argument for the judicial forum."[34] Yet the enforcement of the strict compliance standard does entail consideration of how choices may be narrowed by failure to require early compliance. By focusing upon the environmental harm which the actual filling of the completed reservoir will cause several years hence, the court is implicitly approving the kind of logic, condemned in highway cases, which would have allowed a highway to be built up to both sides of a park before considering the "alternatives" for completing the missing link.[35]

Although we have already discussed agencies' overall compliance for their ongoing programs (see chapter V especially pages 176 *ff.*),

[30] —— F. Supp. at ——, 2 ELR at 20645.

[31] Appendix B.

[32] 350 F. Supp. at 266, 2 ELR at 20547 (W.D. Wash. 1972).

[33] 348 F. Supp. 338, 2 ELR 20620 (W.D. Mo. 1972).

[34] 348 F. Supp. at 354, 2 ELR at 20627. *But see* Sierra Club v. Froehlke (Trinity River–Wallisville Dam), Appendix B.

[35] *See* Citizens to Preserve Overton Park v. Volpe; Named Individual Members of the San Antonio Conservation Society v. Texas Highway Dept., both in Appendix B.

we should stress again that the requirement of early compliance is not changed if the action under consideration was ongoing at the time NEPA was enacted. Compliance for a project in progress must be completed as early as possible. As stated in *Calvert Cliffs'*:

> Although the Act's effective date may not require instant compliance, it must at least require that NEPA procedures, once established, be applied to consider prompt alterations in the plans or operations of facilities approved without compliance.[36]

WHO MUST PREPARE THE STATEMENT

The Act is clear that a "responsible federal official" must prepare the impact statement required by §102(2)(C). Two litigable issues have arisen, however, because of conflicting views on the extent to which the Act allows the task of preparing statements to be passed on to state agencies or to private parties, and the Act's requirements regarding the manner of compliance when several agencies share authority to approve a single action. In each instance one major litigation has been completed,[37] and in each litigation the timing of agency statement preparation was a key factor (see preceding discussion). As in other areas of NEPA implementation, both litigations confirmed that the courts will require compliance "to the fullest extent possible."

Delegation to Private Parties or State Agencies

When the major federal action to be taken is the approval of a private license application, the grant of funds to a private party, or the funding of work carried out by a state, some federal agencies have delegated significant responsibility for statement preparation to the applicant. The Federal Power Commission, the Federal Highway Administration, the Federal Aviation Administration, the Civil Aeronautics Board, the Department of the Interior, and the Interstate Commerce Commission until recently all followed this practice.[38] The guidelines and practices of these agencies should begin to change, however, now that the Supreme Court has refused to

[36] 449 F.2d at 1121, 1 ELR at 20352. Similar considerations governed in Lee v. Resor, 348 F. Supp. 389, 2 ELR 20665 (M.D. Fla. 1972).

[37] Greene County Planning Board v. Federal Power Comm'n, 455 F.2d 412, 2 ELR 20017 (2d Cir.), *cert. denied*, 41 U.S.L.W. 3184 (Oct. 10, 1972); Upper Pecos Ass'n v. Stans, Appendix B.

[38] Comment, *Developments Under the National Environmental Policy Act*, 1 ELR 10022, 10025 *ff*. (February 1971).

grant certiorari in *Greene County Planning Board v. Federal Power Comm'n.*[39]

Greene County and Formal Hearings. In *Greene County* the Power Authority of Southern New York (PASNY) had prepared a preliminary draft statement pursuant to FPC guidelines on a 345-kilowatt transmission line that was to service a pumped storage project near Albany, New York. At the hearing on the contested line, Greene County challenged the commission's reliance upon PASNY's draft, arguing that the commission should have prepared its own, but the hearing examiner denied their request and the commission confirmed. Greene County sought review of the commission's order in the Second Circuit.

The Second Circuit held that NEPA required the commission to prepare its own impact statement prior to any formal hearings conducted by the agency. The court framed the issue in terms of the timing of the commission's statement, noting that the three proposals before it were (1) that the commission not be required to make its own statement until it filed its final decision (the commission's view), (2) that the commission be allowed to prepare its statement on the basis of the hearings before the presiding examiner (the commission's fallback position), and (3) that the commission be required to prepare its statement prior to the hearing (plaintiffs' view). The court accepted the plaintiffs' view.

The principal argument advanced by the commission was that the exception in the CEQ Guidelines which allows agencies to prepare draft statements on the basis of a hearing subject to the Administrative Procedure Act (APA) released the commission from the responsibility of independently preparing an impact statement until its final decision was rendered. The court held, however, that the APA is supplemented by the requirements of NEPA. In the court's view, NEPA's obligations are not limited to adversary proceedings and are not concerned solely with one party's rebutting his adversary's testimony. Rather, NEPA places upon the *agency*, not the contestants of license applications, the burden of representing the public interest in the environment. The court said that even prior to NEPA, §10(a) of the Federal Power Act[40] had changed

[39] Appendix B. *See* Comment, *Delegation of the Drafting of Environmental Impact Statements: Greene County Planning Board v. Federal Power Commission*, 2 ELR 10153 (June 1972); Note, *Environmental Impact Statements—A Duty of Independent Investigation by Federal Agencies*, 44 Colo. L. Rev. 161 (1972).

[40] 16 U.S.C. §803 *et seq.*

the FPC's role from that under the APA of "an umpire blandly calling balls and strikes for adversaries appearing before it,"[41] to providing active, affirmative protection for the public interest. But NEPA "went far beyond" the command of §10(a) to consider environmental factors.[42]

The court then suggested reasons of policy and of congressional intent for rejecting the FPC approach.

> The danger of the procedure, and one obvious shortcoming is the potential, if not likelihood, that the applicant's statement will be based upon self-serving assumptions. . . . intervenors generally have limited resources, both in terms of money and technical expertise, and thus may not be able to provide an effective analysis of environmental factors. It was in part for this reason that Congress has compelled agencies to seek the aid of all available expertise and formulate their own position early in the review process.[43]

Finally, the court in *Greene County* suggested that while it was permissible for the commission to seek comments from other agencies on the basis of the applicant's preliminary impact statement, under NEPA's mandate to consider environmental factors to "the fullest extent possible," it would be preferable to have the agency prepare its own draft for circulation, as is being done by the AEC after the decision in *Calvert Cliffs'*.[44]

The commission has not accepted its NEPA obligations very willingly. With regard to the administrative proceedings in *Greene County*, the court remarked that the "Commission at nearly every turn . . . made it difficult procedurally for the intervenors."[45] After denial of its petition for a rehearing *en banc* before the Second Circuit,[46] the commission took the position, pending action on a petition for certiorari, that the decision was binding at the most for

---

[41] Scenic Hudson I, Scenic Hudson Preservation Conference v. Federal Power Comm'n, Appendix B, 354 F.2d at 620, 1 ELR at 20292.

[42] 455 F.2d at 419, 2 ELR at 20020. Here, of course, the court is referring to the burden of compiling and analyzing information on environmental impacts. Courts are also beginning to notice the problem of the burden of proof in NEPA court cases. The district court in Sierra Club v. Froehlke (Trinity River–Wallisville Dam), Appendix B, stated that the burden of proof that NEPA had been complied with was shifted to the agency after a *prima facie* showing of noncompliance. The chief factor cited by the court was the "part-time" nonexpert character of citizen plaintiffs, who are similar to the poorly funded "intervenors" in Greene County. *See* chapter II, *supra* note 82.

[43] 455 F.2d at 420, 2 ELR at 20020.

[44] 10 CFR Part 50 revised Appendix D, 36 FED. REG. 18071, ELR 46116 (Sept. 9, 1971). *See* §D.1, *as amended* Sept. 29, 1971, ELR 46117.

[45] 455 F.2d at 417 note 12, 2 ELR at 20019 note 12.

[46] *See* Comment, *supra* note 39, at 10155.

only the one contested proceeding and transmission line.[47] After denial of certiorari[48] the commission allowed 15 days only for comments on the procedures which were revised to comply with the *Greene County* decision.[49]

The problem of applying NEPA to the early phases of decision making in the case of agencies which base their decisions on the record developed at trial-type hearings is a persistent one, even after the decision in *Greene County.* The narrow issue concerns the legality of the preparation of a draft statement under §8 of the APA, which prohibits agency participation in the decision-making process until the record is completely developed. The *Greene County* court correctly pointed out, however, that the two senses of "commission" may be confused.

> Certainly no one has suggested that the detailed statement prepared before the hearings must be prepared by the Commission members. It is sufficient for the purposes of NEPA if the statement is prepared by the Commission's staff on the basis of the staff's investigation. Counsel for the Commission conceded at oral argument that this procedure would not violate the APA.[50]

This interpretation of §8, however, does not fully resolve the issue, because it does not clarify the staff statement's relation to final decision making. Under §5 of the APA, the investigative staff may not "participate or advise in the decision," "except as witness or counsel in public proceedings," and the hearing examiner may not consult the investigative staff *ex parte* (without notice to parties and the opportunity for a hearing), "except to the extent required for the disposition of *ex parte* matters as authorized by law."[51] Perhaps the correct answer to this dilemma is that the staff's non-adversarial assessment of environmental impacts under NEPA falls within this last exception, but no court has yet considered the question.

A relevant precedent may be discerned in the operations of the Federal Trade Commission, where staff both conducts studies of market practices and brings suit against alleged violators. A challenge that the commission thus "prejudges the issues" was rejected

---

[47] *In re* Arkansas–Louisiana Gas Co., Before the Federal Power Commission, Docket No. CP70–267, 2 ELR 30007, 30008 (Feb. 18, 1972).

[48] 41 U.S.L.W. 3184 (Oct. 10, 1972).

[49] 37 FED. REG. 23360 (Nov. 2, 1972). Final Regulations at 37 FED. REG. 28412 (Dec. 23, 1972).

[50] 455 F.2d at 422 note 24, 2 ELR at 20021 note 24.

[51] 5 U.S.C. §554(d).

in *FTC v. Cement Institute*[52] on grounds that the FTC was the only tribunal able to hear the case and that the minds of its members were not "irrevocably closed," but open to whatever evidence industry members might present at the ensuing hearing. It would likewise appear that responsibility for a draft or even a final impact statement would not commit a regulatory agency to any one course of action. For example, license applicants would be free to propose mitigation measures not considered by the commission, or to introduce evidence aimed at changing the cost-benefit assessment.

The broader policy issues raised by NEPA's application at the earliest possible moment to quasi-judicial agency decision making are the same as those discussed above in connection with the pesticide program of the Environmental Protection Agency (see chapter IV, page 190). Further, the same issues have been analyzed in an *Environmental Law Reporter* Comment, which concluded by rejecting the view that the many safeguards of the judicial process contained in trial-type hearings ensure that NEPA's goals will not be slighted.

> Agencies have more diverse functions than courts and should not be expected to act as passive umpires presiding over private disputes. Agencies have statutory missions to accomplish, some of which may be inimical to environmental values. The nature of controversies may be—and in the environmental area usually is—"polycentric," i.e., a large number of results may follow and many interests or groups may be affected. Administrative decision-making, therefore, even where trial-type hearings under the APA are held, may be a far cry from the frequently "bipolar" nature of judicial cases. . . .
>
> All of these forces tend to diffuse the rigorous examination of the environmental impacts and alternatives which NEPA requires. By contrast, the earliest possible injection of NEPA procedures into the administrative process could have a tonic effect. A detailed environmental analysis prepared by the agency staff of "environmentalists"—biologists as well as engineers, ecologists as well as systems analysts—who would develop their own expertise, might prevent a later hearing from becoming bogged down in the "wider" environmental issues. This would leave the presiding examiner free to assess the merits of the proceeding, dealing with . . . focused environmental issues [footnote omitted].[53]

Delegation by the Federal Highway Administration. While the *Greene County* case was making its way through the courts, three cases involving the Federal Highway Administration's compliance

---

[52] 333 U.S. 683 (1948).

[53] Comment, *supra* note 39, at 10159. *See* Lon Fuller, *The Forms and Limits of Adjudication* 36 (unpublished mimeograph), cited in B. Boyer, *A Re-evaluation of Trial-type Hearings for Resolving Complex Scientific and Economic*

with NEPA narrowly avoided a ruling on the same point, two more held that FHWA's duties were delegable, while a sixth held that they were not delegable. The issue is raised by FHWA procedures which require states that are recipients of federal trust funds to prepare the impact statements for proposed highway construction.[54] The FHWA attempts to square this procedure with NEPA by having the "responsible federal official" (FHWA division offices) clear draft statements for circulation by signing and dating them and by providing for "review and acceptance" of final statements before they are marked "officially approved."[55] In this way, the FHWA procedures seek to make the federal official responsible for the statement, while allowing the statement to be prepared by the state highway departments.

The first of the cases, *Lathan v. Volpe*,[56] has occasionally been cited as resolving the issue of the legality of FHWA's delegation. The circuit court stated:

> The State defendants are required by Federal Highway Administration regulations . . . to evaluate the environmental consequences of federal-aid highway projects and make a determination that a detailed environmental statement is necessary or that a "negative declaration" of no significant environmental effect will suffice.[57]

However, the decision on remand makes it clear that the question of the proper agency for preparing the statement was not raised at the time. *Lathan*, like many other cases, simply took FHWA procedures at face value in the absence of any specific challenge. When the issue was raised on remand, the district court declined to decide it, holding that it had been introduced too late in the proceedings and was therefore barred.[58]

In *Pizitz, Inc. v. Volpe*,[59] the Fifth Circuit responded to the issue in rather summary form: "We find no merit in the contentions

---

Issues, A Staff Report of the Administrative Conference of the United States 2 (Dec. 1, 1971). *See also* Professor Murphy's suggestions for reform in decision making where quasi-judicial yet "polycentric" environmental agency decision making takes place. A. Murphy, *The National Environmental Policy Act and the Licensing Process: Environmentalist Magna Carta or Agency Coup de Grace?* 72 COL. L. REV. 963, *esp.* 982 *ff.* (October 1972).

[54] PPM 90–1, §6(b), ELR 46106 (Aug. 24, 1971).

[55] *Id.* §§6(j), (k).

[56] Appendix B.

[57] 455 F.2d at 1121, 1 ELR at 20606.

[58] 350 F. Supp. at 267, 2 ELR at 20547.

[59] —— F. Supp. ——, 2 ELR 20378 (M.D. Ala.), *aff'd*, 467 F.2d 208, 2 ELR 20379 (5th Cir.), *modified on rehearing*, 467 F.2d 208, 2 ELR 20635 (5th Cir. 1972).

of appellants that the responsible federal officials could not under the applicable federal statutes accept an environmental impact statement prepared by a state highway department."[60] Shortly after this decision, however, on the motion of *amicus curiae* permitted to intervene out of time, the court issued a supplementary opinion in which it struck this language from its earlier opinion on the ground that it was not necessary to the decision.

Although the question whether the responsible federal officials could, under the applicable federal statutes, accept an environmental impact statement prepared by a state highway department was briefed by the parties on this appeal, it was not necessary to reach this issue. We, therefore, leave that question for decision in a case in which it is essential to the decision.[61]

Undoubtedly, the court should have avoided ruling on the issue, although it was briefed, because the issue was not ruled upon by the district court, which in any case had characterized plaintiffs' NEPA claims as "spurious."[62]

The issue of delegation appeared to be headed for resolution in *Committee to Stop Route 7 v. Volpe*,[63] where the court first held on the authority of *Greene County* that delegation was impermissible, at least as far as preparation of the final statement was concerned.

But as to the final version of the detailed statement required by §102(2) (C), the Court of Appeals ruled unequivocally that this must be prepared by the FPC itself, with the actual work to be done by the agency's own staff. This was held to be a requirement of the statute, and there is no more basis for the statute to be altered by the regulation of the FHWA here than by the FPC's interpretation of the Guidelines of the Council on Environmental Quality in *Greene County*. 455 F.2d at 421.

In fact, the very same danger of self-serving assumptions that concerned the Court in *Greene County* is present here.[64]

The court left open the question of which federal agency should be responsible for preparation; the parties had assumed that it was the FHWA, but the court thought the Department of Transportation itself might also be appropriate, since one of the alternatives sug-

---

60 —— F.2d at ——, 2 ELR at 20379.
61 —— F.2d at ——, 2 ELR at 20635.
62 —— F. Supp. at ——, 2 ELR at 20378.
63 Appendix B.
64 346 F. Supp. at 741, 2 ELR at 20449.

gested was that of rail transit, a topic outside the expertise of the FHWA but within the scope of its parent department.

In further proceedings in *Route 7,* however, the court backed away from affirming that it meant in its earlier judgment to approve the FHWA's procedures for statement preparation. Specifically, federal defendants sought an amended finding of fact that federal approval of the document prepared by the state amounted to federal "preparation" of it. The court refused to rule on the issue on the ground that it was not ripe for adjudication.

A court reviewing whether the manner of preparation of an impact statement complies with the Act should have the benefit of the agency's action in a specific case, supported by the agency's expertise as to why such action complies with the Act. All that is before me at this point is an "administrative intention" that has been "expressed" but which "has not yet come to fruition." *Eccles v. Peoples Bank of Lakewood Village*, 333 U.S. 426, 434 (1948), citing *Ashwander v. Tennessee Valley Authority*, 297 U.S. 288, 324 (1936). Moreover, in carrying out its intention, the federal agency may decide to take innumerable specific steps during the course of the preparation and approval of the final impact statement. It would be premature now to rule on the adequacy of the broad outlines of the procedure.[65]

With regard to its earlier ruling that federal officials had to prepare the final statement, the court remarked that it had intended to do no more than rule broadly that "the particular project at issue in this litigation required compliance with the Act."[66]

The *Route 7* court, like the courts in two of the other highway cases, managed to avoid one of the key remaining issues of NEPA's implementation. Of course courts should decide live controversies and have the benefit of settled facts upon which to rule. But the court in *Calvert Cliffs'* had no difficulty in basing its ruling upon its analysis of the adequacy of the AEC's procedures, although they were no more specific than those of the FHWA. To the extent that FHWA guidelines are already in conflict with NEPA, they can be reached by the courts in an appropriate case. The controversy is a real one, if for no other reason because "an agency is bound by its own regulations."[67]

The fourth case in the series does apply the *Greene County* precedent in order to find, as one alternate ground for its ruling,

[65] —— F. Supp. at ——, 2 ELR at 20612.
[66] —— F. Supp. at ——, 2 ELR at 20612.
[67] Silva v. Romney, Appendix B, 342 F. Supp. at 785, 2 ELR at 20386.

that substantial compliance by a state agency will not satisfy NEPA's requirements. The court in *Northside Tenants' Rights Coalition v. Volpe*[68] held that NEPA's requirements attach to the federal agency, not to the recipient of federal aid, and "it is the federal agency which must prepare the impact statement and balance the project's worth in light of the environmental consequences."[69] The court's seven-line discussion, however, shows that, as in the other three cases, the delegation issue was not in the forefront of the court's mind and was peripheral to the issue of NEPA's applicability to ongoing projects (see chapter V).

The two remaining cases support the position taken by the FHWA and allow delegation of statement preparation to the state agencies. In the first of the cases, *Iowa Citizens for Environmental Quality, Inc. v. Volpe*,[70] the court's reliance on the *unmodified* order of the Fifth Circuit in *Pizitz* is clearly misplaced, while its attempt to distinguish *Greene County* failed to show how the state highway department had less self-interest in building the highway than the license applicant in *Greene County* had in building the power transmission line. The court in *National Forest Preservation Group v. Butz*[71] was slightly more careful. There the court held that the FHWA procedures do not violate the "spirit" of the Act, because FHWA review would be close, the FHWA would take responsibility for the statement, and the statement would be available at all important stages of federal decision making. However, the court had trouble with plaintiffs' argument that the plain meaning of the statute required federal preparation, while it offered only a bald assertion that the factor of self-interest, so cogently set out in *Greene County*, was not present in this case.

The approach adopted by the Second Circuit in *Greene County* applies with equal force to delegation of highway statements. Delegation is appropriate in this area only in a theoretical sense, because FHWA "preparation" by superficial approval of statements prepared by state officials is not likely to be adequate to offset the effect, which is bound to show up in the statements, of the state's interest in getting the project approved. On the other hand, as the following section suggests, having the FHWA bear the entire burden of statement preparation may not be in the best interests of full NEPA

---

[68] Appendix B.
[69] 346 F. Supp. at 248, 2 ELR at 20555.
[70] —— F. Supp. ——, 3 ELR 20013 (S.D. Ia. 1972).
[71] Appendix B.

implementation. Concentration of the entire environmental review on the federal level would divorce that review from the rest of the planning process, which is essentially a state function. A clear benefit to both levels of government would occur if they both had to take environmental factors into account and hire ecologists, biologists, and environmental planners who would bring fresh perspectives to entrenched federal and state bureaucracies.

Sharing the Burdens of Compliance. Although the trend has not been irrevocably established, the courts will probably apply the *Greene County* rationale and require agencies to prepare their own impact statements at an early stage of agency decision making. Yet the holding in *Greene County* does not appear to require the agency to bear all of the additional expense and conduct all the necessary studies; as long as the agency consults with appropriate parties and then prepares a detailed statement that accompanies the proposed action at every distinct stage of agency decision making, it is free to require state governments and private parties to supply information, hire consultants, conduct field studies, and seek other assistance in carrying out its responsibilities. The synthesis and evaluation of such information, however, must be done by the agency.

The difference between requiring applicants to prepare the early drafts and requiring them to submit factual information is not as subtle as it may first seem. Allowing applicants to draft the entire statement puts the opportunity for framing the issues, weighting values, and establishing emphases in the hands of the parties most interested in minimizing the appearance of adverse environmental impacts. The agency thus becomes vulnerable to manipulation in determining which environmental values are important, unless the agency's staff itself examines the suggested project in the field and frames the issues upon which information is needed.

The test of whether the development of data may be safely entrusted to a private applicant should be whether it is of such an essentially factual nature that it may be subsequently validated through independent agency inquiry. . . .

Although the agency cannot avoid hiring adequate staff for the preparation of impact statements, including field research, much can be done within the limits of NEPA to require applicants to conduct or fund factual studies. Fact-finding constitutes the bulk of the time and expense involved in an adequate environmental assessment. If the test suggested above regarding the permissible limits of involvement for applicants is applied, significant costs associated with fact-finding may be shifted to the applicant. As an additional safeguard, agencies might explore the possibility of certifying consulting firms which then might conduct parts of the environmental analysis at the applicant's expense. Or consultants might simply be subject to agency approval. (The chief drawback to this proposal is that consulting firms with adequate skills simply do not yet exist in adequate numbers to meet the need. And there is some reason to look reluctantly upon this necessity mothering the sudden invention of Impact Statements, Inc.)

Other costs might be shifted to the applicants by requiring a license or other fee where an applicant's project is likely to cause environmental harm. Federal fees which help offset the cost to the government of monitoring activities in the private sector are not without precedent; Congress requires the Securities and Exchange Commission to impose both stiff reporting requirements and fees upon firms desiring to register securities for sale in interstate commerce.[72]

The Lead Agency Concept and Multi-Agency Actions

Many major federal actions require multiple federal approvals or the direct participation of several agencies. For example, the various approvals necessary for construction of the Alaskan pipeline required major federal actions involving the Department of the Interior, which issues permits to cross public lands, the Forest Service, which grants permits for locating terminal facilities on national forest lands, and the Corps of Engineers, which issues permits for crossing navigable streams, for taking gravel from streams, and for harbor work. Nor need the action be on the scale of the Alaskan pipeline before involving several agencies. A project to dredge and fill a portion of Hunting Creek, Virginia, for an apartment building required a permit from the Corps of Engineers and approval of the Department of the Interior under the Fish and Wildlife Coordination Act.

Several alternative means of complying with NEPA in circumstances such as these may be imagined. First, each agency might prepare a statement on its specific part of the action. Second, the agencies involved might issue a jointly prepared statement. Third, one agency might be designated to take care of NEPA compliance for all of the agencies involved.

In an apparent attempt to anticipate the difficulty which agencies would have with multiple-agency actions, the CEQ Interim Guidelines of April 30, 1970 opted for the third alternative by requiring the "lead agency" to prepare a single statement on the "cumula-

[72] Comment, *supra* note 39, at 10160.

tively significant impact" of the entire action.[73] The concept was repeated and expanded in CEQ's final guidelines published a year later.[74] While retaining the lead agency concept as its basic recommended approach, CEQ has since given approval to jointly prepared statements,[75] but noted in its third annual report that such statements may be "especially appropriate for new *policy* initiatives formulated at an interagency level [emphasis added],"[76] thus somewhat limiting the scope of the concept.

The attempt by the CEQ to enlarge and refine the lead agency concept, which admittedly is a commendable attempt to deal with a difficult problem, may indicate that the CEQ approach is not yet adequate to the standard of strict compliance set by NEPA. Of the three basic alternatives available to agencies for complying with NEPA, only the lead agency alternative involves the exempted agencies the least in considering the environmental consequences of proposed actions.

The litigation in *Upper Pecos Ass'n v. Stans*[77] illustrates the abuses of the strict compliance standard which may occur when the lead agency concept is adopted. In October 1970 the Economic Development Administration (EDA) of the Department of Commerce made a grant of $3,795,200 to San Miguel County, New Mexico, to finance part of the construction of a road that would pass through the Santa Fe National Forest. No impact statement was prepared. Three months later, the Forest Service, purporting to act as the lead agency, issued a draft statement on the right-of-way easement which it would have to grant. Four-and-a-half more months passed before the final statement was prepared. In the meantime, the county and state had begun to grant funds to an engineering company for plans and specifications on the road.

The district court held that the Forest Service's impact statement satisfied the requirements of NEPA. The Tenth Circuit affirmed 2–1, Judge Murrah dissenting. The majority held that the Forest Service's compliance with NEPA was adequate, because its statement

---

[73] 35 FED. REG. 7390, 1 ELR 46001, §5(b) (April 30, 1970).

[74] Appendix C, §5(b).

[75] *Council on Environmental Quality, Memorandum for Agency and General Counsel Liaison on National Environmental Policy Act (NEPA) Matters, Recommendations for Improving NEPA Procedures*, §4, ELR 46162 (May 16, 1972).

[76] COUNCIL ON ENVIRONMENTAL QUALITY, ENVIRONMENTAL QUALITY, THIRD ANNUAL REPORT 234–35 (August 1972).

[77] Appendix B.

would "provide the basis on which the Forest Service will decide on the issuance of the right-of-way easement" which involves approval of "the location and construction plans of the proposed road."[78]

Plaintiffs sought and were granted certiorari, but before the case was briefed the solicitor general asked the Supreme Court to vacate the Tenth Circuit's judgment and to remand the case to that court with directions that it be dismissed as moot. In the meantime, EDA had frozen the funds previously made available by its grant and had prepared a draft impact statement for the proposed road. The solicitor general assured the Court that EDA would reconsider the grant at every level in light of the environmental information developed. He indicated that the confession of error was made because of the requirements of CEQ's final guidelines and its Memorandum of May 16, 1972.[79] The Court granted his request.[80]

Additional reasons for the confession of error are not difficult to identify. An important component of the overall federal action had proceeded without consideration of the possible environmental consequences. The failure of the Forest Service statement to discuss the impact of the grant and alternatives to it, as contrasted to the impact of the easement and alternatives to it, constituted only part of the problem. Once the grant was made, that fact itself was likely to narrow the Forest Service's environmental review in a manner inconsistent with NEPA's objective of wide-ranging, impartial consideration of federal actions before resources are committed and before alternatives are narrowed. The Forest Service could not realistically be expected to consider alternative uses for the scarce economic development funds once the grant was made. Nor could the Forest Service realistically consider alternative locations and design factors, or whether to build the road at all. The county had obtained a specific budget for a specific road, the detailed plans and specifications for which were being developed at the same time the Forest Service was conducting its environmental review. Options and alternatives had been effectively limited to minor changes in route location and design.

If the decisions of the lower courts had stood, several vital NEPA policy objectives would have thus been defeated. NEPA was designed both to enable and to require development-oriented agencies such as EDA to take environmental factors into account in their decision-

[78] 452 F.2d at 1237, 2 ELR at 20086.
[79] *Supra* note 75.
[80] 93 S. Ct. 458 (1972).

making processes. The lower courts had allowed EDA to escape the action-forcing process which was intended to help ensure that the mandated change takes place. Moreover, NEPA requires consideration of environmental impact well before decisions are made, in time to influence the basic thrust of agency action. Environmental consideration after agency decision making has gone forward, as in *Upper Pecos,* has been specifically disapproved (see pages 251 *ff.*). Nor can agencies withhold their environmental reviews from outside scrutiny. The decisions in *Upper Pecos* relieved EDA from its obligation to consult with other federal, state, and local agencies regarding the impact of the grant.

Admittedly, *Upper Pecos* is an extreme instance of abuse of the lead agency concept, as the government's confession of error indicated. The problem of the timing of the statement's preparation could have been remedied if EDA had awarded the grant after it had considered the impact statement that the Forest Service prepared. The EDA could have actively assisted in defining and preparing the statement, rather than deferring totally to the Forest Service, so that the statement would have included a discussion of the grant as well as of the right-of-way. Both of these possibilities are not only contemplated, but required, by the CEQ Guidelines and especially by its Memorandum of May 16, 1972.

Even after these improvements are made, preparation of statements through the lead agency device leaves something to be desired. It increases the likelihood that the exempted agencies will engage only in *pro forma* consideration of the statement, when they are fortunate enough to find another agency that will undertake the time-consuming chore of identifying potential environmental impacts. It removes from the agency best equipped to know the full effects and future consequences of a course of action the basic responsibility for having identified and discussed them. It allows the agency to "contract out" a task which if done internally would help to bring about the shift in agency values and attitudes that NEPA was intended to initiate. It allows agencies to cumulate impacts, which is desirable, but at the same time it enables each of them to escape focusing individually upon the particular impact of its own part of the overall action.

Of course these criticisms of lead agency statement preparation in no way detract from the strength of criticisms that can be made of alternative techniques for handling multi-agency actions. Preparation of separate agency statements on each part of an overall action ensures the maximum participation of all agencies involved, but at a cost in coordination, sharing of expertise, and consideration of cumulative impacts. Joint statements involve all agencies to a somewhat lesser degree and achieve coordination, sharing of expertise, and consideration of cumulative impacts, but at the risk of poorly defining ultimate responsibility.

On balance, however, of the three basic approaches—separate compliance, joint preparation, and lead agency preparation—the standard of compliance "to the fullest extent possible" would seem to require that the first two alternatives be favored over the last one. Thus the option currently favored under agency and CEQ Guidelines seems the least desirable.

As indicated in this book's conclusion, the preparation of overlapping impact statements is not necessarily wasteful. Thus an option open to agencies is separate preparation of statements, with a joint statement issued to cover areas where cumulative impacts are expected. This suggestion might apply to the new state of affairs in *Upper Pecos*, where separate EDA and Forest Service statements may now require coordination. The effort involved may be greater than that required when only one agency has to prepare a statement, but it does focus each agency's attention on the consequences of its share of the entire action. The additional information on the cumulative impact of the overall action provided by the joint statement completes the information necessary for each agency's decision. In this context one may agree with CEQ that "overview statements will find more extensive use in the future."[81]

CONTENTS OF AN ADEQUATE IMPACT STATEMENT

After the courts spell out the circumstances which require the preparation of impact statements, the locus of litigation shifts to challenges to the adequacy of the statements filed. It is generally held that in order to comply strictly with §102(2) (C), a statement must discuss five listed topics in a "detailed" manner.[82]

It should go without saying that the mere filing of a document labelled "final impact statement" is insufficient to shield an agency from judicial review if the document fails to comply with the standards outlined in NEPA.[83]

---

[81] Third Annual Report, *supra* note 76 at 236.

[82] Appendix A.

[83] SCRAP v. United States, 346 F. Supp. 189, 195 note 8, 2 ELR 20486, 20488 note 8 (D.D.C.), *application for stay pending appeal denied sub nom.* Aberdeen & Rockfish Ry. Co. v. SCRAP, —— U.S. ——, 2 ELR 20491 (1972) (Burger, Circuit Judge), *cert. granted* 41 U.S.L.W. 3339 (Dec. 19, 1972).

The starting point for this discussion is the statutory phrase "detailed statement." One court turned to a dictionary definition of "detailed," finding that "necessarily the EIS must be 'marked by abundant detail or thoroughness in treating small items or parts.' "[84] The general view was better explained by Judge Eisele, who, writing in the Gillham Dam case,[85] looked to the policy of the Act and concluded:

> At the very least NEPA is an environmental full disclosure law. . . . The "detailed statement" required by §102(2) (C) should, at a minimum, contain such information as will alert the President, the Council on Environmental Quality, the public and, indeed, the Congress, to all known *possible* environmental consequences of proposed agency action.[86]

The practice of measuring adequacy by the bench mark of "full disclosure" is now standard. It has been adopted by the courts in at least six cases[87] and is implicit in most cases that discuss adequacy. No competing standard has been suggested.

Useful as it is, the standard of full disclosure raises the question of how full is "full." The same court which introduced the concept qualified it in a subsequent opinion: "Congress, we must assume, intended and expected the courts to interpret the NEPA in a reasonable manner in order to effectuate its obvious purposes and objectives."[88] The "rule of reason" for testing the adequacy of compliance was first suggested by the District of Columbia Circuit in *Natural Resources Defense Council v. Morton.*[89] While it was used there only to delimit the range of alternatives that had to be considered, it may be extended to other aspects of statements.

We now discuss three aspects of the evolving view that a detailed statement should provide full disclosure of the environmental factors involved in a proposed action. The first is the policy justification

---

[84] Environmental Defense Fund v. Corps of Engineers (Tennessee–Tombigbee Waterway), 331 F. Supp. 925, 1 ELR 20466 (D.D.C. 1971), 348 F. Supp. 916, 932, 2 ELR 20536, 20542 (N.D. Miss. 1972).

[85] Environmental Defense Fund v. Corps of Engineers, 325 F. Supp. 728, 1 ELR 20130 (E.D. Ark. 1970–71), —— F. Supp. ——, 2 ELR 20260 (E.D. Ark.), 342 F. Supp. 1211, 2 ELR 20353 (E.D. Ark.), *aff'd*, 470 F.2d 289, 2 ELR 20740 (8th Cir. 1972).

[86] 325 F. Supp. at 759, 1 ELR at 20141.

[87] Environmental Defense Fund v. Tennessee Valley Authority (Tellico Project); Conservation Council of North Carolina v. Froehlke; Sierra Club v. Froehlke (Kickapoo River); Sierra Club v. Froehlke (Trinity River–Wallisville Dam); Allison v. Froehlke; Committee to Stop Route 7 v. Volpe, all cited in Appendix B.

[88] 342 F. Supp. at 1217, 2 ELR at 20355.

[89] Appendix B.

for full disclosure; a statement must inform the reviewing decision makers, the public, and possibly the courts themselves of the issues. The next discusses the standards used in deciding whether challenged statements are adequate. Finally, we will discuss a few special rules that apply to each of the subject areas of §102(2) (C).

The Policy Served by Full Disclosure

Information for Decision Makers. Section 102 was drafted in response to a congressional belief that federal agencies had given inadequate attention to the environmental consequences of their actions. A special report prepared for the Senate Committee on Interior and Insular Affairs traced both the crisis of the cities and that of the natural environment to "an ignorance of and a disregard for man's relationship to his environment."[90] The report focused in particular on the lack of any means for relating knowledge and policy.

> In pending legislation [NEPA] the knowledge assembled through survey and research would be systematically related to official reporting, appraisal, and review. The need for more knowledge has been established beyond doubt. But of equal and perhaps greater importance at this time is the establishment of a system to ensure that existing knowledge and new findings will be organized in a manner suitable for review and decision as matters of public policy.
>
> In summary, to make policy effective through action, a comprehensive system is required for the assembly and reporting of relevant knowledge; and for placing before the President, the Congress, and the people, for public decision, the alternative courses of action that this knowledge suggests.[91]

Calvert Cliffs'[92] was the first case to fill out the legislative framework. As Judge Wright saw it, the purpose of the "detailed statement" was to aid the agencies in their decision making and to alert other interested agencies and the public to the environmental consequences of planned federal action.

> [§§102(2) (C) and (D)] . . . seek to ensure that each agency decision maker has before him and takes into proper account all possible approaches to a particular project (including total abandonment of the project) which would alter the environmental impact and the cost-benefit

---

[90] L. CALDWELL, A SPECIAL REPORT TO THE SENATE COMM. ON INTERIOR AND INSULAR AFFAIRS, A NATIONAL POLICY FOR THE ENVIRONMENT 8, 90th Cong., 2d Sess. (Comm. print July 11, 1968). *Also at* 115 CONG. REC. S. 12126, 12128 (daily ed. Oct. 8, 1969).

[91] *Id.*, at 10–11 [115 CONG. REC. S. 12129 (daily ed. Oct. 8, 1969)].

[92] Appendix B.

balance. Only in that fashion is it likely that the most intelligent, optimally beneficial decision will ultimately be made.[93]

In other decisions, the District of Columbia Circuit has given further indication of the role of the statement in review. It is to be the "environmental source material" for the decision makers,[94] and must be sufficient to enable them to make an "informed choice."[95] In the Tellico Dam case, the district court saw the statement as allowing "those removed from the decision-making process to evaluate and balance the factors on their own."[96] In *Sierra Club v. Froehlke* (Trinity River–Wallisville Dam),[97] the court stated that in view of the uncertainty surrounding Congress' approval of the $1.3 billion Trinity River project, and specifically the $29 million Wallisville Dam, the Corps of Engineers had to submit (or resubmit) the projects to Congress to ensure full congressional awareness in light of NEPA's stated policy of informing decision makers.

It is the intent of this court, in view of the circumstances in this case, that Congress, the appropriate federal agencies and the Council on Environmental Quality pass upon and render decisions on the Wallisville and Trinity Projects as well as other components in the light of NEPA as passed by Congress.[98]

The range of officials who may eventually use the statement and to whom it should be addressed is extremely wide. The officials of sister agencies may make use of the statement. Any executive branch decision may eventually be submitted to presidential review, or Congress may decide to intervene through legislation concerning the project at hand. As the court in *Committee for Nuclear Responsibility v. Seaborg*[99] said:

The statement has significance in focusing environmental factors for informed appraisal by the President, who has broad concern even when

---

[93] 449 F.2d at 1114, 1 ELR at 20348 (D.C. Cir. 1971). More recent cases are substantially in accord: Environmental Defense Fund v. Tennessee Valley Authority (Tellico), 339 F. Supp. 806, 2 ELR 20044 (E.D. Tenn. 1972); Environmental Defense Fund v. Corps of Engineers (Tennessee–Tombigbee), 348 F. Supp. 916, 2 ELR 20536 (N.D. Miss. 1972). Full citations in Appendix B.

[94] Natural Resources Defense Council v. Morton, 458 F.2d at 833, 2 ELR at 20032.

[95] Committee for Nuclear Responsibility v. Seaborg, Appendix B, 463 F.2d at 787, 1 ELR at 20470.

[96] Environmental Defense Fund v. Tennessee Valley Authority (Tellico Dam), Appendix B, 339 F. Supp. at 810, 2 ELR at 20045.

[97] Appendix B.

[98] —— F. Supp. at ——, 3 ELR at ——.

[99] Appendix B.

not directly involved in the decisional process, and in any event by Congress and the public.[100]

Good faith consideration by agency decision makers is the logical culmination of the §102(2) (C) process; detailed disclosure of impacts, without such consideration, is obviously useless to agency decision making. Nevertheless, the courts have had to spell out the requirement. In *Calvert Cliffs'* the AEC was roundly castigated for attempting to write guidelines which would allow the mere circulation of (presumably) detailed statements without properly providing for their actual use:

> The Commission's crabbed interpretation of NEPA makes a mockery of the Act. What possible purpose could there be in the Section 102(2) (C) requirement . . . if "accompany" means no more than the physical proximity—mandating no more than the physical act of passing certain folders and papers, unopened, to reviewing officials along with other folders and papers.[101]

A statement's adequacy, in the end, is measured by its functional usefulness in decision making. The statement must be of a nature and form that enables the decision maker to consider environmental factors in good faith. The decision in *Calvert Cliffs'* goes on to spell out how consideration through a "finely tuned and 'systematic' balancing process" must take place. This requirement is analyzed in detail in chapter VII (see pages 247 *ff.*), where we discuss how the courts have articulated the close relationship which exists between NEPA's action-forcing procedures and its substantive requirements.

In short, as the courts have interpreted it, §102 expresses a congressional policy that a wide range of potential reviewers are to be given and should consider a wide range of information on the environmental consequences of their actions. Actions are not to be taken in ignorance; instead, existing knowledge and new findings are to be organized in a manner suitable for review and decision as matters of public policy.[102]

Information for the Public. Behind the President and the Congress stands the public as ultimate decision maker. NEPA has been interpreted to require both the sharing of information with the public

---

[100] 463 F.2d at 787, 1 ELR at 20470.

[101] 449 F.2d at 1117–18, 1 ELR at 20350. *Accord*, Greene County Planning Board v. Federal Power Comm'n, Appendix B.

[102] *See supra* note 91.

and the involvement of the public in developing information. The legislative history recognizes that NEPA was itself a response to a public demand, with Congress trying to catch up rather than lead.

There is a new kind of revolutionary movement underway in this country. This movement is concerned with the integrity of man's life support system—the human environment. The stage for this movement is shifting from what had once been the exclusive province of a few conservation organizations to the campus, to the urban ghettos, and to the suburbs. . . . S. 1075 is a response by the Congress to the concerns the Nation's youth are expressing. It makes clear that Congress is responsive to the problems of the future.[103]

The specific language of § 102 grants public access to impact statements under the Freedom of Information Act.[104] The role of NEPA in bringing the public into decision making was referred to in the Gillham Dam case.

The Congress, by enacting it [NEPA], may not have intended to alter the then existing decisionmaking responsibilities, or to take away any then existing freedom of decisionmaking, but it certainly intended to make such decisionmaking more responsive and more responsible.[105]

The public information provisions of NEPA were redefined by the CEQ Guidelines, which established the two stages of statement preparation—draft and final. The guidelines express a general policy in favor of public involvement,[106] but the specifics are disappointing. Hearings are left to the discretion of the agency "where appropriate," and the agency must wait only 90 days between submitting the draft and taking action (see discussion, pages 238 *ff.*). In view of the wide variation in project lead times, and the time ordinarily needed for citizen reaction to be conveyed to administrative agencies, the 90-day requirement appears inadequate. It can easily be used to block meaningful public involvement in the NEPA

---

[103] 115 CONG. REC. S. 17452 (daily ed. Dec. 20, 1969) (statement of Mr. Jackson).

[104] 5 U.S.C. §552.

[105] 325 F. Supp. at 759, 1 ELR at 20141. This passage appears to sum up what the district judge in Sierra Club v. Froehlke (Trinity River–Wallisville Dam), Appendix B, recently attempted to spell out precisely with respect to the Corps of Engineers. The Corps' decision-making "responsibilities" and its "freedom of decision making" remain, but its decision making must be made more "responsive" and "responsible" by alteration in its method of evolving project cost-benefit ratios, its evaluation of environmental costs, its use of sister agency expertise, etc.

[106] CEQ Guidelines, § 10(e), Appendix C.

process (see pages 234 *ff.*). Finally, the agencies are allowed to petition the CEQ for suspension of the waiting period.[107]

One case stands directly for the proposition that public involvement in preparation of the final statement is essential to its adequacy. On remand, the court in *Lathan v. Volpe*[108] indicated:

> The public may also raise environmental questions by way of comment to the draft impact statement. Since the final impact statement must respond to these comments, as well as to the comments of government agencies, environmental harm which might have been overlooked by highway officials may be brought to their attention. For this reason, highway officials must give more than cursory consideration to the suggestions and comments of the public in the preparation of the final impact statement. The proper response to comments which are both relevant and reasonable is to either conduct the research necessary to provide satisfactory answers, or to refer to those places in the impact statement which provide them. If the final impact statement fails substantially to do so, it will not meet the minimal statutory requirements [citing *EDF v. Hardin*].[109]

However, *Lathan* may be limited by the fact that it also involved a number of highway law statutes that mandated public hearings.[110]

In *Hanly v. Kleindienst*,[111] the Second Circuit took a strong stand in favor of public participation. The public must be allowed to comment while the agency is deciding whether or not to prepare a statement. Or, in the court's words:

> We . . . hold that before a preliminary or threshold determination of significance is made the responsible agency must give notice to the public of the proposed major federal action and an opportunity to submit relevant facts which might bear upon the agency's threshold decision. . . . The precise procedural steps to be adopted are better left to the agency, which should be in a better position than the court to determine whether solution of the problems faced with respect to a specific major

---

[107] CEQ Guidelines, §10(d). Query whether the CEQ may waive this rule and any other at its sole discretion. Waiver is probably subject to judicial review. Natural Resources Defense Council v. Morton, 337 F. Supp. 170, 2 ELR 20071 (D.D.C. 1972). May a member of the public petition CEQ, rather than an agency, for an extension of the deadline for comments?

[108] Appendix B.

[109] 350 F. Supp. at 265, 2 ELR at 20547.

[110] PPM 90–1, ELR 46106 (Aug. 24, 1971).

[111] Hanly v. Mitchell, —— F. Supp. ——, 2 ELR 20181 (S.D.N.Y.), *rev'd* (Hanly I), 460 F.2d 640, 2 ELR 20216 (2d Cir.), *cert. denied*, 41 U.S.L.W. 3247 (Nov. 7, 1972), *sub nom.* Hanly v. Kleindienst, —— F. Supp. ——, 3 ELR 20016 (S.D.N.Y.), *rev'd* (Hanly II), —— F.2d ——, 2 ELR 20717 (2d Cir. 1972).

federal action can better be achieved through a hearing or by informal acceptance of relevant data.[112]

Certainly the right of the public to submit comments on the statement is, if anything, even greater than the right to be heard while the agency is deciding whether or not to prepare that statement.

Standards of Adequacy

When they come to review the contents of challenged statements, courts have translated the policy of full disclosure into more specific requirements. The judiciary has acted to balance the need for the most complete disclosure possible against the resources which may be devoted to environmental investigation. The court which reviewed the Tennessee–Tombigbee navigation project provides an example:

> In reviewing the sufficiency of an agency's compliance with §102, we do not fathom the phrase "to the fullest extent possible" to be an absolute term requiring perfection. If perfection were the standard, compliance would necessitate the accumulation of the sum total of scientific knowledge of the environmental elements affected by a proposal. It is unreasonable to impute to the Congress such an edict. We preface our consideration of plaintiffs' contentions by declaring that the phrase "to the fullest extent possible" clearly imposes a standard of environmental management requiring nothing less than comprehensive and objective treatment by the responsible agency. . . . Thus, an agency's consideration of environmental matters that is merely partial or performed in a superficial manner does not satisfy the requisite standard.[113]

In determining whether statements are reasonably detailed, the courts have indicated that: (1) statements should be understandable and nonconclusory; (2) they should refer to the full range of knowledge; and (3) they must discuss certain impacts which are typical of some types of action.

Since statements are to be used by both lay reviewers and their scientific advisors, they must be comprehensible and informative for both: "The EIS must be written in language that is understandable to nontechnical minds and yet contain enough scientific reasoning to alert specialists to particular problems within the field of their expertise."[114] One way to accomplish this is to provide the more

[112] —— F.2d ——, 2 ELR 20723.

[113] Environmental Defense Fund v. Corps of Engineers, Appendix B, 348 F. Supp. at 927, 2 ELR at 20540.

[114] *Id.*, 348 F. Supp. at 933, 2 ELR at 20542.

technical aspect of the discussion in appendices to the statement itself.

In a number of cases statements have been rejected on the grounds that they were overly conclusory. These include, in addition to the first Gillham Dam case,[115] *City of New York v. United States*,[116] *SCRAP v. United States*,[117] and a triad of cases challenging construction of I-90 through Washington State, *Daly v. Volpe*,[118] *Brooks v. Volpe*,[119] and *Lathan v. Volpe*.[120] *Brooks*, in particular, found that the statement required under §102(2) (C) could not take the form of a simple "finding." The court reviewed the legislative history of the Act, and concluded that the change in conference from "findings" to "detailed statement" had been made in order to make it more difficult "for administrators to relegate environmental considerations to one more set of forms for signature."[121]

In order that the ensuing review be *informed* review, the cases have also held that the scientific conclusions presented must be supported by references to the relevant literature or to field studies. The court in the Tellico Dam case[122] stated the reason for such a requirement:

> Although comprehensive in scope, the draft statement's cost-benefit analysis consists almost entirely of unsupported conclusions. As a result, a non-expert reader is denied the opportunity to intelligently evaluate TVA's conclusions. In addition, it is impossible to determine the thoroughness of the research upon which TVA based the conclusions, or their relative merit. A lack of careful research and planning is suggested by two statements contained in the statement itself.[123]

Brooks tied this need for sources to the role of the statement as a reviewable record, to be used by the judiciary in determining whether the administrative action had been arbitrary or capricious, and de-

---

[115] Environmental Defense Fund v. Corps of Engineers, Appendix B.

[116] 337 F. Supp. 150, 159, 2 ELR 20275, 20276 (E.D.N.Y.). A revised statement was subsequently accepted. 344 F. Supp. 929, 2 ELR 20688 (E.D.N.Y. 1972).

[117] Appendix B.

[118] Appendix B.

[119] 319 F. Supp. 90, 1 ELR 20045 (W.D. Wash. 1970), 329 F. Supp. 118, 1 ELR 20286 (W.D. Wash. 1971), *rev'd*, 460 F.2d 1193, 2 ELR 20139 (9th Cir.), 350 F. Supp. 269, 2 ELR 20704 (W.D. Wash.), 350 F. Supp. 287 (W.D. Wash. 1972).

[120] Appendix B.

[121] 350 F. Supp. at 274, 2 ELR at 20705.

[122] Environmental Defense Fund v. Tennessee Valley Authority, Appendix B.

[123] 339 F. Supp. at 809, 2 ELR at 20044.

cided that such a determination could not be made in the absence of adequate documentation of the facts underlying the action.[124]

The litigation involving offshore oil leases, *Natural Resources Defense Council v. Morton*,[125] added two corollary propositions to the trend against conclusory statements. The District of Columbia Circuit ruled that an insufficient or overly conclusory statement could not be corrected by the introduction of more evidence at trial or by counsel's exegesis. To do so, it was indicated, would serve the third goal of judicial reviewability at the expense of the other two: "The subject of environmental impact is too important to relegate either to implication or to subsequent justification by counsel. The Statement must set forth the material contemplated by Congress in a form suitable for the enlightenment of the others concerned."[126] In a decision on remand, the district court stated further that addenda prepared by the Interior Department to fill in the gaps left in the initial statement could not themselves be considered part of a "Final Statement" without undergoing the same commenting procedure as was used in producing the initial statement.[127]

In order to avoid rejection as insufficiently detailed or overly conclusory, the statement must set forth not only the views and conclusions of the agency, but also all those brought to its attention during the commenting process, including those claims made by opponents of the undertaking. Even if the agency disagrees violently with a particular point of view, it should include it in the final statement. As the court in *Environmental Defense Fund v. Corps* (Gillham Dam) stated:

> Where experts, or concerned public or private organizations, or even ordinary lay citizens, bring to the attention of the responsible agency environmental impacts which they contend will result from the proposed agency action, then the §102 statement should set forth these contentions and opinions, even if the responsive agency finds no merit in them whatsoever. Of course, the §102 statement can and should also contain the opinion of the responsible agency with respect to all such viewpoints. The record should be complete. Then, if the decisionmakers choose to ignore such factors, they will be doing so with their eyes wide open.[128]

However, there would appear to be some disagreement as to whether all views expressed, or only those supported by "responsible

---

[124] 350 F. Supp. at 276, 2 ELR at 20706.
[125] Appendix B.
[126] 458 F.2d at 836, 2 ELR at 20033.
[127] 337 F. Supp. 170, 2 ELR 20071.
[128] 325 F. Supp. at 759, 1 ELR at 20141.

scientific opinion," need be brought to the attention of the decision makers. In *Committee for Nuclear Responsibility v. Schlesinger,*[129] the D.C. Circuit took the latter position:

> When, as here, the issue of procedure relates to the sufficiency of the presentation in the statement, the court is not to rule on the relative merits of competing scientific opinion. Its function is only to assure that the statement sets forth the opposing scientific views, and does not take the arbitrary and impermissible approach of completely omitting from the statement, and hence from the focus that the statement was intended to provide for the deciding officials, any reference whatever to the existence of responsible scientific opinions concerning possible adverse environmental effects . . . [citing the Gillham Dam case].
>
> Only *responsible* opposing views need be included and hence there is room for discretion on the part of the officials preparing the statement; but there is no room for an assumption that their determination is conclusive [italics in original].[130]

Presumably this limitation was conceived in order to allow mere rumors and groundless fears to be omitted. The time of reviewing decision makers would not be well spent if they had to weed through such notions. Yet the formula would also seem to contain potential for abuse by the agency preparing the statement; the last clause of the court's statement, quoted above, indicates that the judiciary will look quite closely at the decision to exclude proffered views.

The case for restricting statements to "responsible scientific opinion" loses all force when the controversy centers on matters where lay and scientific opinions are equally valid. For example, when a highway is being opposed because of its destructive effect on a community, or when citizens state that their aesthetic or recreational interests would suffer, it would make little sense to omit such views on the ground that they are not "scientific" statements. The place of scientific opinion is in the discussion of ecological effects, while the public itself may prove the best source of information as to the significance of such effects on the quality of the *human* environment.

The *Committee for Nuclear Responsibility* litigation[131] also posited a second limitation on the inclusion of the full range of knowledge in impact statements. This was that dissenting views need not be set forth in full, nor in the words of those who oppose the action.

---

[129] Appendix B.

[130] 463 F.2d at 787, 1 ELR at 20470.

[131] Committee for Nuclear Responsibility v. Schlesinger, Appendix B.

The agency need not set forth at full length views with which it disagrees, all that is required is a meaningful reference that identifies the problem at hand for the responsible official. The agency, of course, is not foreclosed from noting in the statement that it accepts certain contentions or rejects others . . . [citing the Gillham Dam case].[132]

In this view it would be enough if the existence of possible controversy were made known, which may stress the first policy goal of full disclosure at the expense of the second. Superior reviewing officials, such as the President or Congress, can always turn to scientific advisors for details on the scientific merits of the argument. The public has no such recourse, and to permit the case against a project to be thus condensed and rephrased by the agency would deprive opponents of the chance to use the commenting procedure as a forum for amplifying their criticism.

A meaningful reference might be enough to identify a possible *impact*, but more detail would be essential to permit informed examination of an *alternative*. Presumably the scientific advisors could turn to the literature to find out more about the impact, while the mere statement that "X is an alternative" provides no information as to the specific contours of that alternative. The alternative may exist in a number of different variations, and much may depend on the exact size of the alternative project which the agency has in mind. Thus, the district court in *Conservation Council of North Carolina v. Froehlke*[133] was in error when it used the *Committee for Nuclear Responsibility* decision as authority for the proposition that the statement does not have to represent the agency's complete analysis of alternatives. This is particularly true of §102(2) (D), which requires that the agency "study, develop, and *describe* appropriate alternatives," is read in conjunction with the requirement of §102(2) (C) (iii) that there be a detailed statement on "alternatives to the proposed action."[134]

Presumably, under the *Committee for Nuclear Responsibility* rule, the agency's decision to condense views would be carefully scrutinized by the judiciary if a claim arose that such views were distorted or slighted in the paraphrase. The reviewing court would direct itself to the question whether informed review and careful

[132] 463 F.2d at 787, 1 ELR at 20470.

[133] 340 F. Supp. 222, 2 ELR 20155 (M.D. N. Car.), *aff'd*, —— F.2d ——, 2 ELR 20259 (4th Cir. 1972), —— F.2d ——, 3 ELR 20132 (4th Cir. 1973).

[134] This line of argument is suggested by Environmental Defense Fund v. Corps of Engineers (Gillham Dam), Appendix B, in the circuit court opinion, —— F.2d at ——, 2 ELR at 20743.

consideration of alternatives was still possible under these circumstances. Should the agency wish to avoid challenge on such grounds, it would do well to include the actual depositions of project opponents in the statement or as appendices, as was found satisfactory in *Conservation Council.*

Although statements are evaluated on a case-by-case basis, for frequently litigated types of projects the decisions agree that certain recurrent problems must always be discussed in the statement. The trend appears to be that failure to discuss all these aspects, unless specifically justified by the agency, is grounds for invalidating the statement because it excludes an important factor as a matter of law.

Two cases indicate that statements on highway undertakings must examine in depth the resultant air and noise pollution, to the point of requiring that field studies be conducted for the area in question.[135] Statements have also been approved or rejected depending on whether they analyzed the effects of the new road on land use and population distribution, as well as the possibility that traffic on connecting roads will increase.[136]

Water resources projects also suggest *per se* categories.[137] In the Gillham Dam case,[138] the first set of memorandum decisions faulted a Corps statement for, among other reasons, giving insufficient attention to five specific environmental factors: the effects of flow fluctuation on downstream biological productivity and stability, the effects of the intrusion of other fish species, the effects of impoundment on existing food chains and downstream alluvial deposits, and the effect of occasional reservoir "drawdowns" on shoreline vegetation.[139] It would seem that all are essential to a full discussion of impacts, particularly since the court in the later decision *Environmental Defense Fund v. Corps* (Tennessee–Tombigbee)[140] made clear that its acceptance of a statement was due to sufficient

---

[135] Lathan v. Volpe, Appendix B, 350 F. Supp. at 266, 2 ELR at 20547; Keith v. Volpe, —— F. Supp. ——, ——, 2 ELR 20425, 20428–29.

[136] *Id. See also* Pizitz v. Volpe, Appendix B, —— F. Supp. ——, 2 ELR 20378, where a statement which discussed such topics "fully" was accepted.

[137] The most instructive case thus far in this area is the recent opinion in Sierra Club v. Froehlke (Trinity River–Wallisville Dam), Appendix B, where the court provides a lengthy analysis of a myriad of factors related to water resources projects.

[138] Appendix B.

[139] Environmental Defense Fund v. Corps of Engineers, Appendix B, 325 F. Supp. at 747–48, 1 ELR at 20137.

[140] Appendix B.

coverage of project effects on water quality, fauna, items of historical or archaeological value, and the like.[141]

Certain items must be disclosed for all types of actions. The first decision in the Gillham Dam case indicated that any actions of the agency which departed from its responsibilities under statutes other than NEPA (such as the Corps' responsibilities under the Fish and Wildlife Act of 1934) should be set forth in the statement. Similarly, departures from the plan of the project as given in the legislation which authorized it should be acknowledged.[142]

The courts favor discussion of plans for future projects if they might affect a decision on the project at hand. The Second Circuit in *Greene County* stated: "We fail to see how the Commission, if it is to fulfill the demanding standard of 'careful and informed decision-making,' *Calvert Cliffs'* 449 F.2d at 1115, can disregard impending plans for further power development."[143]

In *Sierra Club v. Froehlke* (Trinity River–Wallisville Dam)[144] the court enjoined the long-range $1.3 billion Trinity River Project at the same time it enjoined further work on the Wallisville Dam, a component of the Trinity River Project, until the Corps of Engineers either showed that sufficient independent, localized justification to go ahead with the dam existed, or the larger river development project was justified. This ruling is better reasoned than *Conservation Council of North Carolina v. Froehlke*,[145] where the court held that the relationship to a Corps water project of plans for two atomic power plants should be considered in the *later* statements on the two plants. The opinion appears to assume that when two conflicting actions have been proposed, the one put forth earlier has some sort of vested right to go forward, regardless of the rationality of doing so or the loss of a more desirable way of using resources which had been suggested later. "Overview" impact statements would deal with the overall combination of projects, which in any event would be the subject of separate statements (see chapter VIII pages 290 *ff.*).

There is a split on whether cost-benefit analyses of projects should be included in statements, or at least attached as appendices. Because the issue arises primarily in connection with the role of the

---

[141] 348 F. Supp. at 934–41, 2 ELR at 20543–45.
[142] 325 F. Supp. at 759, 1 ELR at 20141.
[143] Greene County Planning Board v. Federal Power Comm'n, Appendix B.
[144] 455 F.2d at 424, 2 ELR 20022.
[145] Appendix B.

impact statement in the reviewable agency record of decision, it is treated in chapter VII (see pages 252 *ff.*).

Special Problems of Adequacy

Three questions concerning adequacy pose special difficulty, and accordingly they are considered separately. They relate to how unknown or uncertain impacts are to be treated, the range of alternatives that the agency must discuss, and the meaning of §§102(2)(C)(iv) and (v) which relate to short- and long-term use and irretrievable resource commitments.

Uncertain or Unknown Impacts. Since the impact statement is prepared before an action is carried out, its discussion of what will happen once the action is underway is in large measure based on informed guesswork. For some types of action the range of uncertainty may be large, as when the proposed project is the first of its kind. In other cases, the effort required to measure precise effects might be excessively costly. The courts differ in their approach to this problem. Some find sufficient compliance with §102(2)(C) when the statement simply points out the gaps in existing knowledge. Others make the completion of an agency research program a prerequisite to action.

The decisions which find that admission of incomplete knowledge is sufficient for §102(2)(C) seem to balance the utility of more complete disclosure against the costs of greater precision.[146] This approach is illustrated in *Environmental Defense Fund v. Corps* (Gillham Dam).[147] In rejecting on other grounds the first statement filed on the dam, the court noted that an expert witness for plaintiff had testified that a survey of species present in the river was essential in studying the effect of the dam on fish life. Taking into consideration the time required for such a study (six months to a year, plus several more years for a follow-up study), Judge Eisele decided:

---

[146] In Environmental Defense Fund v. Corps of Engineers (Tennessee-Tombigbee), Appendix B, the court jumped to the conclusion that only "probable" effects need be discussed, excluding "mere possibilities unlikely to occur as a result of the proposed activity." 348 F. Supp. at 933, 2 ELR at 20542. The weakness here lies in the failure to see that the magnitude of the risk must also be included; certainly a reviewing decision maker would want to know if there was a chance that the undertaking would be disastrous, even if a disaster were an improbable eventuality. Sierra Club v. Froehlke (Trinity River-Wallisville Dam), Appendix B, indicates that uncertainty undermines the efficacy of decision making to such an extent that under NEPA an agency cannot proceed until it gets the answers, no matter what the cost.

[147] Appendix B.

The Court is not here stating that such a collection or study would be required in order to comply with NEPA. But the opinions of such qualified professionals . . . should be made a part of the impact statement. The decisionmakers can then determine whether to proceed without such a study or to postpone the project while such study is being undertaken.[148]

The court applied the same logic to plaintiffs' claim that defendants had failed to develop methods for considering unquantifiable values [§102(2) (B)]. Since the CEQ had not indicated how this provision was to be implemented, the court stated:

NEPA does not require the impossible. Nor would it require, in effect, a moratorium on all projects which had an environmental impact while awaiting compliance with §102(2) (B). It would suffice if the statement pointed out this deficiency. The decision-makers could then determine whether any purpose would be served in delaying the project while awaiting the development of such criteria.[149]

In a later decision Judge Eisele approved a completely rewritten impact statement, although it allegedly contained some misstatements. He suggested that whatever error there had been was harmless, since "the EIS is sufficient to alert the decision-maker to the problem."[150] The opinion alludes further to the difficulty of composing a perfect statement, and the existence of other ways by which errors might be rectified.

Although the impact statement should, within reason, be as complete as possible, there is nothing to prevent either the agency involved, or the parties opposing the proposed agency action, from bringing new or additional information, opinions, and arguments to the attention of the "upstream" decision-makers even after the final EIS has been forwarded to CEQ. So it is not necessary to dot all the I's and cross all the T's in an impact statement. . . .

It is doubtful that any agency, however objective, however sincere, however well staffed, and however well financed, could come up with a perfect environmental impact statement in connection with any major project. Further studies, evaluations and analyses by experts are almost certain to reveal inadequacies or deficiencies. But even such deficiencies, discovered after the fact, can be brought to the attention of the decision-makers, including, ultimately, the President and the Congress itself.[151]

---

[148] 325 F. Supp. at 760, 1 ELR at 20142.

[149] 325 F. Supp. at 758, 1 ELR at 20141.

[150] 342 F. Supp. at 1216, 2 ELR at 20355.

[151] 342 F. Supp. at 1217, 2 ELR at 20355. *Accord*, Sierra Club v. Froehlke (Kickapoo River), 345 F. Supp. 440, 2 ELR 20307 (W.D. Wisc. 1972).

Similarly, in the Tennessee–Tombigbee case, where spoil disposal was identified as a problem but no solution was proposed, the court accepted the impact statement, since "to rule otherwise would require an agency to compile virtually complete engineering data merely to prepare an impact statement. An undertaking of such magnitude is beyond the scope of §102(2)(C)."[152]

There are several objections to allowing action to continue while further study is carried out. The increased commitment of resources might swing the balance in favor of proceeding with an otherwise undesirable project. Moreover, adverse findings would be diluted, as they trickled in one after another instead of being collected and cogently set forth in one document for reviewers. One solution would be to require the agency to seek out testimony on the range and magnitude of the risks involved in proceeding without specific studies. The suggestion in the Gillham Dam case that similar projects be examined to ascertain likely impacts seems sound.[153]

When the uncertainties are especially large, as when the action is the first of its kind, "full disclosure" would seem to require that the agency establish a system for monitoring the impacts as they occur. The CEQ Guidelines indicate that "it is also important in further action that account be taken of environmental consequences not fully evaluated at the outset of the project or program."[154] The Sixth Circuit recently applied a similar concept in finding that impact statements should be filed annually on ongoing work supported by annual appropriations.

> At least to the extent that a project can still be said to be ongoing, the decision makers are entitled to all information relevant to a determination whether to abandon the project or to alter it. Thus, as long as appropriations are necessary for the continued construction of a project, impact statements should be filed.[155]

The alternative to permitting action to proceed while admitting incomplete knowledge is set out in *Environmental Defense Fund v. Hardin.*[156] There the D.C. District Court turned to §102(2)(A), as well as §102(2)(C), and concluded:

---

[152] 348 F. Supp. at 939, 2 ELR at 20545.

[153] 325 F. Supp. at 748, 1 ELR at 20138.

[154] CEQ Guidelines, §11, Appendix B, referring to actions ongoing as of the passage of NEPA.

[155] Environmental Defense Fund v. Tennessee Valley Authority (Tellico), Appendix B, —— F.2d at ——, 2 ELR at 20734.

[156] 325 F. Supp. 1401, 1 ELR 20207 (D.D.C. 1971).

> [Section 102(2) (A)] . . . makes the completion of an adequate research program a prerequisite to agency action. The adequacy of the research should be judged in light of the scope of the proposed program and the extent to which existing knowledge raises the possibility of potential adverse environmental effects.[157]

The approach is most useful when the program involved is untried, as was the Department of Agriculture's proposal in *Hardin* to spray the pesticide Mirex over a wide area in order to control the fire ant. In this situation, field research might be required, whereas for waterway construction, reference to previous scientific data and literature may be sufficient.[158]

While finding on the facts that Agriculture's research program on Mirex was adequate, the *Hardin* court went on to specify the following condition for adequacy:

> The Act envisions that program formulation will be directed by research results rather than that research programs will be designed to substantiate programs already decided upon. Thus [Section 102(2) (A)] . . . requires a diligent research effort, undertaken in good faith, which utilizes effective methods and reflects the current state of the art of the relevant scientific discipline.[159]

The decision also relates research under §102(2)(A) to the required impact statement. In order for a reviewer to reach a "reasonably accurate decision" on environmental effects, it is necessary to present the research results in the statement.[160] Thus for untried programs, the statement serves to narrow the initially wide range of uncertainty and so achieves one purpose of the Act by preventing decision making in ignorance of the consequences.

The Presentation of Alternatives. The Senate discussion of NEPA clearly indicates that the Act was intended to facilitate policy choices, yet there can be no choice if a decision maker lacks alternatives to the proposed action. The importance of presenting alternatives is emphasized not only in §102(2)(C)(iii), but also in §102 (2)(D), which requires that agencies develop "appropriate alternatives" when there are "unresolved conflicts concerning alternative

---

157 325 F. Supp. at 1403, 1 ELR at 20208.

158 Environmental Defense Fund v. Corps of Engineers (Tennessee–Tombigbee), Appendix B, 348 F. Supp. at 930, 2 ELR at 20541.

159 325 F. Supp. at 1403, 1 ELR at 20208. At least two cases, Lathan v. Volpe, and Keith v. Volpe, both cited in Appendix B, have extended this requirement to proposals for highway construction.

160 325 F. Supp. 1403–04, 1 ELR at 20208.

uses of available resources."[161] These two sections of the statute tell agencies to present reviewers with options other than the one favored by the agency. It is not surprising, then, that the agencies have been challenged on grounds of excluding relevant alternatives from statements and inadequately discussing those presented.

The alternatives of no action at all, and of another action that fully accomplishes the original goal but without any of its objectionable features, constitute the extremes between which discussable alternatives fall. The district court in the Gillham Dam case expressly stated that both extremes must be covered. The possibility of a decision to take no action, i.e., to reject the proposed action, is itself an alternative which must be set forth. The court characterized the Corps' failure to describe fully the consequences of leaving the river alone as a "glaring deficiency" in the statement.[162] In so holding, the case is in full accord with the intent of Congress; the section-by-section analysis presented to the Senate indicated that §102(2)(C)(iii) meant "the alternative ways of accomplishing the objectives of the proposed action and the results of not accomplishing the proposed action."[163] *Keith v. Volpe*[164] related this presentation to the requirement of the consideration of the statement in decision making. There must be good faith consideration of the merits of the project and of the wisdom of abandoning it.[165]

Further, the court in the Gillham Dam case read §102(2)(D) in conjunction with the impact statement requirement, and concluded that the statement should definitely indicate if an alternate "optimum" solution conceivably exists.

> If there exists any alternative which would satisfy all of the competing interests—that is, an alternative in which all of our citizens could "have their cake and eat it too"—this should be made explicit in any "detailed statement" required by NEPA.[166]

In between no action at all and the perfect alternative lies a wide range of more likely choices that offer partial solutions or go only partway toward meeting the original proposal's objectives. *Natural Resources Defense Council v. Morton*[167] makes it plain that these must be discussed.

---

[161] NEPA, §102(2)(D), Appendix A.
[162] 325 F. Supp. at 761, 1 ELR at 20142.
[163] 115 Cong. Rec. S. 40420 (Dec. 20, 1969) (statement of Senator Jackson).
[164] Appendix B.
[165] —— F. Supp. at ——, 2 ELR at 20429.
[166] 325 F. Supp. at 762, 1 ELR at 20143.
[167] Appendix B.

Nor is it appropriate, as Government counsel argues, to disregard alternatives merely because they do not offer a complete solution to the problem. If an alternative would result in supplying only part of the energy that the lease sale would yield, then its use might possibly reduce the scope of the lease sale program and thus alleviate a significant portion of the environmental harm attendant on offshore drilling.[168]

As important as the range of alternatives is the thoroughness with which each alternative must be discussed. Although one district court has stated that the statement need not contain the agency's complete analysis of alternatives,[169] the better standard would appear to be that of *Natural Resources Defense Council v. Morton*, which required that they be discussed in sufficient depth to permit a reviewer to make a reasoned choice.[170] In particular, the circuit court found that the statement should not only mention the alternatives, but attempt to assess the environmental risk of each, in comparison to the main proposal.

A sound construction of NEPA, which takes into account both the legislative history and contemporaneous executive construction, requires a presentation of the environmental risks incident to reasonable alternative courses of action.[171]

The limitation which the decision places on this doctrine is that the effects of alternatives which are merely "remote and speculative possibilities" need not be included, at least when such effects "cannot be readily ascertained."[172]

This same fertile circuit opinion also held that alternatives beyond the power of the agency to implement must be discussed. The court strongly favored having statements discuss solutions which cannot be implemented by the official preparing the report, as long as they are within the power of executive reviewers or the Congress. The *Morton* litigation challenged the Department of the Interior's plan to lease offshore oil tracts, a plan undertaken in response to a presidential message on energy.[173] The message called for an overall program for increasing supply, and the leasing was only one part of this endeavor.

The D.C. Circuit Court reasoned that if alternatives were limited to those which Interior could choose, the more basic question of how

---

[168] 458 F.2d at 836, 2 ELR at 20033.
[169] Conservation Council of North Carolina v. Froehlke, Appendix B.
[170] 458 F.2d at 836, 2 ELR at 20033.
[171] 458 F.2d at 834, 2 ELR at 20032.
[172] 458 F.2d at 838, 2 ELR at 20034.
[173] Text at 117 CONG. REC. S. 8313–17 (June 4, 1971).

responsibility could best be apportioned among the departments would be ignored.

When the proposed action is an integral part of a coordinated plan to deal with a broad problem, the range of alternatives that must be evaluated is broadened. While the Department of the Interior does not have the power to eliminate or reduce oil import quotas such action is within the purview of both Congress and the President, to whom the impact statement goes.[174]

As the court saw it, a broader statement could have been prepared at the time of the presidential message, and, failing this, was required of the agency taking the first step toward implementation.[175] Other statements could, in turn, incorporate this broad initial evaluation by reference. (See the discussion of "multi-tier" statements at pages 290 *ff.*)

A number of other decisions have rejected statements which dealt with only a small part of a broader action, when no alternatives to the overall action were presented. This has happened most frequently in highway cases, where statements typically deal with small sections of road and only ask "is this the best location and/or design," not "should this road be built at all?"[176] The discussion in *Morton*, however, does not illuminate a crucial problem, that of deciding what actually constitutes a "coordinated plan to deal with a broad problem." Had the directive to lease the tracts been sent to Interior in the absence of any presidential message, it might have been far more difficult to prove that some broader action was involved. Such a situation would be similar to that in *Scientists' Institute for Public Information v. Atomic Energy Comm'n*,[177] where the AEC was held not obligated to file a statement on its breeder reactor program until the first demonstration plant had been built. The problem, of course, is that continued commitment of resources to research in this area works to foreclose alternatives; by the time of the first plant, the program might have acquired momentum which could be reversed only with difficulty.

The number of *possible* alternatives threatens to overwhelm any compiler, especially because alternatives may accomplish less than the original proposal contemplated, and they too must be discussed.

---

[174] 458 F.2d at 835, 2 ELR at 20033.

[175] *Id.*

[176] *See* Keith v. Volpe, Committee to Stop Route 7 v. Volpe, and Indian Lookout at Alliance v. Volpe, all in Appendix B.

[177] Scientists' Institute for Public Information v. Atomic Energy Comm'n, —— F. Supp ——, 2 ELR 20642 (D.D.C. 1972).

Some way of reducing the discussion of alternatives to manageable proportions would therefore be desirable. Nevertheless, there are other considerations which militate against allowing the agency the untrammeled discretion to pick the alternatives to be discussed. If it confined itself to suitably undesirable choices, any proposal could be made to look better than the alternatives presented. Similarly, if alternatives are presented in the disjunctive (e.g., *either* build a dam *or* buy title to the floodplain), then many desirable combinations would be overlooked (e.g., it might be cheaper to combine a smaller dam with the purchase of the most threatened land).[178]

For this reason the principal limit on the discussion of alternatives is a "rule of reason" set out in *Natural Resources Defense Council v. Morton,* where the circuit court decided that the alternatives discussed need only be those available in the same time span as the original proposal.

> Since the statement also sets forth that the agency's proposal was put forward to meet a near-term requirement, imposed by an energy shortfall projected for the mid-1970's, the possibility of the environmental impact of long-term solutions requires no additional discussion at this juncture. We say "at this juncture" for the problem requires continuing review. . . .[179]

Another corollary of the "rule of reason" is that only reasonably proximate alternatives are to be discussed, excluding those which involve the repeal of "basic legislation," such as the antitrust laws.[180] This example may have been unfortunate, should it be used by advocates for utilities to foreclose discussion of ways to reduce the demand for power on the grounds that regulatory statutes mandate that capacity be increased to cover such demand. Perhaps it could be pointed out that certain statutes may be ecologically irrational, and cause environmental effects not foreseen at the time they were passed. In such circumstances, these laws are not nearly as remote from the problem nor as free from critical scrutiny as the Sherman and Clayton acts would be.

---

[178] For an example of arbitrary exclusion of alternatives, see Scenic Hudson Preservation Conference v. FPC, 453 F.2d 463; 1 ELR 20496 (2d Cir. 1971), where the commission limited itself to alternative power sources within 100 miles of New York: "In this day of high voltage transmission," asked the dissent, "what is so magic about one hundred miles?" 453 F.2d at 493, 1 ELR at 20512.

[179] 458 F.2d at 837, 2 ELR at 20034.

[180] *Id.*

Short-Term Uses, Long-Term Productivity, and Irreversible Resource Commitments. The other two specific topics mentioned in §§102(2)(C)(iv) and (v) appear to have received little attention from the courts. Since the section as a whole was aimed at implementing the policy goals of NEPA as stated in §§101(a) and (b), these two requirements should be related to the objectives of the Act. In particular, §101(b)(1) states that the nation should "fulfill the responsibilities of each generation as trustee of the environment for succeeding generations."[181] Thus, in stating the "relationship between local short-term uses of man's environment and the maintainance and enhancement of long-term productivity,"[182] the agency should discuss the action from the standpoint of a trustee for future generations, and justify, in writing, any decision to incur lasting losses for the sake of short-term gains. The Senate report on S. 1075 lends some support to this position.

Wherever local, short-term uses of the resources of man's environment are being proposed, a finding must be made that such uses are consistent with the maintainance and enhancement of the long-term productivity of the environment.[183]

As *Brooks v. Volpe*[184] pointed out, the substitution of "detailed statement" for the original bill's "finding" was meant to place a greater burden of justification on the preparing agency.

Judge Oakes' dissent in the *Scenic Hudson* case[185] provides an illustration of the meaning of §102(2)(C)(iv), and of its misinterpretation by an agency. The action involved a license for a pumped storage electric facility at Storm King Mountain which would have permanently impaired the mountain's "scenic grandeur" and "unique beauty." Judge Oakes wrote:

Here the Commission's finding 217 says, incomprehensibly, "any short term adverse impact is more than offset by the enhancement of long term productivity which will result from the project. . . ." I think the finding indicates that the commission did not read the Act very carefully. . . . Not the short-term *impact* on the natural environment, but the short-term *uses* of it in relation to long-term productivity, is the statement required. Here we are considering permanent structures, a long-term and substantial use of an area of great natural beauty . . . involving an

[181] NEPA, §101, Appendix A.
[182] NEPA, §102(2)(C)(iv), Appendix A.
[183] S. Rep. No. 91–269, 91st Cong., 1st Sess. 21 (1969).
[184] Appendix B.
[185] *Supra* note 178.

"irreversible and irretrievable commitment of resources" in the proposed project if licensed.[186]

Other examples of situations where a relationship of short-term use to long-term interests should be discussed include the leasing of offshore oil tracts (as in *Morton*), where the short-term depletion of a resource means that it will be unavailable in the future. Similarly, the short-term gains from Corps water projects should be balanced against the permanent loss of free-flowing streams and the eventual uselessness of a dam as silt accumulates.

Irreversible commitment [§102(2)(C)(v)] may best be considered as complementing discussion of long-term uses and short-term productivity [§102(2)(C)(iv)]. It may be interpreted to require consideration of the permanent loss, not to the environment, but of the resources committed to the project and thus made unavailable for any other use. *Daly v. Volpe*[187] provides an example of application to a specific case; the court decided that the statement on a challenged highway was insufficient, since (among other reasons) it failed to state (1) the cost of land, construction materials, labor, and other economically measurable cost which cannot be retrieved once a highway is constructed; and (2) the resources which may be irretrievably lost, and the nature of each such loss, to which a dollar value cannot be readily assigned—for example, the loss of forested recreational land.[188]

NEPA COMMENTING PROCEDURES

Section 102(2)(C) establishes procedures for obtaining comments and views from outside the agency before major federal action may be taken. The legislative history suggests that the provision was intended to provide coordination in environmental policy making. "The present problem is not simply the lack of a policy. It also involves the need to rationalize and coordinate existing policies."[189] The concluding portion of §102(2)(C) states:

> Prior to making any detailed statement, the responsible Federal official shall consult with and obtain the comments of any Federal agency which has jurisdiction by law or special expertise with respect to any environmental impact involved. Copies of such statement and the comments and

---

[186] 453 F.2d at 492, 1 ELR at 20512.

[187] Appendix B.

[188] 350 F. Supp. at 259, 2 ELR at 20445.

[189] 115 Cong. Rec. S. 12127 (daily ed. Oct. 8, 1969) (from prepared statement by Senator Jackson).

views of the appropriate Federal, State, and local agencies, which are authorized to develop and enforce environmental standards, shall be made available to the President, the Council on Environmental Quality and to the public as provided by section 552 of title 5, United States Code, and shall accompany the proposal through the existing agency review processes.

The Act provides only the statutory skeleton of the commenting process, which has been fleshed out considerably by Executive Order 11514,[190] the CEQ final guidelines, and judicial interpretation of the Act. The CEQ Guidelines set up a comprehensive procedure. "Draft" statements, which should be as complete as the agency can possibly make them, provide the means whereby the agencies are to elicit the comments of sister agencies with environmental expertise, of state and local environmental agencies, and of the public. With the benefit of these comments, the preparing agency completes a "final" statement, which is supposed to accompany the proposal through "existing agency review processes." Buttressed by Executive Order 11514, Section 10 of the CEQ Guidelines[191] puts great stress on public involvement in the process of turning a draft statement into a final one. The Act itself does not mention a public role other than by providing that the public may see the final statement under the Freedom of Information Act ("section 552 of title 5, United States Code").

The Effect of Failure to Obtain and Consider Comments

The courts have not accepted the view that the commenting process is merely ancillary to the central purpose of obtaining an impact statement from the agency proposing action. Rather, by requiring compliance "to the fullest extent possible," the few courts to rule on the issue have held that failure to follow the statutory procedures for obtaining comments will result in rejection of the statement as inadequate.

In *Natural Resources Defense Council, Inc. v. Morton,*[192] the circuit court affirmed a district court holding that the Department of the Interior's final statement inadequately discussed alternatives to the proposed leasing of offshore oil lands. On remand, Interior attempted to comply with the circuit court's decision by supplementing its inadequate statement with an addendum prepared over a weekend, which discussed reasonable alternatives to the proposed action.

[190] Protection and Enhancement of Environmental Quality, Exec. Order No. 11514, 35 FED. REG. 4247, ELR 45003 (March 5, 1970).

[191] CEQ Guidelines, § 10, Appendix C.

[192] Appendix B.

Because the new material had never been circulated for comment as required by §102(2)(C), the district court refused to accept the statement as modified, saying:

If this addendum is to be considered a part of the Final Impact Statement, then it must be subjected to the same comment and review procedures outlined by §4332(2) (C) of NEPA, as was required for the original Final Impact Statement which did not contain the addendum when it was first circulated.[193]

In support of its holding, the district court cited the language in *Calvert Cliffs'* which requires strict compliance with NEPA in spite of delay, cost, and administrative burden. (See chapter III, pages 54 *ff.*) Moreover, the CEQ, which had first defined the draft-to-final process, lacked authority to waive the requirement of circulation, although in this case it attempted to do so. Finally, the court indicated that the procedure must be followed regardless of the likelihood that the comments will be acted upon.

While it is quite conceivable that none of the alternatives to the proposed lease sale are feasible at this time, this fact does not mean that those Federal, State, and local agencies interested should not be given the opportunity to comment on the Addendum to the Final Impact Statement as required by Congress. Whether or not the comments will be valuable in the end is not the question before this court. The court must only determine whether the opportunity for comment as required by Section 4332(2) (C) was afforded.[194]

Another case, *United States v. 247.37 Acres of Land*,[195] indicated that the sponsoring agency must be prepared to prove in court that the commenting procedure was carried out. It had earlier been held that the Corps had to file a statement on an ongoing dam project. When the Corps returned to court with a completed statement and asked that the injunction be lifted, the court rejected the motion.

What this record does not show is what happened thereafter. There is no indication at all whether or not it was commented on by the Environmental Council [sic], whether the public was in any way informed "as provided by Title 5–552," [sic] etc. Either one of two things is perfectly consistent with this record in its present status—(1) That the "agency review process" led to all kinds of environmental advice to the Corps of Engineers, which was thrown in the wastebasket; or (2) That nothing

[193] 337 F. Supp. at 172, 2 ELR at 20072.
[194] *Id.*
[195] —— F. Supp. ——, 1 ELR 20513 (S.D. Ohio 1971), 2 ELR 20154 (S.D. Ohio 1972).

has happened and there is nothing the Corps can do about it to hurry up some reaction to the impact statement.[196]

While these two cases rest on the failure to obtain comments, several others refer to such failure as one factor among others which led to rejection of agency claims of compliance. These decisions give some indication of the approach the courts would be likely to take should the issue be focused solely on the adequacy of compliance with the commenting procedure. In the Gillham Dam case, *Environmental Defense Fund v. Corps of Engineers*,[197] one reason for the ruling against the Corps was:

The evidence of the defendant's consultation with state agencies leads the Court to conclude that in some cases such might have been rather superficial. And this also appears to be the case with respect to their consultation with certain federal agencies.[198]

After trial, the court listed ten instances of noncompliance, three of which concerned the commenting procedure:

(7) The evidence does not indicate that, prior to making the statements, the defendants did "consult with and obtain the comments of" all Federal agencies which have "jurisdiction by law or special expertise with respect to any environmental impact involved."
(8) The statements do not include the "comments and views" of *all* appropriate "State and local agencies which are authorized to develop and enforce environmental standards."
(9) The evidence does not indicate that the "statements and comments and views" of all the appropriate Federal, State, and local agencies did "accompany the proposal through the existing agency review process."[199]

The case of *Akers v. Resor*[200] was brought to enjoin the Corps of Engineers from channelizing tributaries of the West Tennessee River. The court found that the Fish and Wildlife Coordination Act of 1958,[201] which required the Corps to consult "in good faith" with the United States Fish and Wildlife Service and the concerned state agency, and to make "adequate provision" for wildlife conservation, interacted with NEPA to require comment and consultation. Where the Tennessee Fish and Game Commission, the Fish and Wildlife Service, the Bureau of Recreation, the Forest Service, and the Ten-

[196] —— F. Supp. at ——, 2 ELR at 20155.
[197] Appendix B.
[198] 325 F. Supp. at 745, 1 ELR at 20136.
[199] 325 F. Supp. at 758, 1 ELR at 20141.
[200] 339 F. Supp. 1375, 2 ELR 20221 (W.D. Tenn.), —— F. Supp. ——, 3 ELR 20157 (W.D. Tenn. 1972).
[201] 16 U.S.C. §661 *et seq.*

nessee Health Department had stated that the mitigation plan prepared by the Corps was inadequate, the court decided that "in light of the requirements of NEPA," especially §§102(1) and 101(b), "the Act of 1958 must be interpreted to require the Corps to submit a new plan of mitigation to Congress before it proceeds further. . . ."[202]

In rulings in two NEPA highway cases, both released on August 4, 1972, Judge Beeks of the Federal District Court for the Western District of Washington held that it was a fatal procedural defect under NEPA for federal defendants to have failed to give adequate notice or otherwise not to have allowed an opportunity for public comment on the impact statements which they had prepared. In *Daly v. Volpe*,[203] Judge Beeks held not only that the statement at issue remained insufficiently "detailed" but that defendants had given "inadequate advance notice to the public of defendants' intention to favor route E-3,"[204] and had failed to give the public an opportunity "to meaningfully comment" on the impact statement. The court refused to dissolve the injunction issued earlier, but specified steps short of a new location hearing which defendants might take to comply "because of the urgency of the situation in this case."[205] In *Brooks v. Volpe*,[206] Judge Beeks relied on the same two grounds in refusing to dissolve an earlier injunction against further construction of a disputed segment of I-90, located 45 miles from Seattle. He said that the statement was inadequately "detailed," and:

> Finally, defendants failed to give adequate public notice of the existence of the completed impact statement. This is most likely the reason for the absence of comment thereon from the general public. It is shocking that not even the plaintiffs in this lawsuit were so informed [footnote omitted].[207]

Comments must not only be obtained in accordance with §102(2)(C) procedures, they must also be meaningfully employed in preparing the final statement. In both the Gillham Dam case and *Committee for Nuclear Responsibility v. Seaborg*[208] the courts held that "outside" opinion must actually be discussed in the impact statement. While formal §102(2)(C) comments were not specifically

[202] 339 F. Supp. at 1380, 2 ELR at 20223.
[203] Appendix B.
[204] 350 F. Supp. at 260, 3 ELR at 20032.
[205] *Id.*
[206] Appendix B.
[207] 350 F. Supp. at 280, 2 ELR at 20708.
[208] Appendix B.

at issue in the two cases, the rulings still apply to such statements a fortiori. As mentioned earlier (see pages 209 *ff.*), the Arkansas district court appears to state a rule slightly broader than that of the District of Columbia Circuit Court. The Arkansas court said:

> Where experts, or concerned public or private organizations, or even ordinary lay citizens, bring to the attention of the responsible agency environmental impacts which they contend will result from the proposed agency action, then the §102 statement should set forth these contentions and opinions, even if the responsive agency finds no merit in them whatsoever.[209]

The circuit court opinion in *Committee for Nuclear Responsibility* focuses on scientific, not lay, opinion.[210] The rule as expressed by the court of appeals is open to a restrictive reading that would allow agencies preparing statements merely to mention the existence of dissenting viewpoints and comments, without lengthy discussion in the statement itself of the substance of the disagreement. On the other hand, the Arkansas court would not merely require that the impact statement "mention" dissenting opinion; it says that the statement should "contain" and "set forth" dissenting opinion, even nonmeritorious opinion, so that the record will be complete.

Yet a third formulation, developed specifically for comments which the public prepares, implies an even fuller use for comments than would be required by the two leading opinions on the issue. In *Lathan v. Volpe*[211] the court said on remand:

> The public may also raise environmental questions by way of comment to the draft impact statement. Since the final impact statement must respond to these comments, as well as to the comments of government agencies, environmental harm which might have been overlooked by highway officials may be brought to their attention. For this reason, highway officials must give more than cursory consideration to the suggestions and comments of the public in the preparation of the final impact statement. The proper response to comments which are both relevant and reasonable is to either conduct the research necessary to provide satisfactory answers, or to refer to those places in the impact statement which provide them. If the final impact statement fails substantially to do so, it will not meet the statutory requirements.[212]

---

[209] 325 F. Supp. at 759, 1 ELR at 20141.

[210] 463 F.2d at 787, 1 ELR at 20470.

[211] Appendix B.

[212] 350 F. Supp. at 265, 2 ELR at 20547. A dictum imposing research requirements appears in the August 4, 1972 opinion in Daly v. Volpe, Appendix B, 350 F. Supp. at 261, 3 ELR at 20033. Sierra Club v. Froehlke (Trinity River–Wallisville Dam), Appendix B, would appear to endorse this requirement as well. The Trinity River court, however, frames the issue in

EPA's Responsibility for Comments Under the Clean Air Act

Congress added §309 to the Clean Air Act in 1971 to make explicit that the administrator of the Environmental Protection Agency has a duty to comment in writing upon the possible environmental impacts in EPA's area of authority on all agency-originated legislative proposals, all proposed agency guidelines, all new federal construction projects, and all other proposed federal action to which §102(2)(C) of NEPA applies. After his review, the section continues, the administrator must make his comments public and, if he has found the impact of the proposed action to be environmentally "unsatisfactory," he must publish his finding and "refer" the matter to the CEQ.

Sec. 309. (a) The Administrator shall review and comment in writing on the environmental impact of any matter relating to duties and responsibilities granted pursuant to this Act or other provisions of the authority of the Administrator, contained in any (1) legislation proposed by any Federal department or agency, (2) newly authorized Federal projects for construction and any major Federal agency action (other than a project for construction) to which section 102(2) (C) of Public Law 91–190 applies, and (3) proposed regulations published by any department or agency of the Federal Government. Such written comment shall be made public at the conclusion of any such review.

(b) In the event the Administrator determines that any such legislation, action, or regulation is unsatisfactory from the standpoint of public health or welfare or environmental quality, he shall publish his determination and the matter shall be referred to the Council on Environmental Quality.[213]

Only one case has discussed §309 to date. In *National Forest Preservation Group v. Butz*[214] the court stated that plaintiffs had argued that the administrator had to prepare an "environmental impact statement" under the Clean Air Act for inclusion in the NEPA statement that had been prepared by the Forest Service on the contested "Big Sky" land exchange. Presumably, the court was referring to the administrator's obligations under §309. In finding "irregularities in procedure . . . not critical to compliance," the court relied upon three extenuating circumstances. First, the public had had

terms of *deference* to expert commenting agencies. This court would apparently require the agency preparing the statement to accept the facts, opinions, and recommendations for further research of "environmental" agencies as virtually conclusive.

[213] Clean Air Act, 42 U.S.C. §1857 *et seq. as amended.* Section 309 was enacted in the Clean Air Amendments of 1970, Pub. L. No. 91–604 (Dec. 31, 1970).

[214] Appendix B.

ample opportunity to review the Forest Service impact statement before the decisions by the chief forester and the secretary of agriculture, although admittedly the statement had not been prepared in time for use by the forest supervisor and the regional forester. Second, and cryptically, "the E.P.A. Administrator was afforded an opportunity to issue a statement." Third, no objection was made during the administrative procedure to the absence of the administrator's "statement."[215]

The opinion may be criticized in several respects. The NEPA statement was not prepared in time to accompany the proposal through "existing agency review processes." A statement was not present at every phase of decision making. (See chapter IV, page 78.) The court does not appear to have grasped the import of §309, if it was aware of the section at all. If the court was speaking of the administrator's obligations under §309, then it should have required the defendants to obtain a §309 comment from the administrator. The holdings discussed above, especially the one in *Natural Resources Defense Council v. Morton,*[216] appear to require such a result. Finally, the requirement of strict procedural compliance cannot be compromised by allowing late or merely "substantial" compliance. (See chapter III, pages 51 *ff.*) Only the argument that plaintiffs did not exhaust their administrative remedies has some merit. (See chapter II, pages 45 *ff.*)

The dearth of case law on §309 does not obscure the importance of its requirements. For this reason we briefly discuss its legislative history, its scope, and its promise if implemented properly in the future.

The short legislative history suggests that by enacting §309 Congress intended to take full advantage of the debate and differences of opinion that will occur between the "developmental" agencies and "environmental" EPA. The report of the Senate Public Works Committee focused upon the role of the expert "environmental agencies" in the preparation of impact statements, stressing that "it is essential" for mission-oriented agencies to have access to that expertise.[217] Senator Muskie, in submitting the conference report to the Senate, emphasized another aspect of Congress' intent:

---

215 343 F. Supp. at 702, 2 ELR at 20574.

216 Appendix B.

217 S. REP. No. 91–1196, 91st Cong., 2d Sess. 43 (1970) (hereinafter cited as Senate Report).

The conference agreement thus removes the ambiguity concerning the public release of such reviews and comments which has emerged during the debate on appropriations for the supersonic transport. Those comments must be made public when the Environmental Protection Agency completes its review—not when the environmental impact agency decides the public should be informed.[218]

Additional parts of the legislative history indicate that §309 was intended to make EPA more of a "watchdog" than merely a repository of applied expertise. At Administrator Ruckelshaus' confirmation hearings, Senator Muskie said that §309 "makes you a self-starter, whenever you, unilaterally, see an environmental risk. You are given the responsibility to raise the red flag."[219]

The scope of the administrator's authority to comment is as broad as the agency's authority in the areas of air, water, pesticides, radiation, solid wastes, and noise. "Matter relating to" this wide scope of authority will occur in most impact statements. Further, the section-by-section analysis of the Clean Air Act states that the administrator is to comment when the impact in question relates directly *or indirectly* to the secretary's—now the administrator's—authority.[220] In keeping with its wide scope of authority, EPA has indicated that

---

[218] 116 CONG. REC. S. 20602 (daily ed. Dec. 18, 1970). The background of the SST reference is given in a Comment in the ENVIRONMENTAL LAW REPORTER:

> During hearings on the nomination of William Ruckelshaus as Administrator of EPA, Senator Muskie expressed concern that environmental improvement agencies gave oral, unwritten comments to the Department of Transportation on its draft impact statement on the supersonic transport aircraft (SST). Senate Comm. on Public Works, 91st Cong., 2d Sess., *Nomination of William Ruckelshaus* 16 (Comm. Print 1970). Senator Muskie cited §102(2)(C) of NEPA and asserted that the intent of NEPA was to have the comments of "environmental" agencies accompany the developmental agencies' draft impact statements throughout the decision process. Clearly upset that the Senate was being denied the information which §102(2)(C) apparently guaranteed, Senator Muskie pressed Mr. Ruckelshaus to spell out how EPA would interpret its duty to comment on draft statements in the future. Mr. Ruckelshaus pointed out that NEPA was ambiguous about how agency comments should be handled and that the act did not clearly require that comments be made public before the final decision was made (p. 18). Senator Jackson, who was present at the hearing, then confirmed what Mr. Ruckelshaus said about NEPA's ambiguities (p. 19). Later, Senator Muskie cited §309 (then §310) as resolving the ambiguities inherent in §102 of NEPA. Comment, *Section 309 of the Clean Air Act: EPA's Duty to Comment on Environmental Impacts,* 1 ELR 10146, 10147 (September 1971).

[219] Senate Report, *supra* note 217 at 45. Further analysis is contained in the Comment *supra* note 218.

[220] Senate Report, *supra* note 217 at 66.

it is attempting to comment on all statements forwarded to it, the comments increasing in length and complexity with EPA's increasing expertise in the subject matter.[221]

Apparently, three categories of nonmajor federal activities, which do not require impact statements under NEPA, must be the subject of "309 statements," if such actions have an "environmental impact." These three categories are proposed legislation, newly authorized federal construction, and proposed agency regulations. This observation does not derive its importance from the likelihood that a large number of federal actions will be outside NEPA, because the obvious trend is toward allowing very few exceptions. Rather, the additional authority is important because EPA need not wait to receive an agency draft statement before "raising the red flag," since it must comment in any event. Section 309 is all the more crucial in this regard because EPA apparently does not have authority to require a sister agency either to prepare a statement or to produce an existing one. The almost total absence of impact statements in two of the three relevant categories—legislative proposals and proposed guidelines—suggests that "309 statements" may afford the only means in some cases of initiating discussion of potential environmental impacts. Unfortunately, EPA has not yet begun to exercise its authority in this manner.

Nor have EPA and the CEQ made full use of the authority contained in §309(b), which requires the administrator to "refer" proposals to CEQ whenever he determines that the proposed legislation, action, or regulation is "unsatisfactory from the standpoint of public health or welfare or environmental quality."

> Unless §309(b) is to be read as simply declaratory of §102(2) (C) of NEPA, which already requires mere transmittal of comments with the final 102 statements to the CEQ, the "referral" and the "determination" which §309(b) mentions must be read as requiring the Administrator to do something more. Otherwise, §309(b) is nugatory. Section 309(b) apparently requires the Administrator to draw the CEQ's attention to proposed federal action with serious environmental impacts so that the final decision will be made on a higher administrative level than the agency level at which it would ordinarily be made. . . . If an agency disregards the Administrator's finding and proceeds to decide a matter and take action, even if it has waited the full 90 days which CEQ guidelines provide, a negative comment from the Administrator is apparent grounds for restraining the agency until the CEQ has had an

[221] Comment, *supra* note 218 at 10149.

opportunity to arrange for a higher-level administrative or Presidential resolution of the controversy.[222]

After a slow start, EPA published guidelines and began in early 1972 to give notice in the *Federal Register* of the availability of comments prepared under §309. Its comments are ranked on a four-point scale:

Category I. General agreement/lack of objections. EPA has no objection to the action, it suggests only minor changes in the statement or action, or it has no comments to offer.

Category II. Inadequate information. The draft statement does not contain sufficient information for EPA to assess fully the action's environmental impact. EPA requests more information about potential hazards which have been identified in the statement, or it asks that potential hazards not addressed in the draft statement be addressed in the final statement.

Category III. The proposed action, as described in the statement, needs major revisions or major additional safeguards to adequately protect the environment.

Category IV. EPA believes the proposed action is unsatisfactory because of its potentially harmful effect on the environment. Furthermore, the agency believes that environmental safeguards built into the project may not adequately protect the environment. Therefore the EPA recommends that alternative actions, including no action at all, be analyzed further.[223]

Richard Liroff of the Brookings Institution has analyzed how EPA categorized statements received over the nine months between November 1971 and July 13, 1972. He found that 48.1 percent of the statements reviewed fell into category I, indicating general acceptability, while the remainder did not include adequate information upon which to base a proper review. In fact, 7.3 percent of the proposed actions needed to have basic environmental safeguards added, and 0.3 percent were deemed "unsatisfactory." His other findings and conclusions, however, cast further doubt upon this already spotty agency record.[224] The Environmental Protection Agency's review, while it has yet to reveal its full potential, nevertheless shows

---

[222] Comment, *supra* note 218 at 10150.

[223] Comment, *supra* note 218 at 10151. As we go to press, EPA has amended these guidelines to provide for review of final statements with a modified rating scheme.

[224] R. Liroff, *EPA Comments on Environmental Impact Statements: One Indicator of Administrative Response to the National Environmental Policy Act of 1969* (unpublished paper prepared at the Brookings Institution, October 1972).

that it might eventually play a role far more important than that of the CEQ in improving the quality of the impact statement process through substantive, regular criticism "on the record."

The Public and the Commenting Process

While a strict reading of the statute might limit the commenting process to governmental bodies, Executive Order 11514[225] and the CEQ Guidelines[226] evince a desire to allow private individuals and organizations to participate in the procedure. The cases hold that the agencies have an obligation to seek comments from the public and that final impact statements must take these comments into account. Thus, for example, in *Brooks v. Volpe* and *Daly v. Volpe*[227] the court held that NEPA's procedures had not been complied with where federal defendants had failed to give public notice and had otherwise failed to provide an opportunity for public comment on the impact statements which had been prepared. The "right" of the public to comment has even been confirmed in the most regrettable NEPA district court case to date. In its conclusions of law in *Jicarilla Apache Tribe of Indians v. Morton*,[228] the court said that "the public has a right to participate in drafting the final impact statement by submitting comments and environmental information upon any alleged factual or legal matter in the draft impact statement."[229] Several other cases hold that public comments must be appropriately considered and weighed in the final statement. These were discussed earlier in connection with the general duty to take §102(2)(C) comments into account.

The cases suggest that such comment by the general public may be necessary to ensure that impacts and alternatives have been fully explored. Thus *Greene County Planning Board v. Federal Power Comm'n*[230] stated that "Congress has compelled agencies to seek the aid of all available expertise,"[231] while *Scherr v. Volpe*[232] observed that "the Act requires the agencies of the federal government to assemble all of the pertinent information."[233] As discussed above,

[225] *Supra* note 190.
[226] Appendix C.
[227] Appendix B.
[228] Appendix B.
[229] —— F. Supp. at ——, 2 ELR at 20295.
[230] Appendix B.
[231] 455 F.2d at 420, 2 ELR at 20020.
[232] 336 F. Supp. 882, 2 ELR 20068 (W.D. Wisc.), 336 F. Supp. 886, 2 ELR 20068 (W.D. Wisc. 1971), 466 F.2d 1027, 2 ELR 20453 (7th Cir. 1972).
[233] 336 F. Supp. at 889, 2 ELR at 20070.

strong support is provided by the decision on remand in *Lathan v. Volpe*,[234] where the court read into the Act a requirement that public comments be solicited and considered in good faith.

Must hearings be held to ensure public participation in NEPA environmental assessments? The CEQ final guidelines indicate that agency procedures established to ensure "the fullest practicable provision of timely public information . . . in order to obtain the views of interested parties . . . shall include, whenever appropriate, provision for public hearings."[235]

The question of whether to hold hearings is thus delegated to the various agencies. So far no case has interpreted NEPA to require hearings where such would not have to be held under the provisions of other statutes or where they were not provided for in agency procedures. *National Helium Corp. v. Morton*[236] expressed the views of the Tenth Circuit on this point.

In oral arguments the appellees have expressed a desire for extensive administrative proceedings. We do not see any such requirement. This is an intra-department matter in which the Secretary fulfills his obligations by following the mandate of the NEPA. Neither the APA nor the NEPA compels him to appoint an examiner and conduct hearings. Indeed the Department has NEPA procedures in its manual. He ought to at least follow these. These is no indication that Congress in enacting the NEPA intended to impose extensive procedural impediments to Department action.[237]

However, where agencies make a current practice of such hearings, NEPA has been held to require that such hearings extend to environmental issues, and that a statement be prepared in advance of the hearings. In *Greene County*,[238] where plaintiffs challenged FPC implementation procedures, the court ruled against the commission. "We conclude that the Commission was in violation of NEPA by conducting hearings prior to the preparation by *its staff* of its own impact statement."[239] Moreover, the integration of the §102 process with ordinary hearing procedures of the agency must allow the intervenors a reasonable opportunity to comment on the statement, and to cross-examine interested parties.

---

[234] Appendix B, 350 F. Supp. 262, 2 ELR 20545.
[235] CEQ Guidelines, §10(e), Appendix C.
[236] Appendix B.
[237] 455 F.2d at 656, 1 ELR at 20480.
[238] Greene County Planning Board v. Federal Power Comm'n, Appendix B.
[239] 455 F.2d at 422, 2 ELR at 20021.

Since the statement may well go to waste unless it is subject to the full scrutiny of the hearing process, we also believe that the intervenors must be given the opportunity to cross-examine both PASNY and Commission witnesses in light of the statement.[240]

In the August 4, 1972 opinion in *Daly v. Volpe*,[241] however, the court refused to order a new highway location hearing merely because the public had been given inadequate notice of defendants' intention to favor a particular route and because defendants had otherwise failed to give the public an opportunity for meaningful comment on the statement. The court did, however, impose lengthy judge-make notice and participation procedures to ensure that NEPA's purposes were met, even if outside the usual hearing process.[242]

A final case concerning public comments and public hearings deserves to be discussed, although it involves the agency's threshold determination whether or not to prepare a statement at all and although it rests primarily upon an interpretation of §102(2)(B), rather than §102(2)(C). In the Second Circuit's second decision remanding *Hanly v. Kleindienst*,[243] the court stated that the GSA had to show (1) how it had informed the public about its proposed construction project and (2) how it had developed and used procedures for actually receiving and considering public comments on the threshold determinations whether to prepare a statement.[244] The court suggested that the agency consider hearings on the issue. In so ruling, the court relied in part on §102(2)(B), which requires federal agencies to develop methods and procedures for taking nonquantifiable environmental factors into account alongside traditional economic factors.

The general public availability of agency comments is important, not only so that the public's own participation in the impact statement process may be more meaningful, but also so that the NEPA objective of open discussion and debate may be served. Section 102(2) (C) says that the statement and comments "shall be made available to the President, the Council on Environmental Quality and to the Public as provided by section 552 of title 5, U.S.C. [the Freedom of Information Act]." The CEQ Guidelines amplify this requirement in several ways: action is not to be taken ("to the

---

[240] 455 F.2d at 422, 2 ELR at 20021.
[241] Appendix B.
[242] 350 F. Supp. at 260–61, 3 ELR at 20032–33.
[243] Appendix B (Hanly II).
[244] —— F.2d at ——, 2 ELR at 20723.

maximum extent practicable") within 90 days of the availability of the draft and 30 days of the final; agencies are to consult the CEQ should they seek waiver of these periods, as was done by the Interior Department in *NRDC v. Morton*;[245] agencies are to "develop procedures to insure the fullest practicable provision of timely public information . . ., including, whenever appropriate, provision for public hearings," and statements and comments are to be available at the clearinghouses on the state, regional, and metropolitan levels established by OMB Circular 71–1.

The point on which significant litigation has arisen is the applicability of the Freedom of Information Act to statements, and particularly to comments, which the CEQ Guidelines indicate are not to be withheld as "interagency memoranda."[246] This provision was severely tested in the Project Cannikin litigation where the AEC's Amchitka Island nuclear test was challenged on environmental grounds. This case, *Committee for Nuclear Responsibility v. Seaborg*,[247] reached the District of Columbia Circuit on appeal from the district court's grant of summary judgment for defendants. The circuit court reversed, indicating that plaintiffs were entitled to carry on discovery proceedings against the AEC in order to show that the statement that had been issued did not set forth all views or include opposing comments:

> Plaintiffs also alleged the existence of reports by federal agencies recommending against Cannikin specifically because of potential harm to the environment. NEPA clearly indicates that the agency responsible for a project should obtain and release such adverse reports. If these reports exist, and they are not subject to some statutory exemption, plaintiffs must prevail on this contention as well.[248]

The second case in the series came on a government appeal on its claim of executive privilege. In relevant part, this Amchitka opinion provided:

> Normally this balancing process [weighing plaintiffs' interest in the documents against the government's in secrecy] will require an excision, from the factual data in the documents, of material which consists purely of advice, deliberations and recommendations. Certain of the documents, though, may constitute agency comments whose existence prior to the issuance of the impact statement in final form required their inclusion in the statement by virtue of section 102 of NEPA. . . . [The CEQ

[245] 337 F. Supp. 170, 2 ELR 20071.
[246] CEQ Guidelines, §9(f), Appendix C.
[247] Appendix B.
[248] 463 F.2d at 788, 1 ELR at 20470.

Guidelines] specify that the exemption for inter-agency memoranda is not applicable concerning the comments of federal agencies. . . .[249]

The third appeal in the case came once more from plaintiffs, who sought an injunction against the test and the production of some documents withheld by the district judge after *in camera* inspection. Both motions were denied on the ground of lack of time for consideration, and the case was appealed to the Supreme Court. Over three dissents, the Court denied the appeal.[250] Justice Douglas, in dissent, indicated his view of the comment procedure:

> Disclosure of these statements to the public by any federal agency which has "special expertise with respect to any environmental impact involved" is indeed required by §102(C) of the Act. And the courts have consistently held that a defect in the Impact Statement presents a justiciable question and is the basis for equitable relief.[251]

The Short Commenting Period

A final problem with the NEPA commenting process concerns the deadlines set by the CEQ Guidelines on draft and final impact statement preparation. Because litigation has not directly dealt with the issue, our discussion will be brief.

The CEQ's final guidelines allow the federal agencies to take action 90 days after the draft statement is prepared and 30 days after the final. The periods may run concurrently.[252] If the government is determined to act quickly, which it must be admitted has not often been the case, the short commenting period of 30–45 days allows for very little, if any, informed analysis and debate.

The CEQ has proved to be its own best critic in this matter. The third annual CEQ report states:

> Agencies and private groups whose interests and expertise put them frequently in a commenting role on draft 102 statements have complained at times of the difficulty of preparing helpful comments in only 30 to 45 days. For example, the Department of the Interior is asked to comment on hundreds of proposed actions affecting land use and fish and wildlife values. EPA, with its expertise in pollution control, faces a similar situation. EPA's workload is increased by section 309 of the Clean Air Act.[253]

The report goes on to impeach its recommended solution:

---

[249] 463 F.2d at 794, 1 ELR at 20531.
[250] 404 U.S. 917, 1 ELR 20534 (1971).
[251] 404 U.S. at 921, 1 ELR at 20535.
[252] CEQ Guidelines, §10(b), Appendix C.
[253] CEQ Third Annual Report, *supra* note 76 at 237.

One answer to this problem, obviously, is for the commenting entities to add the staff and other resources to handle the commenting task. The opportunity to make Federal decisionmaking better informed and more carefully planned warrants the necessary manpower. However, even with adequate resources, it is often impossible to prepare comments in 30 days that will do justice to a draft statement that may have taken years to prepare.[254]

Yet the CEQ steadfastly insists that the solution is not simply more time:

It is probably impracticable to solve the time problem by an across-the-board extension of the minimum period between circulation of the draft statement and agency action. A significant extension would impose a delay incompatible with the nature of some Government programs.[255]

Possibly there is an answer to CEQ's argument that a longer comment period would cause unjustified delay. While "some" governmental programs might be delayed, the majority would not, as a quick survey of major federal actions on which statements have been filed to date indicates. The majority involve projects that have been years in maturing, not days. Thus if "some" government programs really means "few," then perhaps the proper response should be for CEQ to require various agencies to tailor their guidelines to fit longer review periods, except in the "few" exceptional cases. The CEQ might assist in this process by suggesting categories of action suitable for commenting periods of 90 days, 6 months, and a year or more.

THE CONSEQUENCES OF FAILURE TO COMPLY: JUDICIAL RELIEF UNDER NEPA

Except where there have been extraordinary equities to the contrary, the courts have enjoined projects proceeding in violation of NEPA until the mandates of the Act have been met. Plaintiffs in NEPA cases have usually sought preliminary injunctions, where the courts normally balance an array of factors and grant or deny relief on the overall balance struck. In some NEPA suits, however, the courts have been able to leapfrog the factor of plaintiffs' likelihood of eventual success regarding NEPA's applicability, because even at the hearing on the preliminary injunction the court was virtually certain what its final ruling on the law would be. In other suits, the courts have held that the factors to be balanced have been

---

[254] *Id.*
[255] *Id.*

fundamentally altered by NEPA's requirements or its legislative purposes, so that the traditional balancing of equities on the preliminary injunction becomes heavily weighted by law in favor of granting relief. Thus government or private investment of time, money, or other resources in ongoing projects has not succeeded in tipping the balance toward allowing the projects to continue. Whatever success defendants have had in this regard has usually been in convincing the courts that because of the degree of completion of the projects NEPA did not apply as a matter of law (see chapter V), that substantial compliance had been achieved (see chapter III), or that compliance was only days away (see later discussion in this section).

Once the basic question of law regarding NEPA's applicability has been answered, the majority of the courts have granted the relief requested without discussion, apparently assuming that disposition of the question of law ended the courts' inquiry.[256] To the extent that there is a rule, it may be stated in quite conclusory language. As the court in *Bradford Township v. Illinois State Toll Highway Authority* said, "failure to comply [with NEPA] . . . is basis for an injunction," or, putting it another way, "judicial relief is available to correct failure on the part of a federal agency to follow the procedural requirement under NEPA."[257] Additional cases have explained the grant of relief by observing that further investment of time and money may increase environmental harm, narrow alternatives, and otherwise prevent the decision maker from conducting an unbiased assessment of the project once the impact statement is completed. In *Lathan v. Volpe*,[258] the Ninth Circuit vacated the district court's denial of a preliminary injunction and remanded with explicit instructions regarding compliance before the injunction restraining work on an interchange in the interstate highway system could be lifted. The circuit court observed that if it accepted defendants' contention that a statement did not have to be prepared until the final approval stage, it might well be too late to adjust existing plans to minimize adverse environmental effects. The Ninth Circuit paraphrased the reasoning of the District of Columbia Circuit in *Calvert Cliffs'*:

> In the language of NEPA there is likely to be an "irreversible and irretrievable commitment of resources," which will inevitably restrict the

[256] 463 F.2d 537, 2 ELR 20322 (7th Cir.), *cert. denied*, 41 U.S.L.W. 3313 (Dec. 5, 1972).
[257] 463 F.2d at 539, 2 ELR at 20323.
[258] Appendix B.

[highway officials'] options. Either the [highway planners] will have to undergo a major expense in making alterations in a completed [plan] or the environmental harm will have to be tolerated. It is all too probable that the latter would come to pass.[259]

The Fourth Circuit in *Arlington Coalition on Transportation v. Volpe*,[260] faced with a case involving facts similar to those in *Lathan*, reversed the district court's denial of a preliminary injunction by relying in part on the view that "further investment of time, effort or money . . . would make alteration or abandonment of the route increasingly less wise and, therefore, increasingly unlikely."[261] The Fourth Circuit put great stress upon NEPA's requirement that the decision maker must consider modifying the remainder of the project in light of the facts unearthed by the impact statement. Investment prior to and during such consideration, the court reasoned, would slowly diminish the available options so that if consideration took long enough, they would vanish. For its view that NEPA requires suspension so that options may be preserved, the circuit court cited language in §102(2) (C)—e.g., "proposed action," "should the proposal be implemented," "alternatives to the proposed action"—and the declaration of policy in §101. The court also cited *Lathan* and *Calvert Cliffs'*, and the following language from the Tenth Circuit's opinion in *National Helium Corp. v. Morton*:[262]

The statute does not limit the authority of any governmental agency in any permanent or conclusive manner. It does, however, contain a mandate that action can be taken only following complete awareness on the part of the actor of the environmental consequences of his action and following his having taken the steps required by the Act.[263]

Reasoning similar to that discussed here was employed in *Keith v. Volpe* (the Los Angeles Century Freeway case), *Northside Tenants' Rights Coalition v. Volpe* (citing *Lathan, Calvert Cliffs'*, and *Arlington Coalition*), and *Brooks v. Volpe*.[264] In *Brooks v. Volpe*, the court added to NEPA's purposes that of protecting the environment for future generations in order to justify relief that "sometimes may appear too harsh" when applied to ongoing projects.[265]

---

[259] 455 F.2d at 1121, 1 ELR at 20605.
[260] Appendix B.
[261] 458 F.2d at 1327, 2 ELR at 20162.
[262] Appendix B.
[263] 455 F.2d at 656, 1 ELR at 20480.
[264] All in Appendix B.
[265] 350 F. Supp. at 283, 2 ELR at 20709.

Where courts have felt constrained to explain how the relief granted relates to the requirements for a preliminary injunction, they have either held that the questions of law are determinative, that the circumstances of an injunction under NEPA are so special that the conventional requirements for preliminary relief do not apply, or that when applied to a NEPA suit the requirements are heavily weighted by the Act itself in favor of the injunction.

We have already discussed a number of cases in which the basic legal question of NEPA's applicability was determinative. Why this is so is partially explained in *Sierra Club v. Mason*,[266] where the court pointed out that when "the issues concerning the requirement of an impact statement do not appear to involve any disputed facts, the case may turn out to be in no significantly different posture than when the merits are formally reached,"[267] so that delay until the merits are considered at a full hearing really means delay until the statement is prepared. In these circumstances the question of law may automatically determine the relief granted, because the issue is clear and the relief sought becomes equal to that which a final order might bring. In *Mason* the court did in fact conclude its opinion by stating that the preliminary injunction issued in the case would become permanent in 20 days unless the parties notified the court of any material factual disputes.[268]

One circuit court has ruled that the balancing of equities traditionally used in deciding whether preliminary relief should be allowed is inappropriate in NEPA cases. In the Ninth Circuit decision discussed above in *Lathan v. Volpe*,[269] the court applied the Supreme Court's decision in *United States v. San Francisco*[270] to the "exceptional case" before it.

> We are satisfied that this case does not call for a balancing of equities or for the invocation of the generalities of judicial maxims in order to determine whether an injunction should be issued. . . . The equitable doctrines relied on do not militate against the capacity of a court of equity as a proper forum in which to make a declared policy of Congress effective.[271]

The Ninth Circuit concluded its discussion by saying:

---

[266] 351 F. Supp. 419, 2 ELR 20694 (D. Conn. 1972).
[267] 351 F. Supp. at 426, 2 ELR at 20696.
[268] 351 F. Supp. at 429, 2 ELR at 20698.
[269] Appendix B.
[270] 310 U.S. 16 (1940).
[271] 310 U.S. at 30–31.

This is one of those comparatively rare cases in which, unless the plaintiffs receive *now* whatever relief they are entitled to, there is danger that it will be of little or no value to them or to anyone else when finally obtained.[272]

Northside Tenants' Rights Coalition v. Volpe[273] is in accord, but without discussion.

Other cases take the position that the traditional standards governing preliminary relief apply, but that the value to be accorded various equities has been substantially altered by NEPA itself. As the circuit court in *Natural Resources Defense Council v. Morton*[274] remarked, in charting the role of a court reviewing a lower court's ruling on a preliminary injunction, "not only the avowed forecast as to the probability of success on the merits, but also the analysis of the injury to either or both parties, the public interest, and the balancing of interests, may well come to depend upon an assumption of underlying legal premise."[275]

The Seventh Circuit in *Scherr v. Volpe*,[276] in affirming the grant of a preliminary injunction below, disposed of defendants' contention that in order to obtain a preliminary injunction, plaintiffs would have to show how the highway construction in question would cause irreparable environmental harm, by holding that it was impermissible to shift this burden from government to the private plaintiffs.

What this argument attempts to do is shift the burdens of considering and evaluating the environmental consequences of particular federal actions from the agencies Congress intended to bear them to the public, the beneficiary of this legislation. If these agencies were permitted to avoid their responsibilities under the Act until an individual citizen, who possesses vastly inferior resources, could demonstrate environmental harm, reconsideration at the time by the responsible federal agency would indeed be a hollow gesture [citing *Arlington Coalition* and *Calvert Cliffs'*].[277]

Having rejected the government's attempt to shift the burden to the private parties, the circuit court goes on to describe the kind of irreparable harm that plaintiffs must show before an injunction will issue. Again citing *Calvert Cliffs'*, the court ruled that the "careful

[272] 455 F.2d at 1117, 1 ELR at 20603.
[273] Appendix B.
[274] Appendix B.
[275] 458 F.2d at 832, 2 ELR at 20031.
[276] Appendix B.
[277] 466 F.2d at 1034, 2 ELR at 20456.

and informed decision-making process" mandated by NEPA would be lost forever if plaintiffs were not allowed to prevail after showing a probability of success on the merits.[278]

In *Sierra Club v. Mason*[279] the district court stated that *normally* it would balance the environmental benefits that plaintiffs could show would be protected by the requested injunction against the risks and costs which defendants alleged would be created by enjoining the channel dredging project at issue in the case. However, in view of the requirements of NEPA, and unless "extraordinary equities" existed on the government's side, the court would accept showings of "substantial risk of damage to the environment" and of a "reasonable possibility that adequate consideration of alternatives might disclose some realistic course of action with less risk of damage."[280] Nor did the court overlook the possibility that even if no modifications are in fact possible, an important information function would still be served by requiring compliance.[281]

A number of decisions allowing the preliminary injunction also show leniency toward defendants who would incur considerable losses before compliance could be completed. In four such cases the agencies were allowed a short reprieve to give them time to comply before the injunction went into effect. In *Goose Hollow Foothills League v. Romney*[282] the court stayed its order for 90 days to allow HUD to finish its statement, since no irreparable environmental harm was likely to take place in the interim. In *City of New York v. United States*[283] defendants were also allowed 90 days in which to initiate compliance. Such a reprieve made little difference on the facts, because the railway which was the subject of the abandonment challenged by plaintiffs was kept in operation by another company. Finally, in *Hanly v. Mitchell*,[284] the first decision of the Second Circuit gave the General Services Administration 30 days to make its "preliminary determination" whether NEPA applied.

Only a few cases exist where NEPA was held to apply but the injunction did not issue. In two cases, however, the courts accepted the costs that would be incurred if the project were enjoined as part

---

[278] *Id.*
[279] Appendix B.
[280] 351 F. Supp. at 427, 2 ELR at 20697.
[281] *Id.*
[282] Appendix B.
[283] Appendix B.
[284] Hanly v. Mitchell, Hanly I, Appendix B.

of the test of whether NEPA applied to the federal action at issue. The four-point test for NEPA's applicability to ongoing projects set out in *Environmental Law Fund v. Volpe*[285] included, as a factor to be weighed, the cost to the state were NEPA compliance to be ordered and the project enjoined. *Conservation Society of Southern Vermont v. Volpe*[286] also applied this test. Other cases that are sometimes cited as authority that courts will not enjoin ongoing projects pending NEPA compliance also involve the question whether as a matter of law NEPA applies to the ongoing project. The applicability of NEPA to ongoing projects was discussed in chapter V and is not pursued here.

The well-known case of *Greene County Planning Board v. Federal Power Comm'n*[287] included a situation where NEPA was held to apply but no injunction was issued. Although the court ordered that the commission's approval of one of the three transmission lines at issue had to be preceded by compliance with NEPA, the court refused to undo earlier commission approval of the basic project and the two remaining lines because a final commission order approving the projects had been issued, construction was already far advanced, plaintiffs had failed to make proper objection in the earlier administrative proceedings, and the commission had already complied in part with NEPA's mandate.[288]

The case of *Environmental Defense Fund v. Corps of Engineers* (Truman Dam)[289] also illustrates a court's refusal to enjoin all aspects of an ongoing project pending compliance with NEPA because considerable equities existed on the side of defendants: defendants had agreed to refrain from action which might have environmental impact until the final statement was prepared, no adverse environmental consequences would be caused by the few actions allowed to go forward in the interim, and compliance would be completed within a few months. (The case is criticized on page 185.) These same criticisms apply to the recent district court opinion in *Environmental Defense Fund v. Armstrong*,[290] where the court refused to grant a preliminary injunction against further work on the New Melones Dam because inadequacies in the impact statement could be remedied prior to commencement of actual construction.

---

285 Appendix B.
286 Appendix B.
287 Appendix B.
288 455 F.2d at 425, 2 ELR at 20023.
289 Appendix B.
290 Appendix B.

VII

NEPA and Federal Decision Making

OVER THE past three years the courts have had many opportunities to interpret NEPA. The Act has been involved in 149 separate litigations, some of which have produced several opinions. Yet chapters IV, V, and VI abundantly confirm that the bulk of these interpretations focuses on the one short action-forcing provision set out in §102(2) (C). As important as that section is, its role is nevertheless an essentially subservient one. An archive of disregarded assessments that bring about no real improvement in federal decision making might satisfy §102(2) (C), but would fail NEPA as a whole. Aware of this possibility, the courts have endorsed a wider judicial role in ensuring that final agency decision making actually reflects NEPA's substantive policy.

In addition to interpreting the information and disclosure requirements of §102(2) (C), the courts have spelled out how that information must be "considered" by the agency in making its final decision. The use to which environmental information is put in decision making is fully reviewable, and the impact statement plays an important role as part of the reviewable record of consideration. Furthermore, several courts have held that courts may review agency decisions to determine if they are in accord with NEPA's substantive policy.

The role in federal decision making of action-forcing requirements other than §102(2) (C) has been discussed by the courts, which in a few instances have partially rested injunctive relief on them. The key provisions are §§102(2) (A), (B), and (D), and occasionally (G) and (E). Subsection 102(2) (D)—the "alternatives" section—is much like §102(2) (C) in function, while §§102(2) (A) and (B) impose new and totally different requirements on the agencies. These require a qualitative change in decision-making processes which the courts may find difficult to enforce.

Finally, we consider NEPA when the agencies are not resisting its application but are actually attempting to rely upon it for authority to take an environmentally protective action. In theory NEPA is a more potent force in the hands of government than in the hands of parties suing the government. Its potential in this respect has not yet begun to be realized, but cases like *Zabel v. Tabb*,[1] the first case in which a federal agency sought to assert its NEPA mandate, in the long run may be more important for better federal decision making than the suits against the government.

CALVERT CLIFFS', BALANCING, AND THE REVIEWABLE OBLIGATION TO CONSIDER IMPACTS

The District of Columbia Circuit opinion in *Calvert Cliffs'*[2] held that §102 duties were subject to a strict standard of compliance and that agencies were further obligated to "consider" in good faith the environmental information which they developed. In the court's opinion, "consideration" required a trading off or balancing of environmental factors against economic and technical ones. Other decisions endorse this requirement and go further than *Calvert Cliffs'* in defining the use of the §102 process to create a record for reviewing agency consideration of environmental values. The "balancing analysis," however, has inherent drawbacks, which are discussed in the final portion of this subsection.

Calvert Cliffs'

Before Judge Skelly Wright's opinion for the District of Columbia Circuit in *Calvert Cliffs' Coordinating Committee v. Atomic Energy Comm'n*,[3] courts applying NEPA generally limited themselves to narrow constructions of NEPA's action-forcing requirements. In *Calvert Cliffs'* NEPA received its first comprehensive judicial analysis by a circuit court. Despite certain important modifications and extensions, the *Calvert Cliffs'* interpretation has been accepted as the definitive judicial gloss on NEPA. It has been more frequently cited, analyzed,[4] and relied upon than any other NEPA decision, as the preceding chapters have shown.

---

[1] 430 F.2d 199, 1 ELR 20023 (5th Cir. 1970), *cert. denied*, 401 U.S. 910 (1971).

[2] Calvert Cliffs' Coordinating Committee v. Atomic Energy Comm'n, 449 F.2d 1109, 1 ELR 20346 (D.C. Cir. 1971), *cert. denied*, 404 U.S. 942 (1972).

[3] Appendix B.

[4] B. Cohen and J. Warren, *Judicial Recognition of the Substantive Requirements of the National Environmental Policy Act of 1969*, 13 B.C. IND. & COM. L. REV. 685 (1972); N. Landau, *A Postscript to Calvert Cliffs'*, 13 B.C. IND. & COM. L. REV. 705 (1972); A. Murphy, *The National Environmental Policy Act and the Licensing Process: Environmentalist Magna Carta or Agency Coup de Grace?*, 72 COL. L. REV. 963 (1972); D. Tarlock, *Balancing Environmental Considerations and Energy Demands: A Comment on Calvert Cliffs'*

The specifics of the court's holding are not of principal concern here. They relate to the way in which the Atomic Energy Commission's procedures implementing NEPA failed to meet the Act's requirements.[5] The deficiencies identified in the procedures flowed from the court's finding that the §102 duties were subject to a strict standard of compliance that could only be relaxed if they conflicted

---

Coordinating Committee, Inc. v. AEC, 47 IND. L. J. 645 (1972); Comment, *Landmark Decision on the National Environmental Policy Act in Calvert Cliffs' Coordinating Committee, Inc. v. Atomic Energy Commission*, 1 ELR 10125 (August 1971); Comment, *Calvert Cliffs' Coordinating Committee v. AEC and the Requirement of "Balancing" under NEPA*, 2 ELR 10003 (January 1972); Casenote, 13 B.C. IND. & COM. L. REV. 802 (1972); Casenote, 52 BOSTON U. L. REV. 425 (1972); Casenote, 24 FLA. L. REV. 814 (1972); Casenote, 60 GEO. L. J. 1353 (1972); Casenote, 20 KANS. L. REV. 501 (1972); Casenote, 25 VAND. L. REV. 258 (1972); B. Boyer, *A Re-evaluation of Administrative Trial-type Hearings for Resolving Complex Scientific and Economic Issues*, Staff Report to the Chairman of the Administrative Conference of the United States (Dec. 1, 1971).

[5] The specific holdings of Calvert Cliffs' are set out succinctly in the headnote to the case in the ENVIRONMENTAL LAW REPORTER:

> The regulations promulgated by the AEC to comply with the procedural obligations imposed upon federal agencies by the National Environmental Policy Act of 1969 violate the Act's requirements as follows: (1) the AEC's refusal to independently review the nonradiological environmental impact of nuclear power plant operations upon which state or other federal agencies have already passed conflicts with NEPA's mandate to the relevant agency to assess the complete environmental costs of its action on a case-by-case basis; (2) the AEC's failure to require hearing board review of nonradiological environmental factors unless affirmatively raised by outside parties or staff members violates the Commission's affirmative duty to consider environmental values at every stage of the decision-making process; (3) the AEC's refusal to consider nonradiological environmental factors at hearings officially noticed before March 4, 1971 violates NEPA's mandate that such factors be taken into account by each agency to the fullest extent possible from the time the act went into effect on January 1, 1970; (4) the AEC's refusal to consider alteration of plans, backfitting or construction halts for nuclear facilities which were granted construction permits prior to the effective date of NEPA but for which operating licenses have not yet been granted, so as to allow for interim modifications of these facilities consonant with environmental values, is inconsistent with the Commission's duty to fully consider action which will avoid environmental degradation. The case is therefore remanded to the AEC for proceedings consistent with the court's opinion. NEPA imposes a substantive duty upon every federal agency to consider the effects of each decision upon the environment and to use all practicable means, consistent with other essential considerations of national policy, to avoid environmental degradation. The act also imposes procedural obligations which assure that the substantive duty is properly performed. These procedural duties are not inherently flexible. They require agency compliance to the fullest extent unless there is a clear conflict with the agency's other statutory authority. 1 ELR 20346.

with other agency duties imposed by statute. The opinion caused a furor within the commission and in Congress because of its impact on the commission's ongoing program of nuclear facility licensing.[6] The furor subsided after passage of an amendment to the Atomic Energy Act which allows somewhat speeded-up environmental clearance for already-constructed facilities.[7]

In his opinion, Judge Wright gave full attention to NEPA's substantive requirements. He pointed out that §101(a) sets forth the Act's basic substantive policy: the federal government must "use all practicable means and measures" to protect environmental values. He also cited language from §101(b) which indicated that the substantive policy was a flexible one which "may not require particular substantive results in particular problematic instances." Finally, he contrasted the flexible policy requirements with the inflexible, strict standard of compliance which Congress established for the Act's action-forcing procedures.

With these foundations, the court then relied upon both NEPA's procedural and its substantive sections to elaborate how the fruits of the Act's lengthy §102 procedure must be used to reach a decision that accords with NEPA's substantive policy. Judge Wright stated, "perhaps the greatest importance of NEPA is to require . . . agencies to *consider* environmental issues just as they consider other matters within their mandates."[8] He further remarked:

> Only once—in §102(2) (B)—does the Act state, in terms, that federal agencies must give full "consideration" to environmental impact as part of their decision making processes. However, a requirement of consideration is clearly implicit in the substantive mandate of §101, in the requirement of §102(1) that all laws and regulations be "interpreted and administered" in accord with that mandate, and in the other specific procedural measures compelled by §102(2). The only circuit to interpret NEPA to date has said that "[t]his Act essentially states that every Federal agency shall consider ecological factors when dealing with activities which may have an impact on man's environment." *Zabel v. Tabb*, 5th Cir., 430 F.2d 199, 211 (1970). Thus a purely mechanical

---

[6] JOINT COMM. ON ATOMIC ENERGY, SELECTED MATERIALS ON THE CALVERT CLIFFS' DECISION, ITS ORIGIN AND AFTERMATH, 92D CONG., 1ST SESS. (Joint Comm. Print February 1972); SENATE COMM. ON INTERIOR AND INSULAR AFFAIRS, EFFECT OF CALVERT CLIFFS' AND OTHER COURT DECISIONS UPON NUCLEAR POWER IN THE UNITED STATES, SERIAL NO. 92–28, 92D CONG., 2D SESS. (Comm. Print 1972).

[7] 42 U.S.C. §2242, Pub. L. No. 92–307, 86 Stat. 191 (June 2, 1972). The authority granted to the AEC to issue temporary operating licenses on completion of expedited review procedures expires on October 30, 1973.

[8] 449 F.2d at 1112, 1 ELR at 20347.

compliance with the particular measures required in §102(2) (C) & (D) will not satisfy the Act if they do not amount to full good faith *consideration* of the environment. . . . The requirements of §102(2) must not be read so narrowly as to erase the general import of §§101, 102(1) and 102(2) (A) & (B).[9]

The concept of "consideration" is difficult to define; its use only once in NEPA does not help matters. Interpretations may range from the view that the agency cannot act without painstaking attempts to avoid the impacts in question, to the view that the agency decision maker must simply be aware of the impact which his decision may have. The court in *Calvert Cliffs'* defined the concept in terms of a balancing or trading off of environmental factors against economic and technical ones. Relying primarily upon §§102(2) (A) and (B), the court found that "NEPA mandates a rather finely tuned and 'systematic' balancing analysis."[10] As the court more fully explained:

NEPA mandates a case-by-case balancing judgment on the part of federal agencies. . . . The particular economic and technical benefits of planned action must be assessed and then weighed against the environmental costs; alternatives must be considered which would affect the balance of values. . . . In some cases, the benefits will be great enough to justify a certain quantum of environmental costs; in other cases, they will not be so great and the proposed action may have to be abandoned or significantly altered. . . . The point of the individualized balancing analysis is to ensure . . . that the optimally beneficial action is finally taken.[11]

The role of the courts in reviewing agency consideration of environmental factors is plainly stated in *Calvert Cliffs'* in terms of review of compliance with all of NEPA's §102 provisions, especially §§102(2) (C) and (D). "The requirement of consideration 'to the fullest extent possible' sets a high standard for the agencies, a standard which must be rigorously enforced by reviewing courts."[12] Again, "if the decision was reached procedurally without individualized consideration and balancing of environmental factors—conducted fully and in good faith—it is the responsibility of the courts to reverse."[13] Hence the details of the manner in which an agency conducted its final decision-making process will be subject to

[9] 449 F.2d at 1112, note 5, 1 ELR at 20347, note 5.
[10] 449 F.2d at 1113, 1 ELR at 20348.
[11] 449 F.2d at 1123, 1 ELR at 20353.
[12] 449 F.2d at 1114, 1 ELR at 20348.
[13] 449 F.2d at 1115, 1 ELR at 20349.

close judicial scrutiny under NEPA. The courts may ask to be satisfied that actual consideration of the factors specified in NEPA's substantive policy has taken place.

Further discussing the role of the courts, Judge Wright suggested that even if the procedural duties are properly carried out, the substantive provisions of §101 may enable the courts to review the actual decision made on its merits. Such review would be limited, however, in the *Calvert Cliffs'* view, to whether the decision was arbitrary or clearly gave insufficient weight to environmental values. No further indication was given of how the substantive provisions of §101 might limit agency decision making within these standards of review as a matter of law.

Without purporting to define the content of the reviewable record, the court indicated that the formal §102(2) (C) statement and the description of alternatives required by §102(2) (D) would be the focal items on review. Although the court said that the requirements for an impact statement and a description of alternatives provide "evidence that the mandated decision making process has in fact taken place,"[14] it apparently did not intend to restrict review merely to the record compiled under §102. Other evidence showing actual failure to consider environmental factors, or consideration that in fact showed environmental factors to outweigh competing factors, presumably would still be admissible under the *Calvert Cliffs'* test, as *Daly v. Volpe* illustrates.[15] Also, there is no convincing indication that the court thought that the impact statement should include justifications for the proposed action, so that *all* the factors would appear balanced side by side in a single document (see further discussion on pages 254 *ff.*).

[14] 449 F.2d at 1114, 1 ELR at 20348.

[15] It might be asked what would happen if there was evidence that the statement was *not* considered in good faith in the review process. In Calvert Cliffs' the failure was only potential, lying in the structure of Rule 13. (10 CFR Part 50, App. D, at 249, cited in Calvert Cliffs', 449 F.2d at 1117, 1 ELR at 20350.) One case deals with an actual instance of such failure. Daly v. Volpe, Appendix B, found that inadequate consideration means that NEPA had not been complied with:

> The state's first draft environmental impact statement was inadequately considered by FHWA. Indeed, the decision approving route E-3 came the first business day following receipt of the statement. The statute contemplates more deliberation than the time required to use a rubber stamp. 350 F. Supp. at 259, 2 ELR at 20445.

Daly also indicated that later statements did not correct the deficiency, since the statement is to serve as the basis for the decision, not an after-the-fact justification.

The Record on Review

Other court decisions approve the *Calvert Cliffs'* view that agencies must actually consider environmental factors and that such consideration is judicially reviewable, although they have not added significantly to Judge Wright's interpretation of the balancing analysis. They further confirm that NEPA contributes to the reviewable administrative record of decision making. The attractiveness of the §102 process has in fact created a problem, because some courts would like the impact statement to be able to serve as the *entire* record of final decision making, a role the statement would be hard pressed to play.

In *Ely v. Velde*[16] the Fourth Circuit explained at some length that NEPA required the agency to develop a record which showed that it had complied with relevant procedures and had actually considered the factors specified. Although ultimate decision making was entrusted to the agency, it could not keep its thought process "under wraps." The impact statement was considered an important part of the reviewable record.

> With regard to NEPA, the statutory requirement of a "detailed statement . . . on the environmental impact of the proposed action" places a heavy burden on the LEAA [Law Enforcement Assistance Administration]. To enable a court to ascertain whether there has been a genuine, not a perfunctory compliance with NEPA, the LEAA will be required to explicate fully its course of inquiry, its analysis, and its reasoning.[17]

The Second Circuit first touched upon the issue in *Greene County Planning Board v. Federal Power Comm'n.*[18] However, the court simply assumed that the impact statement constituted part of the reviewable record of agency consideration. In discussing §§102(2) (A), (D), and (E), the court remarked that these sections meant that NEPA's scope extended "far beyond" the requirements that the agency must "consider environmental factors and include those factors in the record subject to review by the courts."[19]

In subsequent opinions in *Hanly v. Kleindienst*,[20] the Second

---

[16] 321 F. Supp. 1088, 1 ELR 20082 (E.D. Va.), *rev'd*, 451 F.2d 1130, 1 ELR 20612 (4th Cir. 1971).

[17] 451 F.2d at 1139, 1 ELR at 20615.

[18] 455 F.2d 412, 2 ELR 20017 (2d Cir.), *cert. denied*, 41 U.S.L.W. 3184 (Oct. 10, 1972).

[19] 455 F.2d at 419, 2 ELR at 20020 (2d Cir. 1972).

[20] Hanly v. Mitchell, —— F. Supp. ——, 2 ELR 20181 (S.D.N.Y.), *rev'd* (Hanly I), 460 F.2d 640, 2 ELR 20216 (2d Cir.), *cert. denied*, 41 U.S.L.W. 3247 (Nov. 7, 1972), *sub nom.* Hanly v. Kleindienst, —— F. Supp. ——, 3 ELR 20016 (S.D.N.Y.), *rev'd* (Hanly II), —— F.2d ——, 2 ELR 20717 (2d Cir. 1972).

Circuit in "*Hanly I*" seized upon review of an agency decision not to prepare an impact statement and remarked that "in the context of an act designed to require federal agencies to affirmatively develop a reviewable environmental record," a perfunctory explanation of the agency's refusal to prepare a statement would not suffice.[21] The agency still had to show through a reviewable record (defined in some detail by *Hanly II*)[22] that it had properly considered environmental factors, although an impact statement might not have to be prepared.

The position of the Eighth Circuit is the same, judging by its conclusions on review of the Gillham Dam case.[23] In deciding that an "arbitrary and capricious" standard applied to review of substantive questions in NEPA cases, the court referred to "the complete record, including the environmental impact statement and the transcript of the proceeding below."[24] The recent opinion of the Eighth Circuit in a similar case, *Environmental Defense Fund v. Froehlke*,[25] confirms this interpretation.

> Finally, the formal impact study supplies a convenient record for the courts to use in reviewing agency decisions on the merits to determine if they are in accord with the substantive policies of NEPA.[26]

Similarly, the three-judge panel in *City of New York v. United States*[27] reviewed an agency impact statement and other materials, specifically finding that compliance did not fall short of "the level of refined and systematic consideration required by NEPA, cf. *Calvert Cliffs'*."[28] A dictum treated the issue of the standard to be applied in reviewing decision making subject to NEPA. Because the APA normally subjects Interstate Commerce Commission decisions regarding railway abandonment such as was at issue in this case to the substantial evidence test, the question arose whether, as part of the record, the commission's environmental determinations would be subject to the same standard. Judge Friendly thought not, for the

[21] 460 F.2d at 647, 2 ELR at 20220.

[22] —— F.2d ——, 2 ELR 20717.

[23] Environmental Defense Fund v. Corps of Engineers, 325 F. Supp. 728, 1 ELR 20130 (E.D. Ark. 1970–71), —— F. Supp. ——, 2 ELR 20260 (E.D. Ark.), 342 F. Supp. 1211, 2 ELR 20353 (E.D. Ark.), *aff'd*, 470 F.2d 289, 2 ELR 20740 (8th Cir. 1972).

[24] 470 F.2d at 301, 2 ELR at 20745.

[25] The Cache River case, —— F.2d ——, 3 ELR 20001 (8th Cir. 1972).

[26] —— F.2d at ——, 3 ELR at 20003.

[27] 337 F. Supp. 150, 2 ELR 20275 (E.D.N.Y.), 344 F. Supp. 929, 2 ELR 20688 (E.D.N.Y. 1972).

[28] 344 F. Supp. at 939, 2 ELR at 20689.

reason that many federal actions subject to NEPA involve informal decision making not subject to the higher APA standard. Nevertheless, he avoided actually deciding the issue, because he found that substantial evidence existed supporting the commission's decision.[29] The dissenter thought the dictum was inadvisable, since it was directed to an issue that was neither briefed nor argued.

The cases are split on the major issue whether the impact statement should be able to serve as the full record of the final agency decision, or whether it is only one component of that record. If it serves as the former, then it must include the data and reasoning which support the agency's action, along with its assessment of possible environmental impacts. The issue has come up in a number of cases regarding agency cost-benefit analyses.

The decision in *Calvert Cliffs'* has been read to imply that the impact statement should include at least some discussion of how costs and benefits were balanced.[30] But as discussed above, the court was using "cost benefit" language loosely and did not seem to mean that cost-benefit analyses or techniques should be transferred into the §102 process. Nor should the decision in *Natural Resources Defense Council v. Morton*[31] be interpreted to enlarge the role of the impact statement in showing how the balancing was done, simply because the court remarked that the impact statement provides "a basis for" evaluation of benefits in light of environmental risks and for comparison of any net environmental risk with the risks entailed by alternatives.[32] Both *Calvert Cliffs'* and *Morton* are consistent with the view that the impact statement should focus on providing information for the environmental risk—and conceivably the *environmental* benefit—portion of the balancing analysis.

Other cases, however, are more explicit and are in direct conflict. The district court in the Gillham Dam case indicated that claims of economic benefit, as well as a critical analysis of these claims by opponents, should be included in order to make the impact statement complete.[33] A similar stance was taken on remand by the district

[29] 344 F. Supp. at 940, 2 ELR at 20690.

[30] Council on Environmental Quality, Environmental Quality, Third Annual Report 245 (August 1972), citing language at 449 F.2d at 1113–14, 1 ELR at 20348.

[31] 337 F. Supp. 165, 2 ELR 20028 (D.D.C.), 337 F. Supp. 167, 2 ELR 20089 (D.D.C. 1971), *motion for summary reversal denied*, 458 F.2d 827, 2 ELR 20029 (D.C. Cir.), *dismissed as moot*, 337 F. Supp. 170, 2 ELR 20071 (D.D.C. 1972).

[32] 458 F.2d at 833, 2 ELR at 20032 (D.C. Cir.).

[33] Environmental Defense Fund v. Corps of Engineers, Appendix B, 325 F. Supp. at 761, 1 ELR at 20142.

court in *Lathan v. Volpe*,[34] which found that cost-benefit analysis was an appropriate part of statements on highway projects.[35] The contrary view has been stated by the court in *Environmental Defense Fund v. Armstrong*:[36] "We find no such requirement in the NEPA that any such cost-benefit analysis be conducted or included in the EIS."[37]

The merits of the point of view expressed in *Armstrong* must be weighed against the merits of the view of the Gillham Dam and *Lathan* district courts. The two main advantages of compiling one "decision document" are that such a document creates a single written reviewable project justification, and that formal cost-benefit analyses, heretofore thought to be beyond the courts' reach, may possibly be examined as part of the impact statement. Against these considerations must be pitted the danger that the inclusion of the analyses will allow project justifications and economic and technical considerations to swallow up environmental impact analysis. This latter result clearly was not wished by Congress. In agencies whose primary expertise will remain oriented to the engineering and economic aspects of development projects for some time to come, such a change in emphasis invites the use of the §102 process for developing the reasons why projects should go forward. NEPA creates a special preliminary procedure, the fruit of which—an environmental impact statement—"shall accompany the proposal through existing agency review processes."[38] This preliminary procedure should remain a one-sided inquiry, whatever the use made of cost-benefit analyses developed for other purposes.[39]

The Council on Environmental Quality, aware of the risk that self-serving project justifications could destroy the focus of the impact statement, nevertheless concludes that §102(2) (C) should include some identification of the nonenvironmental interests favoring the project. A "succinct recital" of these interests would alert the President, Congress, and the public to their nature.[40] But the reviewable record presumably would still include the full project

---

[34] 455 F.2d 1111, 1 ELR 20602 (9th Cir. 1971), *modified on rehearing*, 455 F.2d 1122, 2 ELR 20090 (9th Cir.), 350 F. Supp. 262, 2 ELR 20545 (W.D. Wash. 1972).

[35] 350 F. Supp. at 266, 2 ELR at 20547.

[36] —— F. Supp. ——, 2 ELR 20604 (N.D. Cal.), —— F. Supp. ——, 2 ELR 20735 (N.D. Cal. 1972).

[37] —— F. Supp. at ——, 2 ELR at 20738.

[38] NEPA, §102(2)(C), Appendix A.

[39] Comment, *supra* note 4, at 2 ELR 10003, 10004–05.

[40] CEQ Third Annual Report, *supra* note 30, at 245–46.

justification in whatever form it appears. The difference between an adequate statement and an adequate record for review would continue to be significant under the CEQ view; the CEQ solution apparently is not intended to merge the two.[41]

A Critique of "Balancing"

The judicially enforceable requirement that agencies actually consider environmental impacts when final decisions are made is an important bridge between NEPA's informational requirements and its goal of changing the outcome of agency decisions. Yet on closer examination the *Calvert Cliffs'* "balancing analysis" leaves doubt that it is the kind of directive that will cause agencies to comply with NEPA to the full extent suggested by the Act's substantive requirements and its legislative history.

The court's endorsement of the balancing approach meant at the very least that the agencies have to comply in earnest with Congress' intent to affect final agency decision making. What the endorsement may mean beyond that is more doubtful. The court avoided detailed consideration of how the balancing should be carried out or of how the kinds of techniques described in §§102(2) (A) and (B) might be developed and used. The agencies' obligations, if any, to be specific about how incommensurable costs and benefits were traded off, or how weights were assigned to both quantifiable and nonquantifiable factors, were not discussed. Nor did the court indicate that NEPA's substantive policy required anything more than equal treatment among competing national priorities and goals.

Hence the court apparently intended only to convey the common sense notion of "trading off" when competing interests must be equitably balanced. Its reliance on phrases like "finely tuned," "systematic," and "optimally beneficial" is misleading when applied to such a rough-and-ready process. The analogy to precisely calibrated scales suggests a degree of certainty which the agencies cannot possibly attain without more guidance as to the relative importance of various factors,[42] and without more information of a kind and

[41] Sierra Club v. Froehlke (Trinity River–Wallisville Dam), Appendix B, illustrates another aspect of the problem. The court viewed the impact statement as a "decision document," yet it was only part of the voluminous record, which filled 13 book boxes with 246 separate items ranging from a few pages to several thousand pages each (court's footnote 1).

[42] *E.g.*, the Supreme Court's interpretation of the "parklands statutes," 49 U.S.C. §1653(f) (1966); 23 U.S.C. §138 (1968), in Citizens to Preserve Overton Park v. Volpe, 401 U.S. 402, 1 ELR 20110 (1971) (fully cited in Appendix B).

extent which, as NEPA's legislative history acknowledges, does not yet exist.[43] Furthermore, basic agency missions set priorities and create conflicts which *Calvert Cliffs'* does not help resolve. To point out that Congress added an additional, environmental mandate to existing agency missions does nothing to resolve the head-on clashes which inconsistent mandates create.[44]

In these circumstances, an agency faced with making a decision that is consistent with *Calvert Cliffs'* will most likely resolve uncertainties in favor of the priorities set out in its original grant of authority, at least until Congress or the courts give it more specific directions, or until the longer-run NEPA goals begin to take effect.[45] NEPA may have the effect in the meantime of causing a better record supporting the basic decision to be made and new information about environmental impacts to be assembled, but it may not actually achieve its ultimate purpose of changing the congressionally recognized tendency of federal decision making toward environmental neglect and destruction.

Certainly in light of its reliance on §§102(A) and (B), the court could not have intended to sharpen the balancing process by approving the wholesale importation into the NEPA process of formal cost-benefit analysis techniques, such as those for evaluating water resources projects.[46] Moreover, the possibility of agency manipulation of decisional factors, and of their weights, is as great for these more "precise," economically oriented techniques as it is for the

---

[43] *E.g.*, SENATE COMM. ON INTERIOR AND INSULAR AFFAIRS, NATIONAL ENVIRONMENTAL POLICY ACT OF 1969, S. REP. No. 91–296, 91st Cong., 1st Sess. 3, 9 (research and monitoring functions now transferred to other sections from S. 1075, §201), 10 (CEQ as providing "early warning" through study and analysis; date to be provided in annual report) (July 9, 1969).

[44] NEPA, §§103–105, Appendix A. *See Joint Hearings Before the Senate Comm. on Public Works and the Comm. on Interior and Insular Affairs*, Serial No. 92–H32, 92d Cong., 2d Sess. 394–410 (colloquy between Senators Baker and Buckley and Mr. Cramton), 455–56 (colloquy between Senator Buckley and Mr. Anderson) (March 1, 7, 8 and 9, 1972).

[45] Tarlock, *supra* note 4 at 658.

[46] Cohen and Warren, *supra* note 4 at 698. *See* Hammond, *Convention and Limitation in Benefit-Cost Analysis*, 6 NAT. RES. J. 195 (1966); E. Devine, *The Treatment of Incommensurables in Cost-Benefit Analysis*, 42 LAND ECONOMICS 383 (August 1966); J. Knetsch *et al.*, *Federal Natural Resources Development: Basic Issues in Benefit and Cost Measurement*, Natural Resources Policy Center, George Washington University (May 1969); C. Cicchetti *et al.*, *Benefits or Costs? An Assessment of the Water Resources Council's Proposed Principles and Standards*, Department of Geography and Environmental Engineering, Johns Hopkins University (March 1972). For further bibliography on cost-benefit determinations, *see* A. Priest and R. Turvey, *Cost-Benefit Analysis: A Survey*, 75 THE ECONOMIC JOURNAL 683 (1965).

as-yet unrefined techniques called for by NEPA. (See discussion on pages 265 *ff.*) To its credit, the Atomic Energy Commission has made an energetic attempt to develop better decisional techniques in the wake of *Calvert Cliffs'*, but even a cursory scan of its suggested catalogue of factors shows both the difficulty of putting environmental incommensurables into the decisional balance and the consequent possibilities for manipulation which exist.[47]

The balancing process thus appears to deliver only half of NEPA's loaf. Required to do no more than show that they have engaged in a defensible tradeoff, the agencies can still make environmentally destructive decisions, although they now may do so with a fuller awareness of the destruction which could be caused. As we next discuss, the courts appear uneasy about how NEPA's ultimate purposes may still be frustrated. They may be ready to look further into NEPA's substantive requirement for reviewable standards which further NEPA's legislative intent by more carefully limiting the kinds of decisions which agencies may make.

SUBSTANTIVE DUTIES

The cases confirm that NEPA is more than a full-disclosure law; it was intended to bring about substantive changes in agency decision making. Following *Calvert Cliffs'* lead, the courts have assumed a role in ensuring that such changes occur through judicial review of the agencies' "consideration" of environmental factors. These same cases suggest, however, that the requirement of mere consideration may not define the limit of NEPA's judicially enforceable substantive duties.

The courts may find more "law to apply"[48] as they read the Act closely in light of its legislative history. The substantive provisions of NEPA, especially §§101(a) and (b), present a challenge in this regard because of their broad language. Yet that language is not mere preamble, nor was Congress satisfied to limit its statement of policy to the vague requirement that agencies must use all practicable means consistent with other essential national policies to protect the

---

[47] *See* Comment, *supra* note 4, and AEC documents cited, 2 ELR at 10005.

[48] Citizens to Preserve Overton Park v. Volpe, 401 U.S. 402, 410, 1 ELR 20110, 20112 (1971), construing the Administrative Procedure Act, 5 U.S.C. §701(a)(1) and (2), very narrowly. The APA states that review is foreclosed where "agency action is committed to agency discretion by law." The Supreme Court borrows the phrase "law to apply" from the legislative history of §701. *See* S. REP. No. 758, 79th Cong., 1st Sess. 26 (1945).

environment. In §101(b) Congress specified six detailed goals for the federal government to attain in order to carry out the Act's general policy.[49] When read in conjunction with the legislative history and with the Act's other provisions, particularly the requirement of §102(1) that all other federal policies and laws must be interpreted and administered "to the fullest extent possible" in conformity with NEPA's policy, the six provisions of §101(b) suggest that Congress said more about the freedom of agencies to consider and trade off environmental values than the courts have yet fully recognized.

The court in *Calvert Cliffs'* limited its comments on judicial enforcement of NEPA's substantive requirements to observing:

> The reviewing courts probably cannot reverse a substantive decision on its merits . . . [under §101] unless it be shown that the actual balance of costs and benefits that was struck was arbitrary or clearly gave insufficient weight to environmental values.[50]

Nevertheless, this dictum indicates that in the opinion of the *Calvert Cliffs'* court, §101 was not drawn in such broad terms that it contained no "law to apply." Final agency decision making would not, in this court's view, be committed entirely to agency discretion.

Without citing *Calvert Cliffs'*, the district court in its first opinion in *Akers v. Resor*[51] found that the Corps' decision to proceed with a stream channelization project was reviewable on its merits. Citing the Supreme Court in *Citizens to Preserve Overton Park v. Volpe*,[52] the court said that it was "clear" that the action was reviewable because review had not been prohibited by statute and had not been committed by statute to the Corps' discretion. The decision to proceed with the project could be enjoined if it was arbitrary, not otherwise in accordance with law, or failed to meet statutory procedural requirements. Relying upon §101(b) and §102(1), the court found that the plan for mitigating the adverse impact of part of the project, as submitted to Congress in 1963, was insufficient. The court remarked that the 1963 mitigation plan might satisfy the requirements of the Fish and Wildlife Coordination Act,[53] but it did not satisfy NEPA's requirement that that Act be administered in accordance with NEPA [§102(1)], nor did it satisfy NEPA's requirements [§101(b)] "that all Federal plans and programs be

---

[49] NEPA, §101, Appendix A.

[50] 449 F.2d at 1115, 1 ELR at 20349.

[51] 339 F. Supp. 1375, 2 ELR 20221 (W.D. Tenn.), ——F. Supp. ——, 3 ELR 20157 (W.D. Tenn. 1972).

[52] Appendix B.

[53] 16 U.S.C. §§661–666.

improved to attain environmental objectiveness."[54] Thus apparently the Corps had to request congressional appropriations that would enable it to carry out a modified project which better accomplished NEPA's substantive objectives.

The recent opinion of the Eighth Circuit affirming the district court's decision in the Gillham Dam case fully articulated the view that NEPA imposes judicially reviewable substantive requirements.[55] Although the circuit court affirmed the district court opinion allowing the project to continue, it held that the merits of the Corps' decision to proceed were reviewable under §101 and §102(1) on the arbitrary and capricious standard.

> The language of NEPA, as well as its legislative history, make it clear that the Act is more than an environmental full-disclosure law. NEPA was intended to effect substantive changes in decision making. Section 101(b) of the Act states that agencies have an obligation "to use all practical means, consistent with other essential considerations of national policy, to improve and coordinate Federal plans, functions, programs and resources" to preserve and enhance the environment. To this end, §101 sets out specific environmental goals to serve as a set of policies to guide agency action affecting the environment. . . .[56]
>
> . . . Given an agency obligation to carry out the substantive requirements of the Act, we believe that courts have an obligation to review substantive agency decisions on the merits. Whether we look to common law or the Administrative Procedure Act, absent "legislative guidance as to reviewability, an administrative determination affecting legal rights is reviewable unless some special reason appears for not reviewing. . . ." Here, important legal rights are affected. NEPA is silent as to judicial review, and no special reasons appear for not reviewing the decision of the agency. To the contrary, the prospect of substantive review should improve the quality of agency decisions and should make it more likely that the broad purposes of NEPA will be realized [citation and footnote omitted].[57]
>
> . . . The standard of review to be applied here and in other similar cases is set forth in *Citizens to Preserve Overton Park v. Volpe*. . . . The reviewing court must first determine whether the agency acted within the scope of its authority, and next whether the decision reached was arbitrary, capricious, an abuse of discretion, or otherwise not in accordance with law. In making the latter determination, the court must decide if the agency failed to consider all relevant factors in reaching its decision, or if the decision itself represented a clear error in judgment. Where NEPA is involved, the reviewing court must first determine if the

[54] 339 F. Supp. at 1380, 2 ELR at 20223.
[55] Environmental Defense Fund v. Corps of Engineers, Appendix B.
[56] 470 F.2d at 297, 2 ELR at 20743.
[57] 470 F.2d at 298, 2 ELR at 20744.

agency reached its decision after a full, good faith consideration and balancing of environmental factors. The court must then determine, according to the standards set forth in §§101(b) and 102(1) of the Act, whether "the actual balance of costs and benefits that was struck was arbitrary or clearly gave insufficient weight to environmental values . . . [citations to *Overton Park* and *Calvert Cliffs*' omitted].[58]

The Eighth Circuit interpreted NEPA decisions in three other circuits as supporting its view,[59] while only one circuit was thought be be at variance.[60] It also stated that its conclusion was supported by the "analogous" decision of the Supreme Court in *Overton Park* and by several scattered district court decisions.[61]

Turning to the Corps' decision to proceed with the dam, the circuit court remarked that ordinarily the matter would be remanded to the trial court for review, but that in this case remand would not be necessary because the complete record, including the impact statement and the transcript of the proceeding, was before the court. In a very brief discussion the court rejected plaintiffs' contention that the decision to complete the dam was arbitrary and capricious. The court apparently focused upon §101(b) (4), which requires the agencies to "maintain, wherever possible, an environment which supports diversity, and variety of individual choice," and balanced the sacrifice of diversity against the benefits of flood control. In rejecting plaintiffs' argument, the court said:

We have reached this conclusion after a serious consideration of the arguments in favor of and against completion of the project. In large part this has necessitated a balancing, on the one hand, of the benefits to be derived from flood control, and, on the other, of the importance of a diversified environment.[62]

In holding broadly that NEPA creates substantive duties reviewable under the arbitrary and capricious standard, and that in this

---

[58] 470 F.2d at 300, 2 ELR at 20745.

[59] 470 F.2d at 299, 2 ELR at 20744 note 15. *District of Columbia Circuit:* Calvert Cliffs', Natural Resources Defense Council v. Morton, Committee for Nuclear Responsibility v. Seaborg. *Second Circuit:* Scenic Hudson Preservation Conference v. Federal Power Comm'n (Scenic Hudson II), Hanly v. Mitchell (Hanly I). *Fourth Circuit:* Ely v. Velde [and now one must add Conservation Council of North Carolina v. Froehlke]. All of these cases are fully cited in Appendix B.

[60] The Tenth Circuit in National Helium Corp. v. Morton, 326 F. Supp. 151, 1 ELR 20157 (D. Kan. 1971), *aff'd*, 455 F.2d 650, 1 ELR 20478 (10th Cir. 1971).

[61] Lathan v. Volpe, Brooks v. Volpe, Citizens for Reid State Park v. Laird, Morningside–Lenox Park Ass'n v. Volpe, Environmental Defense Fund v. Corps of Engineers (Tennessee–Tombigbee), all cited in Appendix B.

[62] 470 F.2d at 301, 2 ELR at 20745.

instance the Corps' decision was within the bounds of its discretion, the court failed to construe the applicable NEPA section so that the constraints on the balancing process which it imposed as a matter of law were better articulated. In *Overton Park*, to which the Eighth Circuit analogizes its Gillham Dam opinion, the Supreme Court construed the Parklands Statutes[63] in a manner which rather precisely defined the record which the government would have to show in the district court before it could proceed with construction of the highway link at issue in that case. The secretary of transportation had to balance countervailing factors, but the Court's interpretation of the heavy favor which he had to show by law to parklands preservation radically changed the department's earlier view of the balancing to be performed. In the Gillham Dam case, the particular circumstances which justified a reduction in the diversity of the natural environment remained uncertain. There is "law to apply," the court has said, but the court has not articulated its requirements.

By refusing to remand, the circuit court avoided having to spell out how review of the Corps decision should be conducted. This is unfortunate in light of the district court's observation that "in the instant case it is clear that the damming of the Cossatot will reduce 'diversity and variety of individual choice.' "[64] The circuit court stated that a reviewing court must decide, among other things, whether the agency "clearly gave insufficient weight to environmental values." Yet it does not suggest what the weight accorded the district court's finding must be as a matter of law.

In the Gillham Dam case, the Eighth Circuit sustained the decision of the district court and did not remand. Two weeks later in its opinion reversing and remanding a case involving a stream channelization project along the Cache River,[65] the Eighth Circuit actually did remand with instructions to review the substantive agency decision on its merits to determine if it accorded with NEPA. The court also found that the impact statement which the Corps had prepared was inadequate and that congressional appropriations for the project did not mean that Congress implicitly exempted the project from the requirements of NEPA.

The circuit court rejected plaintiffs' assertions that the Corps' cost-benefit determination was reviewable under 33 U.S.C. §701a of the Flood Control Act of 1934, the Corps' governing "cost-benefit

[63] *Supra* note 42.
[64] 325 F. Supp. at 755, 1 ELR at 20140.
[65] Environmental Defense Fund v. Froehlke (Cache River Project), Appendix B.

statute." But in rejecting plaintiffs' claim, the court made the following observation of utmost significance to subsequent judicial review of public works projects:

> We do not think that the statement of policy in §701a can be used as a vehicle for continuing evaluation of the project by the courts.
>
> We point out, however, that the relief requested by the plaintiffs under §701a is partially available under NEPA. To fully comply with NEPA, the Corps must reappraise the costs and benefits of the project in light of the policies of environmental protection found in NEPA. As we have stated, a decision to proceed with channelization is reviewable in the District Court to determine whether the actual balance of costs and benefits struck by the agency according to the standards of §§101 and 102 of NEPA was arbitrary or clearly gave insufficient weight to environmental factors.[66]

In a brief opinion the Fourth Circuit recently remanded *Conservation Council of North Carolina v. Froehlke*[67] for review under NEPA's substantive provisions, instructing the district court to enjoin any environmentally detrimental activity pending completion of that review. Thus it would appear that at least three circuits—the Fourth, the Eighth, and the District of Columbia—favor review under §§101 and 102(1).

In *Sierra Club v. Froehlke* (Trinity River–Wallisville Dam)[68] a federal district court took the important step of applying NEPA point by point to the Corps of Engineers' decision-making process and to the method which the Corps adopted for determining the cost-benefit ratio for the federal water resources projects at issue. Casting aside views that the cost-benefit ratio was to be reviewed solely by Congress and that the courts lacked jurisdiction to examine it, Judge Bue found that the determination "intertwined" environmental and nonenvironmental factors so that NEPA applied. "When the claimed ratio is composed, in part, of environmental amenities which Congress has required under current law to be given careful attention and consideration, then the courts have an obligation to act, where necessary."[69]

The court did not rest its decision specifically upon deficiencies in the cost-benefit ratio computation. A prima facie case for noncompliance with NEPA had been established on narrower grounds

---

[66] —— F.2d at ——, 3 ELR at 20005.

[67] 340 F. Supp. 222, 2 ELR 20155 (M.D. N. Car.), *aff'd*, —— F.2d ——, 2 ELR 20259 (4th Cir. 1972), —— F.2d ——, 3 ELR 20132 (4th Cir. 1973).

[68] Appendix B.

[69] Text *supra* the court's footnote 346. *See also* cost-benefit materials cited *supra* note 46.

involving inadequacies in the Wallisville Dam impact statement and procedures regarding its circulation for comment. However, the court did say:

> Because the Wallisville impact statement and the record which relates to both projects [Trinity River Project; Wallisville Dam] indicate that the [benefit-cost] balance struck was "arbitrary" and "clearly gave insufficient weight to environmental values," under *Calvert Cliffs*' . . . substantial in-depth revision by the Corps in this area will be required prior to the acceptance of either the Wallisville or the Trinity Project impact statement.[70]

The court was quite specific about the inadequacies of the original cost-benefit determination. Our brief summary cannot do justice to the court's rather lengthy treatment of the issues. In the court's view, selected environmental benefits of the project had been identified by the Corps, quantified in dollars, and added to the benefit side of the calculation, while similar environmental costs or losses if construction were completed had not been quantified or included at all. Some factors quantified as "benefits" seemed to the court to be open to considerable doubt, particularly recreational benefits, which it analyzed in detail. The estimated productive life span of the project, the applicable discount rate, the dimensions of the geographic area over which specified benefits were to be measured, all received skeptical analysis from the court.

Having charged the Corps to do a much better job, the court went on to interpret NEPA as in part specifying the steps which the Corps had to take toward this goal. After intimating that the CEQ perhaps had authority under NEPA to develop a new cost-benefit methodology,[71] and chastising all of the federal government for delaying three years the development of adequate techniques for taking account of difficult-to-quantify environmental factors, the court specified methodological steps to be taken and concluded by requiring the Corps to bring its

> . . . Procedures to the attention of Congress, the appropriate federal agencies, and the Council on Environmental Quality in order that appropriate policy decisions may be made to assist the Corps and the Court in assessing the completeness of environmental impact statements. These will include the determination of proper methods for quantifying and evaluating environmental amenities.[72]

---

[70] Text *infra* the court's footnote 339.

[71] Discussion *supra* court's footnotes 347, 424, and discussion immediately preceding final "Summary and Holding" at end of opinion.

[72] "Summary and Holding of this Court" at end of opinion.

The courts stand on the threshold of an important new chapter in NEPA's judicial interpretation. Perhaps they will be content with a general review of compliance with §101; its broad language may discourage extensive detailed interpretation. On the other hand, *Calvert Cliffs'* vague requirement that agencies "consider" and balance environmental factors may not prescribe a decision-making process that can obtain the rather explicit results desired by Congress when it enacted NEPA. If the courts cannot effectively review agency compliance with NEPA's basic purposes, they may become impatient with repeated reviews of procedural compliance and cursory substantive reviews, and begin to interpret §101 and §102(1) in ways which more precisely define the allowable scope of agency discretion.

In spite of the broad language of §101, it is not too difficult to see where there may be "law to apply." The problem of preserving diversity, for one, was discussed in NEPA's legislative history, and it received close attention in the Gillham district court. For another, §101 strongly suggests a nondegradation policy, which, when interpreted in light of the plain wording of the statute and its legislative history, may restrict agency decision making to the preservation and enhancement of environmental quality, especially where pollution is the degradation threatened. Of course that policy would not require that no tree be cut or that no river be dammed. But it might require that forest management practices that allow slow degradation be discarded, or that an overall river "improvement" project be modified or abandoned. Further, the specific requirements of §101 may be interpreted as altering the way in which developmental benefits are to be weighed. This seems to be the thrust of the Corps cases in the Fourth and Eighth Circuits and of *Sierra Club v. Froehlke* (Trinity River–Wallisville Dam). The requirements of NEPA's §101(b)(3), which calls for the attainment of "the widest range of beneficial uses of the environment without degradation, risk to health or safety, or other undesirable and unintended consequences," together with the requirements of §§101(b)(5) and (6), may become a new "cost-benefit statute" for agency decisions on all manner of federal projects. In this way the courts may provide the specificity which *Calvert Cliffs'* failed to attain.

THE OTHER ACTION-FORCING PROVISIONS

Section 102(2) includes seven other action-forcing provisions besides §102(2)(C). Sometimes these sections are construed as ancillary to §102(2)(C), but already the courts have made clear that they have their own distinct, additional roles to play. This is especially true of §§102(2)(A) and (B), which specify that agency planning and decision-making processes, and the methods and procedures used in them, must be systematic and interdisciplinary and must take unquantified amenities into account. Sections 102(2)(D) and (G) emphasize the fact-finding characteristics of §102(2)(C), rather than planning and decision making, but they, too, have purposes separate from §102(2)(C).

These sections have great significance for agency procedures and methodologies that are not easily challenged in a lawsuit. The state of the art in the "design arts" is still very unclearly defined, even among the key professionals involved, such as architects and planners. Hence the agencies' own willing use of these sections offers the most promise over the long run that the intent behind them will be realized. We have included this brief discussion of judicial interpretations of the sections to show that the courts will probably play the maximum role available to them.

The courts have confirmed that the "major federal action" requirement in §102(2)(C) applies only to the preparation of impact statements. The other seven action-forcing provisions are exempt. This point of view is endorsed in *Hanly v. Kleindienst*,[73] where the court further reasoned that if §102(2)(D) were limited to major federal actions, it would merely duplicate §102(2)(C)(iii). Similar logic is implicit in *Citizens for Reid State Park v. Laird*,[74] where although the action was not large enough for an impact statement, the court specifically found that §§102(2)(A), (B), and (D) were satisfied. A strong dictum to the same effect appears in *City of New York v. United States*.[75]

Sections 102(2)(A) and (B) have been construed in three circuit court decisions. In *Calvert Cliffs'* the court relied upon them to buttress its finding that NEPA required consideration of environmental values through the balancing process (see discussion above). Judge Wright said that §§102(2)(A) and (B) clarified the types of environmental consideration compelled by NEPA, pointing out that "appropriate" consideration does not give agencies discretion to undervalue environmental factors in decision making. NEPA requires consideration "appropriate" to the problem of protecting the envi-

[73] Hanly II, Appendix B, —— F.2d at ——, 2 ELR at 20723.
[74] 336 F. Supp. 783, 2 ELR 20122 (D. Me. 1972).
[75] Appendix B, 337 F. Supp. at 158, 2 ELR at 20276.

ronment, not "appropriate" to the whims of federal agencies. The court quoted Senator Jackson's floor remarks on §102(2)(B): "Subsection 102(2)(B) requires the development of procedures designed to insure that all relevant environmental values and amenities are considered in the calculus of project development and decision making."[76] Finally, the court suggested that compliance with §§102(2)(A) and (B) must precede compliance with §102(2)(C).[77]

The Second Circuit's second decision in *Hanly v. Kleindienst* (*Hanly II*)[78] also stated that §102(2)(B) applies before the threshold determination whether to prepare a statement is made under §102(2)(C). The court found that the agency had satisfied §102(2)(A) by employing architects familiar with the design requirements of the project area and by attempting to harmonize the facility with existing structures. The court found that §102(2)(B) had not been satisfied, however, because that section required the agency to give adequate public notice of the proposed action and an opportunity to present facts relevant to the §102(2)(C) threshold determination, neither of which the agency had done. While the court noted that a public hearing did not have to be held, it did suggest that a formal hearing would be advisable in many instances. By using the §102(2)(B) requirements as a vehicle for deciding whether an impact statement had to be prepared, the court did not mean to imply that that section was merely ancillary to §102(2)(C). The court specifically stated that the agency had to comply with §102(2)(B) in any event.

The earlier Second Circuit decision in *Scenic Hudson II*[79] briefly considered §§102(2)(A) and (B), but not with the same sympathetic reception generally accorded them. In *Scenic Hudson II* the court held that the interdisciplinary requirement of §102(2)(A) was satisfied by the FPC's public hearings which had been conducted after the previous remand. During these hearings testimony was taken from a variety of disciplines, but no impact statement was prepared. As Justice Douglas stated in his dissent from the denial of certiorari, the onus for implementing the interdisciplinary approach fell upon

---

[76] 449 F.2d at 1113 note 9, 1 ELR at 20348 note 9.

[77] 449 F.2d at 1114, 1 ELR at 20348.

[78] Appendix B.

[79] Scenic Hudson Preservation Conference v. Federal Power Comm'n (Scenic Hudson I), 354 F.2d 608, 1 ELR 20292 (2d Cir. 1965), *cert. denied sub nom.* Consolidated Edison Co. of New York v. Scenic Hudson Preservation Conference, 384 U.S. 941 (1966); Scenic Hudson II, 453 F.2d 463, 1 ELR 20496 (2d Cir. 1971), *cert. denied*, 407 U.S. 926, 2 ELR 20436 (1972).

the intervenors and private parties opposing the project.[80] But the court was faced with a case which it had remanded five years previously for consideration of environmental values under the Federal Power Act, §10(a), a precursor of NEPA. Under such circumstances, the majority may have intended to ratify the results of the hearings but not necessarily the procedures utilized. Future projects might possibly be held to a stricter standard under §102(2)(A). Indeed, reliance on intervenors or parties for compliance with NEPA seems now to be precluded by the Second Circuit's decision in *Greene County*[81] (see discussion pages 186 *ff*.).

In addition to the three circuit court decisions, several district court opinions have considered the requirements of §§102(2)(A) and (B). In the Gillham Dam case,[82] the court rejected the argument that §102(2)(A) applied to new projects only. Despite the fact that Gillham Dam was authorized in 1958, and that it was 63 percent completed, the court thought that it was not too late to utilize an interdisciplinary approach when preparing the impact statement.[83] Further, the court applied §102(2)(B) in requiring the Corps to include additional factors in the impact statement on the "negative" or "loss" side. A statement of water quality loss resulting from destruction of the free-flowing stream must be included, the court stated, at least where the Corps claimed benefits of enhanced water quality from construction of the dam.[84] Similarly, criticism of the Corps' economic claims by those opposing the project also belonged in the impact statement.

The court in *Akers v. Resor*[85] focused on the benefit side of the cost-benefit analysis carried out in that case. Because increased crop production was included in *Akers* as a benefit of flood protection, the Department of Agriculture, which often pays farmers *not* to cultivate existing lands, should have been consulted.[86] Hence an "interdisciplinary approach" under §102(2)(A) must include comments from agencies having jurisdiction over areas affected by a project. The *Akers* court also held that consideration of unquantified values must be based upon "reasonable" assumptions. The Corps made an unreasonable assumption when it concluded that the value of lost

---

[80] 407 U.S. at 932, 2 ELR at 20437.
[81] Greene County Planning Board v. Federal Power Comm'n, Appendix B.
[82] Environmental Defense Fund v. Corps of Engineers, Appendix B.
[83] 325 F. Supp. at 756–57, 1 ELR at 20140–41.
[84] 325 F. Supp. at 761, 1 ELR at 20142.
[85] Appendix B.
[86] —— F. Supp. at ——, 3 ELR at 20158.

opportunities for hiking and bird watching could be ignored because most of the land affected was privately owned.

The court in the Tennessee–Tombigbee case[87] stated that §§102 (A) and (B) were designed to ensure that the full cost—social, economic, and environmental—of agency action will be known. Quoting legislative history, the court reasoned that §§102(2)(A) and (B) must be applied in early stages of planning. The court held that §102(2)(B) does not necessitate the use of particular techniques, such as computers, but is satisfied if the methodology used "effectively measures life's amenities in terms of the present state of the art." Here, a six-man scientific team conducting a rigorous examination of the project satisfied §§102(2)(A) and (B). The team was led by a civil engineer with a master's degree in water resource management and included two sanitary engineers, a civil engineer who was also a biologist, and two additional biologists. The team gathered data from the Corps and outside sources, consulted sixty persons or agencies, circulated a draft impact statement to twenty state and federal agencies, and analyzed the comments received. However, the Corps did not conduct field studies and relied solely on available scientific data and literature. Yet this was sufficient, the court held, in the absence of any showing by plaintiff that such data did *not* provide an adequate basis for assessment. Similarly, lacking any evidence of bias, the court found that the fact that all members of the team were from the Corps was insignificant. The court could not require more, considering the difficulty of enforcing a methodological approach to a project which began 16 years ago and which was substantially completed at the time of suit.

The decision in *Environmental Defense Fund v. Hardin*[88] interpreted §§102(2)(A) and (G) broadly, although the court did not have to solve the difficult problem of enforcing its interpretation. In this case the court held that the research conducted by the Department of Agriculture on the use of the pesticide Mirex to control fire ants satisfied §§102(2)(A) and (G). Section 102(2)(A) makes the completion of an "adequate" research program a prerequisite to agency action, adequacy being determined by the scope of the proposed project and by the extent potential adverse effects are indicated by existing knowledge. NEPA, the court continued, "envisions" that

[87] Environmental Defense Fund v. Corps of Engineers, 331 F. Supp. 925, 1 ELR 20466 (D.D.C. 1971), 348 F. Supp. 916, 2 ELR 20536 (N.D. Miss. 1972).

[88] 325 F. Supp. 1401, 1 ELR 20207 (D.D.C. 1971).

the evolution of projects will be the result of research efforts rather than that research will be utilized to rationalize projects already proposed.[89] Section 102(2)(A) requires a research effort which reflects the "current state of the art" of relevant scientific inquiry, and the research results, together with adequate documentation, belong in the impact statement. Finally, the court indicated that §102(2)(G) directs agencies to undertake research of a "broader scope" than was traditionally within their jurisdiction.[90]

We now turn to the judicial glosses which have been put on §102(2)(D). These have been largely concerned with the relationship of that section to the similar requirement imposed by §102(2)(C)(iii). Does §102(2)(D) require more extensive analysis of alternatives than §102(2)(C)(iii), less extensive analysis, or the same analysis? What affirmative agency action is required by §102(2)(D), either alone or in conjunction with §(C)(iii)?

In both *Committee to Stop Route 7 v. Volpe*[91] and *Natural Resources Defense Council v. Morton*,[92] §102(2)(D) was utilized in conjunction with §102(2)(C)(iii) to require a more extensive discussion of alternatives than was originally undertaken by the agencies. In *Route 7* the district court held that an impact statement on a small segment of highway was not an adequate consideration of alternatives as required by §§102(2)(C)(iii) and (D).[93] The D.C. Circuit in *Morton* utilized §102(2)(D) and §102(2)(C)(iii) to require the Department of the Interior to consider a broad range of alternatives which could satisfy short-term energy requirements in lieu of offshore drilling. The court rejected Interior's argument that both sections did not require a discussion of environmental consequences or that the department need consider only alternatives that it had the authority to adopt.

While neither *Morton* nor *Route 7* differentiates §102(2)(C)(iii) from §102(2)(D), the "reasonableness" standard of *Morton* was applied exclusively to §102(2)(D) by the Eighth Circuit in the Gillham Dam case. Citing *Morton* favorably, the Eighth Circuit stated that §102(2)(D) requires more extensive treatment of alternatives than §102(2)(C)(iii), although the court said that this did not mean an analysis under §102(2)(D) should not be included in the impact

---

[89] 325 F. Supp. at 1404, 1 ELR at 20208.

[90] *Id.*

[91] 346 F. Supp. 731, 2 ELR 20446 (D. Conn.), *motions to amend the judgment denied,* —— F. Supp. ——, 2 ELR 20612 (D. Conn. 1972).

[92] Appendix B.

[93] 346 F. Supp. at 740, 2 ELR at 20449.

statement.[94] The court concluded that a 37-page discussion of alternatives in a 200-page impact statement satisfied the broader requirement.

The district court decision in *Conservation Council of North Carolina v. Froehlke*[95] also suggested that the required discussion of alternatives under §102(2)(D) is broader than under §102(2)(C)(iii). The court held that §102(2)(C)(iii) does not necessarily have to represent the complete analysis of alternatives, noting that while §102(2)(D) requires agencies to "study, develop and describe" alternatives, §102(2)(C)(iii) only requires agencies to include a meaningful reference to alternatives that will identify problems for the responsible federal official.[96]

ZABEL V. TABB: AGENCY RELIANCE ON NEPA

A revealing fact about NEPA's three years in the courts is that federal agencies have rarely relied upon it to defend their actions in taking environmentally protective measures. The agencies have not creatively used NEPA's substantive requirements as authority for promulgating environmentally protective guidelines, for denying permits and licenses, and for proposing changes in legislative authority. Yet NEPA not only binds the agencies, it also frees them to take action. "[A federal agency] . . . is not only permitted, but compelled, to take environmental factors into account."[97]

The case of *Zabel v. Tabb*[98] is an exception which proves the rule. In *Zabel* the plaintiff had applied to the Corps of Engineers for a dredge-and-fill permit in 1966, had seen it denied in 1967, and had convinced the federal district court to compel issuance of the permit in 1969. In 1970 the Fifth Circuit reversed. The decisive issue was whether the Corps had authority to deny the permit on any ground other than interference with navigation. The court held that the Fish and Wildlife Coordination Act[99] and NEPA authorized the Corps to condition or deny a permit on environmental grounds alone. Further, the court held that "although the Congressional command was not in existence at the time the permit in question was denied, the

[94] 470 F.2d at 296, 2 ELR at 20743.
[95] Appendix B.
[96] 340 F. Supp. at 228, 2 ELR at 20157.
[97] Calvert Cliffs', Appendix B, 449 F.2d at 1112, 1 ELR at 20347.
[98] Appendix B. *See* Comment, *National Environmental Policy Act of 1969: A Mandate to the Corps of Engineers to Consider Ecological Factors. Zabel v. Tabb, 430 F.2d 199 (5 Cir. 1970),* 50 Bos. Univ. L. Rev. 616 (1970).
[99] 16 U.S.C. §§661–666.

correctness of that decision must be determined by the applicable standards of today."[100]

The Corps had relied upon the Fish and Wildlife Coordination Act to deny the permit. The Fifth Circuit observed that the enactment of NEPA gave "added impetus" to the Fish and Wildlife Coordination Act's requirements for consultation and consideration. The government's memorandum on the significance of NEPA to the case, however, gave NEPA an extremely narrow reading, one which ignored the expansion of basic agency mandates which Congress intended.[101] The circuit court gave the Act a more liberal reading than the government apparently wished.

To our knowledge, *Zabel* has virtually no progeny in which the government has defended federal action on the ground that NEPA enabled a measure to be taken. Reliance on NEPA has occurred in administrative adversarial proceedings, although we do not know the full extent of such reliance, which appears to be very slight. The two instances given here, therefore, must be viewed as illustrative, not exhaustive.

In *In re Archer*,[102] the Board of Land Appeals of the Department of the Interior relied upon NEPA in turning back a challenge to restraints that Interior had imposed on mineral prospectors, requiring them to protect and restore prospecting sites. The board said, "such requirements are reasonably related to the environmental ethic of this department and to the obligation of this Department under . . . [NEPA]."[103] A subsequent board opinion, however, seems to assume that in a ruling adverse to the environment, no obligation exists to reconcile the result with NEPA.[104]

In *Burlington Northern, Inc. Abandonment*[105] Burlington Northern Railroad sought Interstate Commerce Commission approval to abandon an 11-mile section of track lying mostly within the city of

---

[100] 430 F.2d at 213, 1 ELR at 20030. *See also* Bankers Life and Casualty Co. v. Village of North Palm Beach, Florida, —— F.2d ——, 2 ELR 20528 (5th Cir. 1972).

[101] ELR Dig. No. [40], Doc. I.

[102] *In re* J. D. Archer, Before the Department of the Interior, Board of Land Appeals, 1 ELR 30035 (May 26, 1971).

[103] 1 ELR at 30035.

[104] *In re* Alfred E. Koenig, Before the Department of the Interior, Board of Land Appeals, 2 ELR 30002 (Oct. 26, 1971). *See* Comment, *Interior Board of Land Appeals Confirms Right of Access to Mining Claims Across Public Lands*, 2 ELR 10015 (February 1972).

[105] Burlington Northern, Inc., Abandonment between Freemont and Kenmore, King Co., Wash., Before the Interstate Commerce Comm'n, 2 ELR 30034 (October 5, 1972).

Seattle. Intervenors sought to block the piecemeal disposition of the right-of-way, desiring that it be preserved as a park for a bicycle and pedestrian path. The commission found that NEPA required it to consider environmental effects along with traditional considerations of public convenience and necessity in reviewing the request for abandonment. The commission found that it could not require the company to dedicate the land to public use without working a confiscatory result, but it did delay permission to abandon for 90 days in order to allow any interested party to purchase the property at a price less than that which would be paid were condemnation proceedings carried out. The commission cited *City of New York v. United States,*[106] in which the commission was required to comply with NEPA in an earlier abandonment proceeding. One may speculate that the earlier litigation compelling the commission to comply with NEPA had an effect on its administrative decision in *Burlington.*

Administrative cases where the agencies have refused to read NEPA broadly are more typical of general agency attitudes. A particularly bad instance of such an attitude is *In re El Paso Natural Gas Co.,*[107] where the Federal Power Commission found that it did not have to comply with NEPA in approving a natural gas curtailment plan. A curtailment plan establishes priorities for denying gas to customers during a shortage. In the commission's view §102(2)(C) was inapplicable because the plan might lie dormant for many years or never be used. Moreover, because the impacts which might occur under the plan lie in the future, the commission could not say what environmental consequences might ensue.[108]

Attorneys need hardly be reminded that a judicial order requiring an administrative agency to exercise its discretion in a particular way is virtually impossible to obtain. Yet the kind of judicial relief which would stimulate affirmative federal use of NEPA is of this nature. In one administrative proceeding the Natural Resources Defense Council tried to evade this rule by asking the Civil Aeronautics Board to comply with §103 of NEPA, which, NRDC argued, required the CAB to spell out how it might use its statutory authority to set rates and approve mergers and air routes in a way which furthered the

[106] Appendix B.

[107] *In re* El Paso Natural Gas Co., Order Denying Motion to Terminate Proceeding and to Require Staff to Prepare and Circulate Environmental Impact Statement, Before the Federal Power Comm'n, 2 ELR 30025 (Aug. 22, 1972).

[108] 2 ELR at 30026.

national environmental policy. In dismissing the petition, the board put the burden right back on NRDC:

> In short, the Board believes that its approach to environmental problems is in full compliance with the policies of NEPA. While we are always receptive to suggestions for improvement, the instant complaint falls far short of the type of concrete proposals that would assist us in refining our procedures.[109]

Formal but nonadversarial reliance on NEPA as a basis for federal action has occurred in a few instances. Executive orders establishing the Refuse Act Permit Program, regulating the use of off-road vehicles on public lands, and barring the use of poisons in federal predator control programs have drawn on NEPA as part of their statutory authority.[110] The Forest Service has modified the manner in which it will implement the multiple use concept, relying in part on NEPA.[111] It has also relied in part on NEPA in issuing proposed regulations governing mining on national forest lands.[112] Similar reliance has been attributed by the CEQ to the Corps of Engineers for its dredge-and-fill permit rules, and to the Atomic Energy Commission for its rules which require licensees to account more carefully for long-term environmental costs.[113]

The instances in which federal agencies affirmatively assert their power under NEPA to take environmentally protective actions will probably be more important in the long run than the instances in which NEPA is forced upon them in citizens' lawsuits. Courts in general pay special attention to the agency's interpretation of its power under an enabling statute. Although NEPA is a special case because it applies to all agencies, nevertheless the courts would probably be impressed with an agency's claim to act where that claim is not in its obvious self-interest. Hence it is especially discouraging that to date the government has made very little affirmative use of the legal authority provided it by NEPA.

---

[109] Complaint of the Natural Resources Defense Council, Before the Civil Aeronautics Board, 1 ELR 30045 (July 26, 1971).

[110] CEQ Third Annual Report, *supra* note 30 at 227.

[111] Forest Service, Emergency Directive No. 1 (Nov. 9, 1971), Title 2100, Forest Service Manual.

[112] Released to selected recipients March 23, 1971. *See* Comment, *supra* note 104, 2 ELR at 10016.

[113] CEQ Third Annual Report, *supra* note 30 at 227.

VIII

The Impact Statement in Perspective

No one who has read the preceding chapters really needs to be convinced that the requirement for impact statements has dominated NEPA's interpretation by the courts. Almost every word of §102(2)(C) has been construed several times, with important variations in shades of meaning which have sometimes produced inconsistent results. Yet it should also be clear that the courts have not stopped at the limits of §102(2)(C); they have begun to review compliance with the broader NEPA goals set out in its other sections, as chapter VII explained. This chapter first summarizes the judicial interpretation of NEPA. It is intended to restore the perspective which may have been lost in the detailed discussion of specific holdings which dominates the preceding chapters.

SUMMARY

NEPA's legislative history has presented certain problems to the courts (see chapter I). In enacting NEPA, Congress' main objectives were to enlarge the federal agencies' basic mandates through enactment of the national environmental policy, to establish action-forcing procedures for the implementation of that policy, to create the Council on Environmental Quality, to foster the development of information and indices of environmental quality, and to provide for an annual progress report. During the past three years, Congress' intentions with regard to the action-forcing requirements have proved to be much more important from the courts' point of view than Congress' intentions with respect to its other objectives. Yet in the legislative history Congress gave practically all of its attention to the lack of reliable information, the need for a policy, and the need for new federal institutional responses. The critical action-forcing requirements received short shrift.

In NEPA's meager legislative history Congress nevertheless succeeded in establishing that it meant the action-forcing procedures to be taken seriously. The requirement of compliance "to the fullest extent possible" was interpreted by Senator Jackson to mean that only a conflict of statutory authority would excuse an agency from carrying out all of the specific procedures of §102 (see chapter III). Further, and in spite of an ambiguity which still persists regarding actions taken for environmental protection, every agency was to comply fully. The groundwork was sparse but adequate.

The courts' role in implementing NEPA has resulted more from what Congress did not say than from what it did. Congress did not discuss how (or whether) the agencies were to be policed as they complied with NEPA. Judicial review was not directly discussed, while the roles of the CEQ and the Office of Management and Budget (OMB) received only passing attention. These oversights are strange in view of the role played in obtaining NEPA's passage by political scientists and other students of the administrative process. Stranger still, Congress' many lawyers did not seem interested in the complex questions involved in achieving compliance with NEPA's novel action-forcing requirements. The OMB apparently was to play a key role, although the promulgation of an executive order shortly after NEPA's passage greatly enhanced CEQ's claim to the title of executive overseer of NEPA's implementation. When enforcement from the Executive Office lagged, the courts moved to fill the vacuum and began to review issues such as whether an impact statement had to be prepared or whether it was adequate. The courts willingly applied CEQ and agency guidelines when they become available, but not before the courts had established a commanding presence as interpreter of the multitude of "procedural" questions of law presented by the legislation.

Whether active judicial review would generally occur was an open question for some time after NEPA was enacted (see chapter II). A variety of complex factors undoubtedly enhanced the willingness of courts to hear NEPA cases. Chief among them, perhaps, was the way in which NEPA prescribed administrative reform through a set of requirements and procedures that strongly reinforced the traditional goals of judicial review. Although NEPA focuses on agency decision making which affects the environment, it prescribes standards for such decision making which are very much like those applied by reviewing courts in general. Both NEPA and (increasingly) the canons of review call for the establishment of procedures for principled decision making, for articulation of the reasons behind decisions, for the elaboration of the risks which proposed action entails, for discussion and consideration of alternatives as a test of the soundness of decisions taken, for a wider view of the public interest under relevant legislation, and for increased public participation. The earlier passage of several statutes enlarging agency decision-making responsibilities along these lines where the environment was threatened, and the concurrent tightening up of judicial review in general, further enhanced NEPA's sympathetic reception.

Whatever the reasons for undertaking review, in three years the courts have actively examined agency compliance with NEPA in a number of key areas. First, the courts have reviewed preliminary agency determinations whether or not to prepare an impact statement, usually applying the arbitrary and capricious standard. The frequency of remands, however, suggests that a stricter standard is in fact being applied. A few courts explicitly apply a higher standard that appears to amount to a *de novo* redetermination. Second, the Act's policy of full disclosure strongly influences the courts to review carefully the level of thoroughness of statement preparation. Usually such review takes the form of a *de novo* determination whether or not as a matter of law particular factors required to be discussed by the detailed subsections of §102(2)(C) have been omitted from the agency's statement or have been given only cursory treatment. Third, courts have begun to review the actual consideration given environmental factors by the decision makers. Fourth, courts have begun to review proposed decision making on its merits to see if it complies with NEPA's substantive policies. Together, these forms of review ensure judicial oversight from beginning to end as agencies implement the Act.

The citizens' suit under NEPA, like the citizens' suit under other legislation, benefits from the expansion of the requirements for standing which began in the mid-1960s. Practically without exception, the NEPA cases allow standing. Moreover, NEPA arguably creates new legal interests, so that harm to the public's right to know, to participate, and to have the interests of future generations protected may constitute injury in fact that can be alleged to support standing. It remains unclear whether NEPA suits brought by industry to compel the preparation of impact statements by the Environmental Protection Agency rest upon the type of "injury in fact" which arguably lies within the "zone of interests" protected by NEPA, so that the Supreme Court's two-part test for standing is satisfied. An industry plaintiff's alleged economic harm from environmental regulation by EPA would not appear to lie within NEPA's zone of interests, nor would it seem that the types of injury in fact which do lie within that zone (environmental, aesthetic, etc.) can constitute injury in fact to an industrial corporation. In view of the possibility that such suits are being brought in large part merely to delay the implementation of EPA's programs, the courts may examine more closely whether such plaintiffs possess a sufficient adversarial interest in seeing NEPA effectively implemented.

NEPA suits by and large have sought the preparation of an adequate impact statement under §102(2)(C). The courts have construed that section's requirements so that it has broad application to a very wide range of federal activities and agencies (see chapter IV). Regarding the degree of federal participation necessary before §102(2)(C) applies, if the section's other requirements are met, the cases implicitly confirm that the section applies where federal agencies provide services and carry out direct construction programs, as it does also where agencies permit, license, certify, or otherwise regulate private conduct. Only a few cases have been decided which involve a private request to a federal agency for use of a federally controlled area, but those cases either assumed that §102(2)(C) applied or affirmed that it did with the briefest mention. Section 102(2)(C) has also been applied without undue difficulty to federal contracts, grants, and loans to private parties.

Federal aid to states has posed more serious problems. Section 102(2)(C) has been applied easily to the grant of funds where the federal agency participated additionally in the project in a significant manner. But block grants and revenue-sharing pose an important test for the widespread applicability of §102(2)(C). The *Ely* case is a strong precedent for its application to block grants, and the logic of that case might apply to revenue sharing as well.

The federal-state matching grant program for highway construction has led to rather extensive litigation regarding whether the "federal" presence is sufficient to require compliance with the impact statement requirement, which has been resisted on the grounds that the state will carry out the highway project in question solely with state funds and that statement preparation would be premature because federal funding or approval has not yet been sought. The cases are divided on the issue, but the opinions finding various kinds of federal involvement in overall highway construction sufficient to require a statement appear to have the edge, both in logic and authority, over those which hold the other way. The basis for finding that an impact statement must be prepared although the "federal presence" has not yet been fully manifested is that later compliance, after some alternatives have been foreclosed as a practical matter, is unlikely to give the federal decision maker the full choice of options which NEPA contemplates. The evasion of the federal requirement by the careful pooling of state funds in challenged projects, while closely related projects receive federal approval and funding, has not been allowed.

The courts have broadly construed the key phrases "major action" and "significantly affecting" to include the maximum number of federal actions, as the relative magnitudes of the projects involved in the cases show. While shying away in general from ready-to-use formulas for applying the phrases, the courts have used NEPA's broad language and the CEQ Guidelines to reinforce their determination that §102(2)(C) applies to projects of modest size. Agency procedures have also played a role in some cases by defining a lower threshold for NEPA's application than the court might have set without the agency's guidance. Agencies have also helped to lower the threshold of §102(2)(C)'s applicability by taking steps which the courts have interpreted as a concession that a statement should be prepared. Finally, by preparing statements on projects of modest size which are challenged on other NEPA grounds than magnitude (e.g., "retroactivity," adequacy, timing), agencies appear to concede that the action is "major" or "significantly" affects the environment. Such instances may then be used later to illustrate the lower limits of §102(2)(C).

In order to enhance the importance of a proposed action so that §102(2)(C) applies, courts in a few cases have looked at the broader implications of the action, its potential cumulative effects, or its place in a larger scheme of development. An emerging judicial tendency to define categories of federal action to which §102(2)(C) applies *per se* may also lower the "major" federal action requirement. Particular federal activities are identified in NEPA's legislative history as prime environmental offenders (e.g., highway projects, water resources projects, electrical generating facilities). The courts may simply endorse statement preparation for entire categories of activities, without examining them on a case-by-case basis.

An additional trend in interpretation of §102(2)(C) can be traced to the suggestion in the CEQ Guidelines that highly controversial actions should always be covered in impact statements. In spite of some contrary authority, it appears that the cases which allow citizen opposition to qualify as "controversy" are the better reasoned ones. It may be necessary to develop a test of good faith opposition, but that alternative is preferable to abandoning the insightful CEQ requirement that recognized the value of citizen opinion about environmental degradation.

Finally, the courts have engaged in a surprisingly limited number of direct comparisons of the magnitude of various projects. Yet with an increasing number of cases in which the magnitude or impact of projects is critical, such comparisons will undoubtedly increase.

A literal reading of §102(2)(C) suggests that two standards have to be met before an action is important enough to merit coverage in an impact statement. First, actions may have to be found to be "major." Then they must be found to have a "significant" effect on the environment. The courts which have explicitly considered the issue are divided but, surprisingly, favor a two-test standard. The balance is partially restored by the many courts which resolve the question by simply assuming that NEPA must cover all actions which may have significant environmental impact. Moreover, the adoption by some courts of a two-test standard has never resulted in a determination that an action with important potential environmental impact need not be covered in a statement because it is not "major." The issue thus appears to be of importance only because courts have paid lip service to a semantic two-part standard. The word "major" has not been used to shield agency action in instances where Congress plainly contemplated the application of NEPA's procedures.

Judicial review of the agency determination whether or not to prepare a statement for a specific action is one of the courts' most important roles in enforcing NEPA. Early in NEPA's history the courts began to stand at the gates to the §102 process, refusing to allow them to be closed except to the most inconsequential actions. A bare majority of courts have favored review under the arbitrary and capricious standard, although several courts have endorsed higher standards which seem to amount to *de novo* review. The courts which state that they are reviewing under the arbitrary and capricious standard, however, have frequently found that the agencies did in fact abuse their discretion.

The reality of close judicial scrutiny of these "threshold determinations" suggests that review *de novo* is more appropriate. The absence of systematic public review by CEQ, OMB, or EPA also argues for strong judicial oversight. Deference to administrative expertise is appropriate where a focused statute pinpoints an ill which the agency specializes in treating, but NEPA applies across the board to all agencies and calls upon them to act against their self-interest. Thus the circumstances are congenial to close scrutiny by "generalist" courts which can help ensure NEPA's uniform implementation throughout the federal government.

By its terms, NEPA applies to *all* agencies of federal government.

An ambiguous exchange between Senators Jackson and Muskie in the legislative history suggested, however, that federal environmental protection activities were exempt. Yet "environmental" agencies other than the Environmental Protection Agency have routinely prepared statements, and debate has now focused on EPA's responsibilities under the Act. The few cases which even mention the issue are divided; only one district court opinion has discussed the question in any detail.

Congress recently became involved in the issue by passing §511 (C)(1) of the Federal Water Pollution Control Amendments of 1972. That section exempts EPA from preparing impact statements on all aspects of the water quality program, except the issuance of discharge permits for new sources and the grant of funds for publicly owned waste treatment works. Congress apparently has opted for a program-by-program approach to the problem, starting with water pollution. Yet it may well be the courts which first decide how two other EPA programs will be handled. About thirty lawsuits seeking impact statements on actions taken under the Clean Air Act have been filed, and one of these has already resulted in a district court decision requiring EPA to comply. At least one litigation involving EPA's pesticide program is pending.

Regarding whether NEPA should apply to EPA's programs, we take the position that long-term benefit to EPA's overall environmental role would occur if it did. Short-term dislocations would be compensated by the increased likelihood that EPA decisions to further one environmental goal would be taken in awareness of the possible impact on other environmental concerns; by the improvement NEPA would bring to planning and interagency coordination and sharing of expertise; by the benefits of public participation through the comment process; and by the *Calvert Cliffs'* requirement for a balanced weighing of costs and benefits. The arguments against compliance are strong, but they are not persuasive that EPA's performance could not be improved through the NEPA procedures.

One of the most neglected aspects of the §102(2)(C) requirement is its application to legislative proposals originating within the agencies. Two of the four cases interpreting the requirement applied it to appropriations requests; a third refused to do so. The remaining case construed the requirement narrowly and avoided its application to a statutorily required report on transportation needs, including legislative needs. No case has yet specifically dealt with the problem of the relief to be granted when a legislative proposal is not accompanied by a statement. Can the agency be required to withdraw the legislation? The dilemma of appropriate relief suggests a larger role for Congress, OMB, and CEQ in policing the requirement, yet all of these have been noticeably remiss in calling for compliance.

A few categories of actions and agencies would appear to be eligible on one ground or another for total exemption from the requirement of NEPA §102(2)(C). However, the courts have been amazingly stingy in dispersing exemptions. Only one agency—the Price Commission—has been exempted as an agency from NEPA requirements, while some actions (not all) affecting national security are the only kinds of actions exempted. The Price Commission was set up specifically to take speedy action adjudged to be inconsistent with NEPA's requirements for deliberation. Yet for other, more permanent agencies which take temporary or emergency action, the *SCRAP* opinion endorses the case-by-case approach, which appears also to have evolved for national security matters.

The application of §102(2)(C) to projects that were not completely approved or constructed by the date of NEPA's enactment has been an issue in over fifty decided cases. A summary cannot do the cases justice, because they weave what is clearly the most tangled web of NEPA law (see chapter V). Nevertheless, it is fair to say that the courts have once again applied NEPA to the fullest extent possible, stinting only where reappraisal of the basic course of action was virtually impossible. Projects in which up to 70 percent of the funding had been obligated have been held to require a statement, as have projects very near completion on the ground.

We discern three important categories of cases. The first involves the federal "action" of committing funds to a private party, regulating his conduct, or authorizing certain behavior. In this area the courts have tended to apply a "critical action" approach, looking for a legal formality which exempts the agency from statement preparation if the action was concluded before NEPA's effective date and if no significant federal involvement lingers.

The second set of cases focuses upon a different meaning of "action." These cases look to the progress of projects on the ground, and if substantial federal "action" remains to be completed, they apply NEPA to that incremental action. Finally, the third group of cases falls between the first two and constitutes a hybrid category of its own. Exemplified by the highway suits, these cases involve continued federal presence in actions that are taken by private parties or by the states. The courts have caused considerable confusion in this area by failing to distinguish systematically between formal administrative action and action on the ground.

After the preliminary issues regarding NEPA's applicability have been resolved (see chapters IV and V), the agency must settle down to the task of impact statement preparation by fulfilling specific procedures and requirements of §102(2)(C). The basic issues as they have developed in litigation concern the timing of preparation, who is to prepare the statement, what must be included, and how comments on statements are to be obtained (see chapter VI). As always, the standards governing the preparation and content of statements are conditioned by the requirement of compliance "to the fullest extent possible" (see chapter III).

The requirement of strict compliance has been interpreted to mean that statements must be prepared at the earliest possible time for every distinct stage of agency decision making. The main reason for the requirement is that the coordinated advance planning and consideration of options which NEPA requires may be curtailed if commitments of money and prestige are made before the NEPA review is conducted. This is as true for the remaining portions of ongoing projects as it is for new federal initiatives. Yet some cases seem to approve the initiation of §102(2)(C) review after early and conceivably quite important planning actions have already been taken. Although these cases do not say that NEPA's general requirement for environmental awareness in all federal activities does not apply to early initiatives, they do seem willing to delay somewhat formal §102(2)(C) reviews.

Conflicting views have developed about the extent to which NEPA allows the task of preparing statements to be passed on to state agencies or private parties. The Act is clear that a "responsible federal official" must prepare the statement; however, several agencies have attempted to redefine "preparation" to mean cursory "approval," at least as far as draft statements are concerned. The Second Circuit's decision in *Greene County* indicated that the courts will probably not permit broad delegation of this responsibility in the future, even in agencies which must decide on the record in quasi-judicial proceedings. Courts in several challenges to highway projects have already considered the issue, but no trend has yet emerged. In our opinion, the intent of the Act to place the basic responsibility for "preparation," i.e., synthesis and evaluation of information, on the "responsible federal official" should be reaffirmed, for the reasons set out in *Greene County.*

Many major federal actions require approvals or direct participation by more than one agency. Such actions raise a problem regarding how §102(2)(C) should be complied with in these circumstances.

The CEQ has endorsed the lead agency concept, by which the agency best able to prepare the statement undertakes to compile one comprehensive statement for all agencies involved. The danger is that the exempt agencies will not have brought home to them the full extent of their new environmental obligations, as the leading case in this area, *Upper Pecos,* illustrates. We think that of the three main alternatives for handling multi-agency actions—lead agency, joint preparation, and separate compliance—the lead agency approach is the least desirable.

The courts have measured the adequacy of the statement itself by the standard of "full disclosure," which itself is partially qualified by a "rule of reason" governing the length and detail of the agency's investigations. Full disclosure enables the agency decision makers, the Congress, the President, and the public to review the decision made. With respect to the agency decision makers, who have the first direct opportunity to use the information developed, the courts have required that they actually "consider" the contents of the statement and engage in a systematic tradeoff of competing factors (see chapter VII). Judicial interpretation of the public's role in developing adequate statements and in reviewing complete statements has given the public a larger role than might be expected from a reading of the statute alone.

When actually reviewing the contents of challenged statements, the courts have elaborated the policy of full disclosure into more specific requirements. They have said that statements must be understandable and nonconclusory, that they must refer to the full range of knowledge, and that they must discuss certain impacts which are typical of some types of action. Especially important is the use made in the statement of responsible dissenting scientific opinion, as illustrated by comment offered on the statement in the Cannikin litigation. Further, even appropriate references to supporting information may be inadequate where important reasons exist for fuller disclosure of underlying logic (e.g., public need, lack of specificity about alternatives, etc.).

The impact statement's discussion of what will actually happen once the proposed action is under way in some instances must be based on guesswork. The cases differ over whether the statement must simply point out the gaps in knowledge or whether research must be conducted to reduce the uncertainty which the decision maker faces. The need for the information must be balanced against the cost of obtaining it; however, the courts have been quite willing to read in NEPA a requirement that agencies not only *disclose* what they can easily learn, but that they affirmatively *develop* additional information for inclusion in the statement.

One of the most important—and controversial—of the detailed requirements of §102(2)(C) calls for a discussion of alternatives to the proposed action which might cause less environmental damage. The requirement is buttressed by §102(2)(D), which employs similar language. The courts have required a full discussion of alternatives in order to satisfy Congress' intent to affect federal policy choices through NEPA. The range of alternatives to be discussed at least includes no action at all; an equivalent action with few, if any, of the environmental side effects of the original proposal; modified action on a reduced scale that concomitantly reduces adverse impacts; and mitigation to reduce the proposal's impact. The depth of discussion of each of these must be sufficient for a reviewer to make a reasoned choice, in the view of the leading case on alternatives, *Natural Resources Defense Council v. Morton.* This same fertile circuit opinion also held that alternatives beyond the power of the agency to implement must be discussed for the benefit of other agencies, the President, and the public. As in other aspects of adequacy, the discussion is tempered by the "rule of reason."

The concluding portion of §102(2)(C) establishes procedures for obtaining comments from outside the agency in time to affect final agency decision making. Judicial interpretation has played an important role in fleshing out these procedures by requiring that they be followed as a condition of the statement's adequacy. Comment must be integrated meaningfully into the statement itself, including comment from the lay public and from responsible dissenting scientists. The court in *Lathan v. Volpe* even suggested that "relevant and reasonable" doubts raised by the public might have to be resolved by research before the final statement could be held to be adequate.

The Environmental Protection Agency has a separate statutory obligation under §309 of the Clean Air Act requiring it to comment on impact statements. Should EPA's administrator find the impact of proposed agency action to be environmentally "unsatisfactory," he must publish his finding and "refer" the matter to the CEQ. Only one case has applied §309 to date, but it is at best ambiguous and is of questionable relevance to the course subsequent litigation involving §309 might take. The legislative history suggests that §309 might be used to obtain interdepartmental or Executive Office review of controversial actions where conflict has developed.

Except where there have been some extraordinary equities to the contrary, the courts have enjoined projects proceeding in violation of NEPA until the mandates of the Act were met. Plaintiffs in NEPA cases usually seek preliminary injunctions, but in some NEPA suits the courts appear to have leapfrogged to the merits of a permanent injunction, because at the hearing on the preliminary injunction the court was virtually certain what its final ruling on the law would be. In other cases, courts have held that the factors to be balanced have been altered by NEPA's requirements and legislative history, so that the equities are heavily weighted by law in favor of granting relief. The few cases denying an injunction rest upon the grounds that substantial compliance had been achieved, that compliance was only a few days away, or that because the project was begun before NEPA was enacted and was nearly complete, the impact statement need not be prepared as a matter of law.

Practically the entire foregoing summary is concerned with the specifics of judicial interpretation of §102(2)(C). Yet that statutory requirement is useful only in the service it performs to force action on NEPA's substantive policy. That policy seeks to reform agency decision making through the enlargement of agency mandates, so that agencies consider environmental factors where relevant in all of their activities (see chapter VII). In addition to interpreting the information and disclosure requirements of §102(2)(C), *Calvert Cliffs'* and other opinions have spelled out how the agencies must use that information in reaching final decisions. The consideration which they give environmental factors has been held to be fully reviewable, with the impact statement playing an important role as part of the reviewable record. Further, several courts have held that agency decisions are reviewable on their merits to determine if they are in accord with NEPA's substantive policy. These cases indicate the probable principal area of judicial concern with NEPA in the coming months. Together with the few judicial holdings that interpret the action-forcing sections other than §102(2)(C), and with the very limited reliance on NEPA by the government, these additional NEPA holdings help put the impact statement to work in proper perspective.

CONCLUSION

What have three years of litigation under NEPA accomplished? The answer varies depending upon one's expectations. If the standard is the extent to which litigation has achieved NEPA's ultimate goal of a better environment through better federal decision making, then apparently the cases that we have discussed have not accomplished very much. If a lesser standard is acceptable, litigation has accomplished a great deal. The courts have strictly enforced NEPA's procedural requirements and have gone a long way toward ensuring that adequate agency decision-making machinery exists so that better substantive results may be obtained in the future.

Very few definite instances of NEPA's having materially altered a federal program or project have been offered,[1] and no one has yet come forward with more than a subjective impression that NEPA has begun to cause the fundamental change which Congress intended. Chairman Russell Train of CEQ has stated in Senate testimony that NEPA's implementation is iceberglike; the showy, controversial tip that involves public attention or litigation is complemented by the massive bulk of change that lies beneath the surface, hidden from view but nevertheless fundamentally altering the way the federal government responds to environmental problems.[2] This point of view is more or less endorsed in CEQ's third annual report.[3] The problem with this perspective is that it rests more upon the amount of attention that NEPA has received in the agencies, which is admittedly great, than upon the results it has obtained. We think this because it is our impression that at present most agencies are anxious to make public any instances in which they in fact relied on NEPA in changing projects.[4] Aside from CEQ, most other reports and analyses are considerably more skeptical.[5]

We can add only our own subjective opinion, based on the

---

[1] COUNCIL ON ENVIRONMENTAL QUALITY, ENVIRONMENTAL QUALITY, THIRD ANNUAL REPORT at 226–27.

[2] Joint Hearings, *supra* note 44, chapter VII at 29.

[3] CEQ Third Annual Report, *supra* note 1 at 257.

[4] The "in-depth" interview of Corps of Engineers personnel of the Wilmington, North Carolina district by Bob Boyd, "The 'Corps' Shows Ecology Soft Spot," THE CHARLOTTE OBSERVER (January 8, 1973), illustrates the point. The Corps' district engineer welcomed the favorable publicity for the Corps' decision to deny a permit for pier construction to serve a recreational and golfing development on Bald Head Island. The Corps relied on NEPA and Zabel v. Tabb, Appendix B, and labeled the project "contrary to the public interest." The article goes on to cite other instances supporting the district engineers' enthusiastic assertions that the Corps had turned over a new leaf.

[5] *E.g.*, Comptroller General of the United States, ADEQUACY OF SELECTED ENVIRONMENTAL IMPACT STATEMENTS PREPARED UNDER THE NATIONAL ENVIRONMENTAL POLICY ACT OF 1969, REPORT TO THE SUBCOMM. ON FISHERIES AND WILDLIFE CONSERVATION OF THE HOUSE COMM. ON MERCHANT MARINE AND FISHERIES, B–170186 (November 27, 1972). *See also* A. Tarlock, *Balancing Environmental Considerations and Energy Demands: A Comment on Calvert Cliffs' Coordinating Committee, Inc. v. AEC*, 47 INDIANA L. J. 645, 671–72 (1972); Note, *The National Environmental Policy Act: A Sheep in Wolf's Clothing?*,

decided cases. The impression we have is that with few exceptions the agencies have not yet begun to take NEPA's substantive mandate very seriously. In general they have complied by preparing superficial analyses of environmental impact, usually after basic proposals were well along in agency review processes. Thus we fear that NEPA litigation has been primarily successful in stimulating after-the-fact rationalizations which are examined less by agency decision makers than by agency lawyers, whose job it is to ensure that the agency's environmental review can survive legal challenge. Again, our perspective is limited by what we can infer from the decided cases, which focus on situations where things have gone wrong, not on situations where they have gone right. Yet in order to review particular decisions, courts may probe entire agency processes for complying with NEPA, so that the insight afforded by the decided cases is valuable, even if its value cannot compare to direct study, agency by agency, of the use and abuse to which NEPA has been put.

NEPA's slow progress in bringing about the vast changes Congress intended does not mean that the Act is a failure. To the contrary, three or four years may be about right for swinging NEPA's systematic, sometimes cumbersome procedures into place. That process has been slow because Congress did not spell out the details of the §102 requirements, necessitating much time-consuming litigation for interpretation. Further, the agencies have resisted implementation of the Act, because it is a "reform statute" that deliberately cuts against the grain of established agency values and practices.

The accomplishments of NEPA litigation that fall short of the

---

37 BROOKLYN L. REV. (1970); R. Gillette, *National Environmental Policy Act: How Well Is it Working?*, 176 SCIENCE 146 (April 14, 1972). Richard N. L. Andrews, in his Ph.D. thesis cited *supra* note 6 in chapter I, concludes that the "instruments of administrative change" built into the law, even when coupled with vigorous judicial review, have been generally insufficient for the attainment of NEPA's basic purposes, at least as far as the four federal programs studied were concerned: the civil works program of the Corps of Engineers; the Small Watersheds Program of the Soil and Conservation Service; the water resource development program of the Tennessee Valley Authority; and the nuclear reactor licensing program of the Atomic Energy Commission. In his unpublished paper entitled "EPA Comments on Impact Statements: One Indicator of Administrative Response to the National Environmental Policy Act of 1969," Richard A. Liroff of the Brookings Institution found that in EPA's published classifications of impact statements for the period November 1971 to July 13, 1972, the agency had classified 51.9 percent of the statements reviewed as at least inadequate.

ultimate goal of a better environment are considerably more encouraging. These accomplishments have helped ensure that the machinery for proper agency review of actions is in place, even if it has not yet been extensively used to achieve better substantive results. The summary shows that over the past three years the courts have broadly interpreted and applied a variety of key NEPA provisions. "All" agencies had to comply "to the fullest extent possible." In cases like *Calvert Cliffs'* and *Greene County*, the courts directly examined agency procedures for implementing NEPA and invalidated them where they did not meet the Act's strict requirements. The courts faced and largely overcame the difficult problem of applying NEPA to the welter of federal projects and activities that antedated the Act's passage. They also began to evolve an array of public rights to participate in NEPA reviews, to receive environmental information prepared by the agencies, and to sue to vindicate these rights if necessary. In opening decision making further to the light of day, the courts enforced NEPA's provisions for interagency and intergovernmental comment, playing on the fruitful tensions which Congress deliberately created. Through review of the adequacy of impact statements, the courts exercised a large measure of control over the thoroughness and competency of the environmental assessments made. If the foregoing accomplishments did not ensure different agency decisions, they at least established that facts which might support different decisions had to be disclosed and discussed.

The key to these accomplishments is something of an accomplishment in itself. Vigorous judicial review, as we have already pointed out, was not a foregone conclusion when NEPA became law. As the cases began to come down, however, it became clear that the courts would review agency compliance with NEPA's provisions from start to finish. At the risk of repetition and oversimplification, the key reviewable agency actions under NEPA are the following:

1. The courts will review how the agencies organize (and the methods used) to decide on a course of action subject to NEPA.
2. The courts will closely review whether the agency must prepare an impact statement.
3. The courts will review *de novo* whether the information disclosed is adequate.
4. The courts will review how the agencies considered the information they developed.
5. The courts will review whether the final decision made is substantively correct under §101.

In setting the scope of this book, we decided not to formulate sweeping recommendations based on our analysis of the cases. At this point in NEPA's interpretation, and with the time available to us, we thought that a close analysis of the patterns and trends in NEPA litigation would be useful enough in itself. During our analysis, however, as the preceding chapters show, we did occasionally endorse a particular line of cases, disapprove the course that other litigation had taken, and even suggest major changes in NEPA's administrative interpretation. These included endorsement of *de novo* review of threshold determinations whether or not to prepare an impact statement; approval of the preparation of impact statements by EPA; disapproval of agency delegation of statement preparation; concern over the handling of multi-agency actions through the "lead agency" approach; encouragement to EPA to use §309 of the Clean Air Act as a basis for systematically monitoring the quality of impact statements; disagreement with the CEQ guideline allowing only 90 and 30 days for comment on draft and final statements, respectively; doubt that *Calvert Cliffs*' "balancing" can be made viable; disapproval of the use of the impact statement as the entire record of decision; and, finally, approval of judicial review of compliance with the substantive duties of §101.

Intermittently, we endorsed what we believe to be a more rational approach to impact statement preparation through overlapping "tiers" of statements. (See pages 108, 122, 177, and 220.) We think this suggestion is important enough to merit a final word here. Under this approach, the first statement prepared would cover pending legislation or broad, new federal policies; statements to follow would be prepared as each distinct initiative in implementing the legislation or policy was formulated. The later statements would cover increasingly specific programmatic initiatives and impacts, and would refer back to the wider, policy-oriented statements for their treatment of far-ranging alternatives and basic federal policy.

Numerous advantages could be obtained through this approach. It would establish a record of least-cost, gradually circumscribed decision making without subjecting the agency to reconsideration of basic principles each time a specific action was contemplated. For instance, the alternative of flood-plain zoning could be exhaustively considered in an early comprehensive statement on the best way to manage a river basin; that alternative need not then be comprehensively reconsidered in statements on particular projects if the decision is made to construct a series of dams or river levees. The latter statements could focus on localized impacts, without the agency's having failed to give the comprehensive, early environmental review called for by NEPA.

The approach would have a large impact in areas such as EPA's pesticide registration program, where thousands of federal permits and licenses are issued, with NEPA apparently requiring a statement on each one. Comprehensive statements on rule making, on entire programs, and on parts of programs would make it possible to give brief, focused attention to a specific license's impact (if significant) without lengthy discussion of the wider issues covered in earlier statements.

Litigation under NEPA has focused on particular projects that implement policies formulated much earlier in the federal decision-making process. For these specific initiatives, the courts have been very thorough in requiring compliance with NEPA, and in a sense this book is the chronicle of their accomplishments. Yet at the same time strict implementation for specific, eminently litigable projects lacks the "leverage" which NEPA's application at a much earlier time could exert to bring about sound environmental planning. NEPA's legislative history often states that Congress wanted federal planners to take environmental factors into account at the moment they began to formulate their ideas and proposals, so that environmental awareness and responsibility would be infused into the very fabric of the federal government. The action-forcing requirements obviously could not guarantee the desired change and could not legislate a new federal ethic, but their success clearly was intended to be measured by the extent to which they pushed federal officials toward this result. In view of Congress' intention, the multiple-tier approach to statement preparation could bring about a great improvement.

While it is possible to criticize NEPA litigation because it has focused attention on the late phases of decision making, the court-imposed requirements for statement adequacy may also be interpreted as putting agencies on notice that if they do not consider environmental impacts during the earliest possible phases of policy formulation and decision making, they may have to back up and consider them in impact statements which are broader in scope than might be required had such early consideration actually taken place. The leading case of *Natural Resources Defense Council v. Morton*, for instance, is clearly open to this interpretation.[6] By having the

[6] Appendix B. *See* Comment, *NEPA and Federal Policy-Making: NRDC v. Morton, Legislative Impact Statement, and Better NEPA Procedures*, 2 ELR 10038 (April 1972).

agencies comply late in the process, the courts have arguably imposed unrealistic requirements for adequate and comprehensive consideration of environmental values. Perhaps the vigorous, court-imposed requirements would cause less difficulty if earlier steps to implement NEPA were taken, such as the preparation of policy and program statements, as recommended here.

One may only speculate why the courts have not been asked to hear cases calling for statements on early, policy-oriented agency actions. Citizens' suits are often local in focus and come into being only when a specific environmental threat, such as a highway or a dam, materializes. Federal policy making also may be conducted behind closed doors, so that basic commitments have been made before possible plaintiffs realize their interest. As we pointed out in regard to legislative impact statements, finding a suable defendant may be difficult because the Executive's freedom to formulate broad policy is unamenable to suit. Finally, as we have mentioned, some case law takes a narrower view of federal "action" than is necessary. A federal proposal may be both distinct and comprehensive, although formal steps have not been taken toward its realization.

To sum up, if a summary of an undertaking such as this is at all useful, NEPA's first three years have been marked primarily by a focus on procedural compliance with the Act. The procedures initiated by §102 at their best have caused more adequate disclosure of possible environmental impacts and, in some instances, have led to useful debate and discussion about them. A beginning has also been made toward NEPA's ultimate goal of better federal decison making, so that in the next few years the promise of the litigation discussed in this book could conceivably be realized.

APPENDIX A

*The National Environmental Policy Act of 1969**

An Act to establish a national policy for the environment, to provide for the establishment of a Council on Environmental Quality, and for other purposes.

Be it enacted by the Senate and House of Representatives of the United States of America in Congress assembled, That this Act may be cited as the "National Environmental Policy Act of 1969."

PURPOSE

Sec. 2. The purposes of this Act are: To declare a national policy which will encourage productive and enjoyable harmony between man and his environment; to promote efforts which will prevent or eliminate damage to the environment and biosphere and stimulate the health and welfare of man; to enrich the understanding of the ecological systems and natural resources important to the Nation; and to establish a Council on Environmental Quality.

---

* 42 U.S.C. §4321 *et seq.*, 83 Stat. 852, Pub. L. 91–190. The United States Code sections correspond to the section numbers of the Public Law in the following manner:

Section 2 is 42 U.S.C. §4321
Section 101 is 42 U.S.C. §4331
Section 102 is 42 U.S.C. §4332
Section 103 is 42 U.S.C. §4333
Section 104 is 42 U.S.C. §4334
Section 105 is 42 U.S.C. §4335
Section 201 is 42 U.S.C. §4341
Section 202 is 42 U.S.C. §4342
Section 203 is 42 U.S.C. §4343
Section 204 is 42 U.S.C. §4344
Section 205 is 42 U.S.C. §4345
Section 206 is 42 U.S.C. §4346
Section 207 is 42 U.S.C. §4347

Alphabetical and numerical subsections are the same in the Public Law and the United States Code. Only Public Law citations are given in the text.

TITLE I

Declaration of National Environmental Policy

SEC. 101. (a) The Congress, recognizing the profound impact of man's activity on the interrelations of all components of the natural environment, particularly the profound influences of population growth, high-density urbanization, industrial expansion, resource exploitation, and new and expanding technological advances and recognizing further the critical importance of restoring and maintaining environmental quality to the overall welfare and development of man, declares that it is the continuing policy of the Federal Government, in cooperation with State and local governments, and other concerned public and private organizations, to use all practicable means and measures, including financial and technical assistance, in a manner calculated to foster and promote the general welfare, to create and maintain conditions under which man and nature can exist in productive harmony, and fulfill the social, economic, and other requirements of present and future generations of Americans.

(b) In order to carry out the policy set forth in this Act, it is the continuing responsibility of the Federal Government to use all practicable means, consistent with other essential considerations of national policy, to improve and coordinate Federal plans, functions, programs, and resources to the end that the Nation may—

(1) Fulfill the responsibilities of each generation as trustee of the environment for succeeding generations;

(2) Assure for all Americans safe, healthful, productive, and esthetically and culturally pleasing surroundings;

(3) Attain the widest range of beneficial uses of the environment without degradation, risk to health or safety, or other undesirable and unintended consequences;

(4) Preserve important historic, cultural, and natural aspects of our national heritage, and maintain, wherever possible, an environment which supports diversity, and variety of individual choice;

(5) Achieve a balance between population and resource use which will permit high standards of living and a wide sharing of life's amenities; and

(6) Enhance the quality of renewable resources and approach the maximum attainable recycling of depletable resources.

(c) The Congress recognizes that each person should enjoy a healthful environment and that each person has a responsibility to contribute to the preservation and enhancement of the environment.

SEC. 102. The Congress authorizes and directs that, to the fullest extent possible: (1) the policies, regulations, and public laws of the United States shall be interpreted and administered in accordance with the policies set forth in this Act, and (2) all agencies of the Federal Government shall—

(A) Utilize a systematic, interdisciplinary approach which will insure the integrated use of the natural and social sciences and the environmental design arts in planning and in decisionmaking which may have an impact on man's environment;

(B) Identify and develop methods and procedures, in consultation with the Council on Environmental Quality established by title II of this Act, which will insure that presently unquantified environmental amenities and values may be given appropriate consideration in decisionmaking along with economic and technical considerations;

(C) Include in every recommendation or report on proposals for legislation and other major Federal actions significantly affecting the quality of the human environment, a detailed statement by the responsible official on—

(i) The environmental impact of the proposed action,

(ii) Any adverse environmental effects which cannot be avoided should the proposal be implemented,

(iii) Alternatives to the proposed action,

(iv) The relationship between local short-term uses of man's environment and the maintenance and enhancement of long-term productivity, and

(v) Any irreversible and irretrievable commitments of resources which would be involved in the proposed action should it be implemented.

Prior to making any detailed statement, the responsible Federal official shall consult with and obtain the comments of any Federal agency which has jurisdiction by law or special expertise with respect to any environmental impact involved. Copies of such statement and the comments and views of the appropriate Federal, State, and local agencies, which are authorized to develop and enforce environmental standards, shall be made available to the President, the Council on Environmental Quality and to the public as provided by section 552 of title 5, United States Code, and shall accompany the proposal through the existing agency review processes;

(D) Study, develop, and describe appropriate alternatives to recommended courses of action in any proposal which involves unresolved conflicts concerning alternative uses of available resources;

(E) Recognize the worldwide and long-range character of environmental problems and, where consistent with the foreign policy of the United States, lend appropriate support to initiatives, resolutions, and programs designed to maximize international cooperation in anticipating and preventing a decline in the quality of mankind's world environment;

(F) Make available to States, counties, municipalities, institutions, and individuals, advice and information useful in restoring, maintaining, and enhancing the quality of the environment;

(G) Initiate and utilize ecological information in the planning and development of resource-oriented projects; and

(H) Assist the Council on Environmental Quality established by title II of this Act.

SEC. 103. All agencies of the Federal Government shall review their present statutory authority, administrative regulations, and current policies and procedures for the purpose of determining whether there are any deficiencies or inconsistencies therein which prohibit full compliance with the purposes and provisions of this Act and shall propose to the President not later than July 1, 1971, such measures as may be necessary to bring their authority and policies into conformity with the intent, purposes, and procedures set forth in this Act.

SEC. 104. Nothing in section 102 or 103 shall in any way affect the specific statutory obligations of any Federal agency (1) to comply with criteria or standards of environmental quality, (2) to coordinate or consult with any other Federal or State agency, or (3) to act, or refrain from acting contingent upon the recommendations or certification of any other Federal or State agency.

SEC. 105. The policies and goals set forth in this Act are supplementary to those set forth in existing authorizations of Federal agencies.

TITLE II

Council on Environmental Quality

SEC. 201. The President shall transmit to the Congress annually beginning July 1, 1970, an Environmental Quality Report (hereinafter referred to as the "report") which shall set forth (1) the status and condition of the major natural, manmade, or altered environmental classes of the Nation, including, but not limited to, the air, the aquatic, including marine, estuarine, and fresh water, and the terrestrial environment, including, but not limited to, the forest, dryland, wetland, range, urban, suburban and rural environment; (2) current and foreseeable trends in the quality, management and utilization of such environments and the effects of those trends on the social, economic, and other requirements of the Nation; (3) the adequacy of available natural resources for fulfilling human and economic requirements of the Nation in the light of expected population pressures; (4) a review of the programs and activities (including regulatory activities) of the Federal Government, the State and local governments, and nongovernmental entities or individuals with particular reference to their effect on the environment and on the conservation, development and utilization of natural resources; and (5) a program for remedying the deficiencies of existing programs and activities, together with recommendations for legislation.

SEC. 202. There is created in the Executive Office of the President a Council on Environmental Quality (hereinafter referred to as the "Council"). The Council shall be composed of three members who shall be appointed by the President to serve at his pleasure, by and with the advice and consent of the Senate. The President shall designate one of the members of the Council to serve as Chairman. Each member shall be a person who, as a result of his training, experience, and attainments, is exceptionally well qualified to analyze and interpret environmental trends and information of all kinds; to appraise programs and activities of the Federal Government in the light of the policy set forth in title I of this Act; to be conscious of and responsive to the scientific, economic, social, esthetic, and cultural needs and interests of the Nation; and to formulate and recommend national policies to promote the improvement of the quality of the environment.

SEC. 203. The Council may employ such officers and employees as may be necessary to carry out its functions under this Act. In addition, the Council may employ and fix the compensation of such experts and consultants as may be necessary for the carrying out of its functions under this Act, in accordance with section 3109 of title 5, United States Code (but without regard to the last sentence thereof).

SEC. 204. It shall be the duty and function of the Council—

(1) To assist and advise the President in the preparation of the Environmental Quality Report required by section 201;

(2) To gather timely and authoritative information concerning the conditions and trends in the quality of the environment both current and prospective, to analyze and interpret such information for the purpose of determining whether such conditions and trends are interfering, or are likely to interfere, with the achievement of the policy set forth in title I of this Act, and to compile and submit to the President studies relating to such conditions and trends;

(3) To review and appraise the various programs and activities of the Federal Government in the light of the policy set forth in title I of this Act for the purpose of determining the extent to which such programs and activities are contributing to the achievement of such policy, and to make recommendations to the President with respect thereto;

(4) To develop and recommend to the President national policies to foster and promote the improvement of environmental quality to meet the conservation, social, economic, health, and other requirements and goals of the Nation;

(5) To conduct investigations, studies, surveys, research, and analyses relating to ecological systems and environmental quality;

(6) To document and define changes in the natural environment, including the plant and animal systems, and to accumulate necessary data and other information for a continuing analysis of these changes or trends and an interpretation of their underlying causes;

(7) To report at least once each year to the President on the state and condition of the environment; and

(8) To make and furnish such studies, reports thereon, and recommendations with respect to matters of policy and legislation as the President may request.

SEC. 205. In exercising its powers, functions, and duties under this Act, the Council shall—

(1) Consult with the Citizens' Advisory Committee on Environmental Quality established by Executive Order No. 11472, dated May 29, 1969, and with such representatives of science, industry, agriculture, labor, conservation organizations, State and local governments and other groups, as it deems advisable; and

(2) Utilize, to the fullest extent possible, the services, facilities and information (including statistical information) of public and private agencies and organizations, and individuals, in order that duplication of effort and expense may be avoided, thus assuring that the Council's activities will not unnecessarily overlap or conflict with similar activities authorized by law and performed by established agencies.

SEC. 206. Members of the Council shall serve full time and the Chairman of the Council shall be compensated at the rate provided for Level II of the Executive Schedule Pay Rates (5 U.S.C. 5313). The other members of the Council shall be compensated at the rate provided for Level IV of the Executive Schedule Pay Rates (5 U.S.C. 5315).

SEC. 207. There are authorized to be appropriated to carry out the provisions of this Act not to exceed $300,000 for fiscal year 1970, $700,000 for fiscal year 1971, and $1 million for each fiscal year thereafter.

Approved January 1, 1970.

APPENDIX B

Table of NEPA Cases and Index to Citations

Aberdeen & Rockfish Ry. v. SCRAP (See *SCRAP v. United States*).

Akers v. Resor, 339 F. Supp. 1375, 2 ELR 20221 (W.D. Tenn.), —— F. Supp. ——, 3 ELR 20157 (W.D. Tenn. 1972). 150, 226, 259–60, 268–69.

Alabama ex rel. Baxley v. Woody, —— F.2d ——, 3 ELR 20122 (5th Cir. 1973).

Allison v. Froehlke, —— F. Supp. ——, 2 ELR 20357 (W.D. Tex.), *aff'd*, —— F.2d ——, 3 ELR 20011 (5th Cir. 1972). 150, 201.

Anaconda Co. v. Ruckelshaus, —— F. Supp. ——, 3 ELR 20024 (D. Colo. 1972). 10, 39, 114–16, 122.

Arlington Coalition on Transportation v. Volpe, 332 F. Supp. 1218, 1 ELR 20486 (E.D. Va. 1971), *rev'd*, 458 F.2d 1323, 2 ELR 20162 (4th Cir.), *cert. denied sub nom. Fugate v. Arlington Coalition on Transportation*, 41 U.S.L.W. 3249 (Nov. 7, 1972). 44–45, 59, 67–68, 71, 107, 124–25, 143, 151, 154, 163, 166, 170–76, 180–81, 182–83, 241, 243.

Atlanta Gas Light Co. v. Southern Natural Gas Co., 338 F. Supp. 1039 (N.D. Ga. 1972). 47

Bankers Life and Casualty Co. v. Village of North Palm Beach, Florida, —— F.2d ——, 2 ELR 20528 (5th Cir. 1972). 146, 272.

Berkson v. Morton, —— F. Supp. ——, 2 ELR 20659 (D. Md. 1971). 78.

Billings v. Camp, —— F. Supp. ——, 2 ELR 20687 (D.D.C. 1972). 78.

Boston Waterfront Residents' Ass'n v. Romney, 343 F. Supp. 89, 2 ELR 20359 (D. Mass. 1972). 47, 86, 153–55.

Kings County Economic Community Development Ass'n v. Hardin, 333 F. Supp. 1302 (N.D. Cal. 1971), —— F. Supp. ——, 2 ELR 20151 (E.D. Cal. 1972), —— F.2d ——, 3 ELR 20335 (9th Cir. 1973). 29.

Kisner v. Butz, 350 F. Supp. 310, 2 ELR 20709 (N.D. W. Va. 1972). 79, 101, 104.

Kitchen v. Federal Communications Comm'n, 464 F.2d 801, 2 ELR 20534 (D.C. Cir. 1972). 62.

La Raza Unida v. Volpe, 337 F. Supp. 221, 1 ELR 20642 (N.D. Cal. 1971), —— F. Supp. ——, 2 ELR 20691 (N.D. Cal. 1972). 31, 65–66, 70, 125, 168–69, 184–85.

Lathan v. Volpe, 455 F.2d 1111, 1 ELR 20602 (9th Cir. 1971), *modified on rehearing*, 455 F.2d 1122, 2 ELR 20090 (9th Cir.), 350 F. Supp. 262, 2 ELR 20545 (W.D. Wash. 1972). 59, 65, 70, 107, 160–61, 168, 180, 185, 191, 206, 208, 212, 217, 228, 234–35, 240–41, 242–43, 254–55, 261, 285.

Lee v. Resor, 348 F. Supp. 389, 2 ELR 20665 (M.D. Fla. 1972). 50, 176–78, 186.

Life of the Land v. Volpe, —— F. Supp. ——, 3 ELR 20180 (D. Hawaii 1972).

Live in a Favorable Environment v. Volpe, —— F. Supp. ——, 3 ELR 20039 (E.D. Va. 1972).

Lloyd Harbor Study Group v. Seaborg, —— F. Supp. ——, 1 ELR 20188 (E.D.N.Y. 1971). 47.

Maddox v. Bradley, 345 F. Supp. 1255, 2 ELR 20404 (N.D. Tex. 1972). 35, 79, 95–96, 151–52.

Maryland-National Capital Park & Planning Comm'n v. United States Postal Service, C.A., 349 F. Supp. 1212, 2 ELR 20656 (D.D.C. 1972). 78, 91, 99.

McLean Gardens Residents Ass'n v. National Capital Planning Comm'n, ——F. Supp. ——, 2 ELR 20659 (D.D.C.), *motion for stay of injunction and summary reversal denied*, ——F.2d ——, 2 ELR 20662 (D.C. Cir. 1972). 58, 63–64, 78, 86, 94.

McQueary v. Laird, 449 F.2d 608, 1 ELR 20607 (10th Cir. 1971). 134–35.

Mink v. Environmental Protection Agency, 464 F.2d 742, 1 ELR 20527 (D.C. Cir. 1971), *rev'd*, U.S. ——, 3 ELR 20057 (1973). 108.

Minnesota Environmental Control Citizen's Ass'n v. Atomic Energy Comm'n, —— F. Supp. ——, 3 ELR 20034 (D. Minn. 1972).

Monroe County Conservation Ass'n v. Hansen, —— F. Supp. ——, 1 ELR 20362 (W.D.N.Y. 1971).

Monroe County Conservation Council v. Volpe, —— F. Supp. ——, 2 ELR 20015 (W.D.N.Y.), *rev'd*, —— F.2d ——, 3 ELR 20006 (2d Cir. 1972).

Montgomery County v. Richardson, —— F. Supp. ——, 2 ELR 20140 (D.D.C. 1972). 77.

Northeast Area Welfare Rights Organization v. Volpe, —— F. Supp. ——, 1 ELR 20186 (E.D. Wash. 1970). 65, 68–69.

Northside Tenants' Rights Coalition v. Volpe, 346 F. Supp. 244, 2 ELR 20553 (E.D. Wis. 1972), ——F. Supp. ——, 3 ELR 20154 (E.D. Wis. 1973). 173–74, 182–83, 194, 241, 242–43.

Pennsylvania Environmental Council v. Bartlett, 315 F. Supp. 238, 2 ELR 20752 (M.D. Pa. 1970), *aff'd,* 454 F.2d 613, 1 ELR 20622 (3rd Cir. 1971). 44–45, 60, 123–24, 144–45, 149, 153, 161–63, 165, 167, 175.

People of Enewetak v. Laird, —— F. Supp. ——, 2 ELR 20739 (D. Hawaii 1972), —— F. Supp. ——, 3 ELR 20190 (D. Hawaii 1973). 136.

Petterson v. Resor, 331 F. Supp. 1302, 2 ELR 20013 (D. Ore. 1971), *sub nom. Petterson v. Froehlke,* —— F. Supp. ——, 2 ELR 20747 (D. Ore. 1972). 146.

Pizitz, Inc. v. Volpe, —— F. Supp. ——, 2 ELR 20378 (M.D. Ala.), *aff'd,* 467 F.2d 208, 2 ELR 20379 (5th Cir.), *modified on rehearing,* 467 F.2d 208, 2 ELR 20635 (5th Cir. 1972). 42, 91, 99–100, 175, 191–92, 194, 212.

Port of New York Authority v. Interstate Commerce Comm'n, 451 F.2d 783, 2 ELR 20105 (2d Cir. 1971). 139–40.

Public Service Comm'n of New York v. Federal Power Comm'n, 463 F.2d 824, 2 ELR 20213 (D.C. Cir. 1972).

Ragland v. Mueller, 460 F.2d 1196, 2 ELR 20320 (5th Cir. 1972). 123, 151–52, 174–75.

San Antonio Conservation Society v. Texas Highway Department (See *Named Individual Members of the San Antonio Conservation Society v. Texas Highway Department*). 67–68, 170.

St. Joseph Historical Society v. Land Clearance & Redevelopment Authority of St. Joseph, Missouri, —— F. Supp. ——, 2 ELR 20749 (W.D. Mo. 1972).

San Francisco Tomorrow v. Romney, 342 F. Supp. 77, 2 ELR 20273 (N.D. Cal. 1972), *aff'd in part, rev'd in part,* —— F.2d ——, 3 ELR 20124 (9th Cir. 1973). 35, 153–55.

Save Our Ten Acres v. Kreger, —— F. Supp. ——, 2 ELR 20305 (S.D. Ala.), *rev'd,* —— F.2d ——, 3 ELR 20041 (5th Cir. 1973). 23, 87, 94, 95, 100, 150.

Save the Dunes Council v. Froehlke, —— F. Supp. ——, 2 ELR 20356 (N.D. Ind. 1972). 47.

Scenic Hudson Preservation Conference v. Federal Power Comm'n (*Scenic Hudson I*), 354 F.2d 608, 1 ELR 20292 (2d Cir. 1965), *cert. denied sub nom. Consolidated Edison Co. of New York v. Scenic Hudson Preservation Conference,* 384 U.S. 941 (1966); *Scenic Hudson II,* 453 F.2d 463, 1 ELR 20496 (2d Cir. 1971), *cert. denied,* 407 U.S. 926, 2

Students Challenging Regulatory Agency Procedures v. United States (See *SCRAP v. United States*).

Tanner v. Armco Steel, 340 F. Supp. 532, 2 ELR 20246 (S.D. Tex. 1972).

Texas Committee on Natural Resources v. Resor, —— F. Supp. ——, 1 ELR 20466 (E.D. Tex. 1971). 47.

Texas Committee on Natural Resources v. United States, —— F. Supp. ——, 2 ELR 20574 (W.D. Tex. 1970), *vacated*, 430 F.2d 1315 (5th Cir. 1970). 15, 58–59, 78, 85, 86, 152–53.

Thompson v. Fugate, —— F. Supp. ——, 1 ELR 20369 (E.D. Va.), *injunction expanded*, —— F.2d ——, 1 ELR 20370 (4th Cir.), *injunction expanded, aff'd in part and rev'd in part*, 452 F.2d 57, 1 ELR 20599 (4th Cir. 1971), 347 F. Supp. 120, 2 ELR 20612 (E.D. Va. 1972). 64–65, 173.

Transcontinental Gas Pipeline Corp. v. Hackensack Meadowlands Development Comm'n, 464 F.2d 1358, 2 ELR 20495 (3rd Cir. 1972). 79, 80, 84, 158.

United States v. Town of Brookhaven, —— F. Supp. ——, 1 ELR 20377 (E.D.N.Y. 1971).

United States v. 247.37 Acres of Land, —— F. Supp. ——, 1 ELR 20513 (S.D. Ohio 1971), —— F. Supp. ——, 2 ELR 20154 (S.D. Ohio 1972). 77, 81, 85, 148, 225–26.

United States v. 2,606.84 Acres of Land in Tarrant Co., Texas, 309 F. Supp. 887 (N.D. Tex. 1969), *rev'd*, 432 F.2d 1286 (5th Cir. 1970), *cert. denied*, 1 ELR 20155 (1971).

Upper Pecos Ass'n v. Stans, 328 F. Supp. 332, 1 ELR 20228 (D.N.M.), *aff'd*, 452 F.2d 1233, 2 ELR 20085 (10th Cir. 1971), *vacated*, 93 S. Ct. 458 (1972). 29, 77, 83, 184, 197–200, 284.

Virginians for Dulles v. Volpe, 344 F. Supp. 573, 2 ELR 20360 (E.D. Va. 1972). 78, 80, 90, 151.

Ward v. Ackroyd, 344 F. Supp. 1202, 2 ELR 20405 (D. Md. 1972). 32–33, 45, 125, 166, 172.

West Virginia Highlands Conservancy v. Island Creek Coal Co., 441 F.2d 232, 1 ELR 20160 (4th Cir. 1971). 29, 47, 58.

Wilderness Society v. Hickel, 325 F. Supp. 422, 1 ELR 20042 (D.D.C. 1970), *sub nom. Wilderness Society v. Morton*, 463 F.2d 1261, 2 ELR 20250 (D.C. Cir.), —— F. Supp. ——, 2 ELR 20583 (D.D.C. 1972), *rev'd*, —— F.2d ——, 3 ELR 20085 (D.C. Cir. 1973). 32, 47, 58, 76.

Willamette Heights Neighborhood Ass'n v. Volpe, 334 F. Supp. 990, 2 ELR 20043 (D. Ore. 1971). 168, 175.

Zabel v. Tabb, 430 F.2d 199, 1 ELR 20023 (5th Cir. 1970), *cert. denied*, 401 U.S. 910 (1971). 145–46, 170, 180–81, 247, 249–50, 271–72, 287–88.

Zlotnick v. Redevelopment Land Agency, —— F. Supp. ——, 2 ELR 20235 (D.D.C. 1972). 35, 41–42, 77, 150.

APPENDIX C

*Guidelines Council on Environmental Quality April 1971**

STATEMENTS ON PROPOSED FEDERAL ACTIONS AFFECTING THE ENVIRONMENT

1. *Purpose.* This memorandum provides guidelines to Federal departments, agencies, and establishments for preparing detailed environmental statements on the proposals for legislation and other major Federal actions significantly affecting the quality of the human environment as required by section 102(2)(C) of the National Environmental Policy Act (Public Law 91–190) (hereafter "the Act"). Underlying the preparation of such environmental statements is the mandate of both the Act and Executive Order 11514 (35 F.R. 4247) of March 4, 1970, that all Federal agencies, to the fullest extent possible, direct their policies, plans and programs so as to meet national environmental goals. The objective of section 102(2)(C) of the Act and of these guidelines is to build into the agency decision-making process an appropriate and careful consideration of the environmental aspects of proposed action and to assist agencies in implementing not only the letter, but the spirit, of the Act. This memorandum also provides guidance on implementation of section 309 of the Clean Air Act, as amended (42 U.S.C. 1857 *et seq.*).

2. *Policy.* As early as possible and in all cases prior to agency decision concerning major action or recommendation or a favorable report on legislation that significantly affects the environment, Federal agencies will, in consultation with other appropriate Federal, State, and local agencies, assess in detail the potential environmental impact in order that adverse effects are avoided, and environmental quality is restored or enhanced, to the fullest extent practicable. In particular, alternative actions that will minimize adverse impact should be explored and both the long- and short-range implications to man, his physical and social surroundings, and to nature, should be evaluated in

* 36 FED. REG. 7724–29 (Apr. 23, 1971) ELR 46049. The Council on Environmental Quality proposed amendments to its guidelines just before this book went to press. *See* 38 FED. REG. 10856 (May 2, 1973).

order to avoid to the fullest extent practicable undesirable consequences for the environment.

3. *Agency and OMB procedures.* (a) Pursuant to section 2(f) of Executive Order 11514, the heads of Federal agencies have been directed to proceed with measures required by section 102(2)(C) of the Act. Consequently, each agency will establish, in consultation with the Council on Environmental Quality, not later than June 1, 1970 (and, by July 1, 1971, with respect to requirements imposed by revisions in these guidelines, which will apply to draft environmental statements circulated after June 30, 1971), its own formal procedures for: (1) Identifying those agency actions requiring environmental statements, the appropriate time prior to decision for the consultations required by section 102(2)(C), and the agency review process for which environmental statements are to be available, (2) obtaining information required in their preparation, (3) designating the officials who are to be responsible for the statements, (4) consulting with and taking account of the comments of appropriate Federal, State, and local agencies, including obtaining the comment of the Administrator of the Environmental Protection Agency, whether or not an environmental statement is prepared, when required under section 309 of the Clean Air Act, as amended, and section 8 of these guidelines, and (5) meeting the requirements of section 2(b) of Executive Order 11514 for providing timely public information on Federal plans and programs with environmental impact including procedures responsive to section 10 of these guidelines. These procedures should be consonant with the guidelines contained herein. Each agency should file seven (7) copies of all such procedures with the Council on Environmental Quality, which will provide advice to agencies in the preparation of their procedures and guidance on the application and interpretation of the Council's guidelines. The Environmental Protection Agency will assist in resolving any question relating to section 309 of the Clean Air Act, as amended.

(b) Each Federal agency should consult, with the assistance of the Council on Environmental Quality and the Office of Management and Budget if desired, with other appropriate Federal agencies in the development of the above procedures so as to achieve consistency in dealing with similar activities and to assure effective coordination among agencies in their review of proposed activities.

(c) State and local review of agency procedures, regulations, and policies for the administration of Federal programs of assistance to State and local governments will be conducted pursuant to procedures established by the Office of Management and Budget Circular No. A–85. For agency procedures subject to OMB Circular No. A–85 a 30-day extension in the July 1, 1971, deadline set in section 3(a) is granted.

(d) It is imperative that existing mechanisms for obtaining the views of Federal, State, and local agencies on proposed Federal actions be utilized to the extent practicable in dealing with environmental matters. The Office of Management and Budget will issue instructions, as necessary, to take full advantage of existing mechanisms (relating to procedures for handling legislation, preparation of budgetary materials, new procedures, water resource and other projects, etc.).

4. *Federal agencies included.* Section 102(2)(C) applies to all agencies of the Federal Government with respect to recommendations or favorable reports on proposals for (i) legislation and (ii) other major Federal actions significantly affecting the quality of the human environment. The phrase "to the fullest extent possible" in section 102(2)(C) is meant to make clear that each agency of the Federal Government shall comply with the requirement unless existing law applicable to the agency's operations expressly prohibits or makes compliance impossible. (Section 105 of the Act provides that "The

policies and goals set forth in this Act are supplementary to those set forth in existing authorizations of Federal agencies.")

5. *Actions included.* The following criteria will be employed by agencies in deciding whether a proposed action requires the preparation of an environmental statement:

(a) "Actions" include but are not limited to:

(i) Recommendations or favorable reports relating to legislation including that for appropriations. The requirement for following the section 102(2)(C) procedure as elaborated in these guidelines applies to both (i) agency recommendations on their own proposals for legislation and (ii) agency reports on legislation initiated elsewhere. (In the latter case only the agency which has primary responsibility for the subject matter involved will prepare an environmental statement.) The Office of Management and Budget will supplement these general guidelines with specific instructions relating to the way in which the section 102(2)(C) procedure fits into its legislative clearance process;

(ii) Projects and continuing activities: directly undertaken by Federal agencies; supported in whole or in part through Federal contracts, grants, subsidiaries, loans, or other forms of funding assistance; involving a Federal lease, permit, license, certificate, or other entitlement for use;

(iii) Policy, regulations, and procedure-making.

(b) The statutory clause "major Federal actions significantly affecting the quality of the human environment" is to be construed by agencies with a view to the overall, cumulative impact of the action proposed (and of further actions contemplated). Such actions may be localized in their impact, but if there is potential that the environment may be significantly affected, the statement is to be prepared. Proposed actions, the environmental impact of which is likely to be highly controversial, should be covered in all cases. In considering what constitutes major action significantly affecting the environment, agencies should bear in mind that the effect of many Federal decisions about a project or complex of projects can be individually limited but cumulatively considerable. This can occur when one or more agencies over a period of years puts into a project individually minor but collectively major resources, when one decision involving a limited amount of money is a precedent for action in much larger cases or represents a decision in principle about a future major course of action, or when several government agencies individually make decisions about partial aspects of a major action. The lead agency should prepare an environmental statement if it is reasonable to anticipate a cumulatively significant impact on the environment from Federal action. "Lead agency" refers to the Federal agency which has primary authority for committing the Federal Government to a course of action with significant environmental impact. As necessary, the Council on Environmental Quality will assist in resolving questions of lead agency determination.

(c) Section 101(b) of the Act indicates the broad range of aspects of the environment to be surveyed in any assessment of significant effect. The Act also indicates that adverse significant effects include those that degrade the quality of the environment, curtail the range of beneficial uses of the environment, and serve short-term, to the disadvantage of long-term, environmental goals. Significant effects can also include actions which may have both beneficial and detrimental effects, even if, on balance, the agency believes that the effect will be beneficial. Significant adverse effects on the quality of the human environment include both those that directly affect human beings and those that indirectly affect human beings through adverse effects on the environment.

(d) Because of the Act's legislative history, environmental protective regulatory activities concurred in or taken by the Environmental Protection Agency are not deemed actions which require the preparation of environmental statements under section 102(2)(C) of the Act.

6. *Content of environmental statement.* (a) The following points are to be covered:

(i) A description of the proposed action including information and technical data adequate to permit a careful assessment of environmental impact by commenting agencies. Where relevant, maps should be provided.

(ii) The probable impact of the proposed action on the environment, including impact on ecological systems such as wildlife, fish, and marine life. Both primary and secondary significant consequences for the environment should be included in the analysis. For example, the implications, if any, of the action for population distribution or concentration should be estimated and an assessment made of the effect of any possible change in population patterns upon the resource base, including land use, water, and public services, of the area in question.

(iii) Any probable adverse environmental effects which cannot be avoided (such as water or air pollution, undesirable land use patterns, damage to life systems, urban congestion, threats to health or other consequences adverse to the environmental goals set out in section 101(b) of the Act).

(iv) Alternatives to the proposed action (section 102(2)(D) of the Act requires the responsible agency to "study, develop, and describe appropriate alternatives to recommended courses of action in any proposal which involves unresolved conflicts concerning alternative uses of available resources"). A rigorous exploration and objective evaluation of alternative actions that might avoid some or all of the adverse environmental effects is essential. Sufficient analysis of such alternatives and their costs and impact on the environment should accompany the proposed action through the agency review process in order not to foreclose prematurely options which might have less detrimental effects.

(v) The relationship between local short-term uses of man's environment and the maintenance and enhancement of long-term productivity. This in essence requires the agency to assess the action for cumulative and long-term effects from the perspective that each generation is trustee of the environment for succeeding generations.

(vi) Any irreversible and irretrievable commitments of resources which would be involved in the proposed action should it be implemented. This requires the agency to identify the extent to which the action curtails the range of beneficial uses of the environment.

(vii) Where appropriate, a discussion of problems and objections raised by other Federal, State, and local agencies and by private organizations and individuals in the review process and the disposition of the issues involved. (The section may be added at the end of the review process in the final text of the environmental statement.)

(b) With respect to water quality aspects of the proposed action which have been previously certified by the appropriate State or interstate organization as being in substantial compliance with applicable water quality standards, the comment of the Environmental Protection Agency should also be requested.

(c) Each environmental statement should be prepared in accordance with the precept in section 102(2)(A) of the Act that all agencies of the Federal Government utilize a systematic, interdisciplinary approach which will insure the integrated use of the natural and social sciences and the environmental design arts in planning and decisionmaking which may have an impact on man's environment.

(d) Where an agency follows a practice of declining to favor an alternative until public hearings have been held on a proposed action, a draft environmental statement may be prepared and circulated indicating that two or more alternatives are under consideration.

(e) Appendix 1 prescribes the form of the summary sheet which should accompany each draft and final environmental statement [*omitted*].

7. *Federal agencies to be consulted in connection with preparation of environmental statement.* A Federal agency considering an action requiring an environmental statement, on the basis of: (i) A draft environmental statement for which it takes responsibility, or (ii) comparable information followed by a hearing subject to the provisions of the Administrative Procedure Act, should consult with, and obtain the comment on the environmental impact of the action of, Federal agencies with jurisdiction by law or special expertise with respect to any environmental impact involved. These Federal agencies include components of (depending on the aspect or aspects of the environment):

Advisory Council on Historic Preservation
Department of Agriculture
Department of Commerce
Department of Defense
Department of Health, Education, and Welfare
Department of Housing and Urban Development
Department of the Interior
Department of State
Department of Transportation
Atomic Energy Commission
Federal Power Commission
Environmental Protection Agency
Office of Economic Opportunity

For actions specifically affecting the environment of their geographic jurisdictions, the following Federal and Federal-State agencies are also to be consulted:

Tennessee Valley Authority
Appalachian Regional Commission
National Capital Planning Commission
Delaware River Basin Commission
Susquehanna River Basin Commission

Agencies seeking comment should determine which one or more of the above listed agencies are appropriate to consult on the basis of the areas of expertise identified in appendix 2 to these guidelines [*omitted*]. It is recommended: (i) That the above listed departments and agencies establish contact points, which often are most appropriately regional offices, for providing comments on the environmental statements, and (ii) that departments from which comment is solicited coordinate and consolidate the comments of their component entities. The requirement in section 102(2)(C) to obtain comment from Federal agencies having jurisdiction or special expertise is in addition to any specific statutory obligation of any Federal agency to coordinate or consult with any other Federal or State agency. Agencies seeking comment may establish time limits of not less than thirty (30) days for reply, after which it may be presumed, unless the agency consulted requests a specified extension of time, that the agency consulted has no comment to make. Agencies seeking comment should endeavor to comply with requests for extensions of time of up to fifteen (15) days.

8. *Interim EPA procedures for implementation of section 309 of the Clean Air Act, as amended.* (a) Section 309 of the Clean Air Act, as amended, provides:

SEC. 309. (a) The Administrator shall review and comment in writing on the environmental impact of any matter relating to duties and responsibilities granted pursuant to this Act or other provisions of the authority of the Administrator, contained in any (1) legislation proposed by any Federal department or agency, (2) newly authorized Federal projects for construction and any major Federal agency action (other than a project for construction) to which section 102(2)(C) of Public Law 91–190 applies, and (3) proposed regulations published by any department or agency of the Federal Government. Such written comment shall be made public at the conclusion of any such review.

(b) In the event the Administrator determines that any such legislation, action, or regulation is unsatisfactory from the standpoint of public health or welfare or environmental quality, he shall publish his determination and the matter shall be referred to the Council on Environmental Quality.

(b) Accordingly, wherever an agency action related to air or water quality, noise abatement and control, pesticide regulation, solid waste disposal, radiation criteria and standards, or other provisions of the authority of the Administrator if the Environmental Protection Agency is involved, including his enforcement authority, Federal agencies are required to submit for review and comment by the Administrator in writing: (i) Proposals for new Federal construction projects and other major Federal agency actions to which section 102(2)(C) of the National Environmental Policy Act applies, and (ii) proposed legislation and regulations, whether or not section 102(2)(C) of the National Environmental Policy Act applies. (Actions requiring review by the Administrator do not include litigation or enforcement proceedings.) The Administrator's comments shall constitute his comments for the purposes of both section 309 of the Clean Air Act and section 102(2)(C) of the National Environmental Policy Act. A period of 45 days shall be allowed for such review. The Administrator's written comment shall be furnished to the responsible Federal department or agency, to the Council on Environmental Quality and summarized in a notice published in the Federal Register. The public may obtain copies of such comment on request from the Environmental Protection Agency.

9. *State and local review.* Where no public hearing has been held on the proposed action at which the appropriate State and local review has been invited, and where review of the environmental impact of the proposed action by State and local agencies authorized to develop and enforce environmental standards is relevant, such State and local review shall be provided as follows:

(a) For direct Federal development projects and projects assisted under programs listed in attachment D of the Office of Management and Budget Circular No. A–95, review of draft environmental statements by State and local governments will be through procedures set forth under part 1 of Circular No. A–95.

(b) Where these procedures are not appropriate and where a proposed action affects matters within their jurisdiction, review of the draft environmental statement on a proposed action by State and local agencies authorized to develop and enforce environmental standards and their comments on the environmental impact of the proposed action may be obtained directly or by distributing the draft environmental statement to the appropriate State, regional, and metropolitan clearinghouse unless the Governor of the State involved has designated some other point for obtaining this review.

10. *Use of statements in agency review processes; distribution to Council on Environmental Quality; availability to public.* (a) Agencies will need to identify at what stage or stages of a series of actions relating to a particular matter the environmental statement procedures of this directive will be applied. It will often be necessary to use the procedures both in the development of a national program and in the review of proposed projects within the national program. However, where a grant-in-aid program does not entail prior approval by Federal agencies of specific projects, the view of Federal, State, and local agencies in the legislative process may have to suffice. The principle to be applied is to obtain views of other agencies at the earliest feasible time in

the development of program and project proposals. Care should be exercised so as not to duplicate the clearance process, but when actions being considered differ significantly from those that have already been reviewed pursuant to section 102(2)(C) of the Act an environmental statement should be provided.

(b) Ten (10) copies of draft environmental statements (when prepared), ten (10) copies of all comments made thereon (to be forwarded to the Council by the entity making comment at the time comment is forwarded to the responsible agency), and ten (10) copies of the final text of environmental statements (together with all comments received thereon by the responsible agency from Federal, State, and local agencies and from private organizations and individuals) shall be supplied to the Council on Environmental Quality in the Executive Office of the President (this will serve as making environmental statements available to the President). It is important that draft environmental statements be prepared and circulated for comment and furnished to the Council early enough in the agency review process before an action is taken in order to permit meaningful consideration of the environmental issues involved. To the maximum extent practicable no administrative action (i.e., any proposed action to be taken by the agency other than agency proposals for legislation to Congress or agency reports on legislation) subject to section 102(2)(C) is to be taken sooner than ninety (90) days after a draft environmental statement has been circulated for comment, furnished to the Council and, except where advance public disclosure will result in significantly increased costs of procurement to the Government, made available to the public pursuant to these guidelines; neither should such administrative action be taken sooner than thirty (30) days after the final text of an environmental statement (together with comments) has been made available to the Council and the public. If the final text of an environmental statement is filed within ninety (90) days after a draft statement has been circulated for comment, furnished to the Council and made public pursuant to this section of these guidelines, the thirty (30) day period and ninety (90) day period may run concurrently to the extent that they overlap.

(c) With respect to recommendations or reports on proposals for legislation to which section 102(2)(C) applies, the final text of the environmental statement and comments thereon should be available to the Congress and to the public in support of the proposed legislation or report. In cases where the scheduling of congressional hearings on recommendations or reports on proposals for legislation which the Federal agency has forwarded to the Congress does not allow adequate time for the completion of a final text of an environmental statement (together with comments), a draft environmental statement may be furnished to the Congress and made available to the public pending transmittal of the comments as received and the final text.

(d) Where emergency circumstances make it necessary to take an action with significant environmental impact without observing the provisions of these guidelines concerning minimum periods for agency review and advance availability of environmental statements, the Federal agency proposing to take action should consult with the Council on Environmental Quality about alternative arrangements. Similarly, where there are overriding considerations of expense to the Government or impaired program effectiveness, the responsible agency should consult the Council concerning appropriate modifications of the minimum periods.

(e) In accord with the policy of the National Environmental Policy Act and Executive Order 11514, agencies have a responsibility to develop procedures to insure the fullest practicable provision of timely public information and understanding of Federal plans and programs with environmental impact in order to obtain the views of interested parties. These procedures shall include, whenever appropriate, provision for public hearings, and shall provide the public with relevant information, including information on alternative courses of action. Agencies which hold hearings on proposed administrative actions or legislation should make the draft environmental statement available to the public at least 15 days prior to the time of the relevant hearings except where the agency prepares the draft statement on the basis of a hearing subject to the Administrative Procedure Act and preceded by adequate public notice and information to identify the issues and obtain the comments provided for in sections 6 to 9 of these guidelines.

(f) The agency which prepared the environmental statement is responsible for making the statement and the comments received available to the public pursuant to the provisions of the Freedom of Information Act (5 U.S.C., sec. 552), without regard to the exclusion of interagency memoranda when such memoranda transmit comments of Federal agencies listed in section 7 of these guidelines upon the environmental impact of proposed actions subject to section 102(2)(C).

(g) Agency procedures prepared pursuant to section 3 of these guidelines shall implement these public information requirements and shall include arrangements for availability of environmental statements and comments at the head and appropriate regional offices of the responsible agency and at appropriate State, regional, and metropolitan clearinghouses unless the Governor of the State involved designates some other point for receipt of this information.

11. *Application of section 102(2)(C) procedure to existing projects and programs.* To the maximum extent practicable the section 102(2)(C) procedure should be applied to further major Federal actions having a significant effect on the environment even though they arise from projects or programs initiated prior to enactment of the Act on January 1, 1970. Where it is not practicable to reassess the basic course of action, it is still important that further incremental major actions be shaped so as to minimize adverse environmental consequences. It is also important in further action that account be taken of environmental consequences not fully evaluated at the outset of the project or program.

12. *Supplementary guidelines, evaluation of procedures.* (a) The Council on Environmental Quality after examining environmental statements and agency procedures with respect to such statements will issue such supplements to these guidelines as are necessary.

(b) Agencies will continue to assess their experience in the implementation of the section 102(2)(C) provisions of the Act and in conforming with these guidelines and report thereon to the Council on Environmental Quality by December 1, 1971. Such reports should include an identification of the problem areas and suggestions for revision or clarification of these guidelines to achieve effective coordination of views on environmental aspects (and alternatives, where appropriate) of proposed actions without imposing unproductive administrative procedures.

RUSSELL E. TRAIN
Chairman

Index